THE *NEW* PESTICIDE USER'S GUIDE

Bert L. Bohmont

Professor
and
Agricultural Chemicals Coordinator

College of Agricultural Sciences

Colorado State University

RESTON PUBLISHING COMPANY, INC.
A Prentice-Hall Company
Reston, Virginia

Library of Congress Cataloging in Publication Data

Bohmont, Bert L.
    The new pesticide user's guide.

    Includes index.
    1. Pesticides--Handbooks, manuals, etc. 1. Title
SB951.B584 1983   632'.95   82-21497
ISBN 0-8359-5538-9

Other publications by Dr. Bohmont:

Pesticides: Why and How They are Used

A correspondence course available through:

    Colorado State University
    Division of Continuing Education
    Rockwell Hall
    Fort Collins, Colorado 80523
    (303) 491-5288

10 9 8 7 6 5

Printed in the United States of America.

Table Of Contents

Chapter 8 - PESTICIDE SAFETY (continued)

Foreword

This book is a revision of The New Pesticide User's Guide published in 1981. Additional material has been added to chapters 6, 7, 8, 9, 11, 12 and 14. Some materials have been moved for better continuity and an Index and Appendix have been included for ease of use and reference.

You will note that this book is printed in a larger foremat than the 1981 version and, of course, is bound in hard cover for durability.

This publication should be a valuable source of information to those who are new to pesticides or are wanting to have an understanding of the importance of pesticides in our society. High school and college teachers should find it very useful in conveying to their students that pesticides are highly regulated chemicals that deserve respect in their use and in their role in food and fiber production.

Recognizing that pesticides are an essential tool in helping to control most pests, the information in this book is oriented toward those applying pesticides. It should prove especially useful to commercial pesticide applicators as well as employees of city, county, state and federal agencies. Pesticide dealers, salesmen and consultants should also find it helpful in their work.

The science of pesticide use has become a highly specialized field. The federal laws require that individuals applying certain designated pesticides be able to substantiate or demonstrate that they have knowledge and capabilities to use the materials safely and effectively. This book can help supply the necessary information in order for pesticide applicators to use pesticides in a responsible manner. Important information is presented throughout the 15 chapters as well as the glossary and appendix.

Information or suggestions for the use of specific pesticides for specific pest control problems is NOT included in this book. Specific control measures and recommendations have been intentionally omitted because they are subject to change and may soon become obsolete. Only current recommendations should be used along with making sure that you are using the latest pesticide label. Pest control guides for the control of most pests are available through the State University Cooperative Extension Service in every state.

The information contained in this publication is supplied with the understanding that there is no intended endorsement of a specific product or practice, nor is discrimination intended toward any product or practice included in/or omitted from this book.

Acknowledgments

Information and illustrations for this book were drawn from
many sources including publications from State Cooperative Extension
Services in California, Kansas, North Carolina, the Northeast Region,
Oklahoma, the Southeast Region, Washington and Wyoming, as well as
Colorado. Use of these materials is gratefully acknowledged.

Specimen labels are from 1982 label books from CIBA-GEIGY
Chemical Corporation, Hopkins Agricultural Chemical Company, and
Union Carbide Corporation. Use of the sample labels is acknowledged
and appreciated.

I acknowledge with appreciation the information drawn from Dr.
Major L. Boddicker in the "Vertebrate Pest" chapter. Some of the
pest illustrations are from "Furbearers of Colorado" published by
the Colorado Division of Wildlife in cooperation with the Colorado
Department of Education. Weed illustrations in the "Weeds" chapter
are reductions from USDA Handbook Number 366 "Selected Weeds of the
United States".

Special appreciation and acknowledgement is extended to Dr. W. K.
Hock, The Pennsylvania State University, for permission to use numerous
illustrations and information from the 1975 publication "Pest
Management and Environmental Quality" and to Dr. M. S. Stimman,
University of California, for permission to use illustrations and
information from the 1977 publication "Pesticide Application and
Safety Training".

I have tried to acknowledge all use of illustrations and
materials, but may have missed an original source because of the
extensive interchange of materials by Cooperative Extension Service
writers, sometimes making the original source uncertain.

A special debt of gratitude is due to my wife, Kathleen, for
her patience and understanding during the preparation of this manu-
script and for her encouragement that enabled me to see it through
to completion.

1
INTRODUCTION
TO
PESTICIDES

Introduction To Pesticides

History records that agricultural chemicals have been used since ancient times; the ancient Romans are known to have used burning sulfur to control insects. They were also known to have used salt to keep the weeds under control. The Ninth Century Chinese used arsenic mixed with water to control insects. Early in the 1800's, pyrethrin and rotenone were discovered to be useful as insecticides for the control of many different insect species. Paris green, a mixture of copper and arsenic, was discovered in 1865 and subsequently used to control the Colorado Potato Beetle. In 1882, a fungicide known as Bordeaux mixture, made from a mixture of lime and copper sulfate, was discovered to be useful as a fungicide for the control of downy mildew in grapes. Mercury dust was developed in 1890 as a seed treatment, and subsequently, in 1915, liquid mercury was developed as a seed treatment to protect seeds from fungus diseases. Mercury was banned from use in 1970, primarily because of the adverse national publicity concerning an accidental exposure of a family to mercury through the ingestion of pork that had been produced from seed screenings contaminated with a mercury fungicide.

Pesticides Protect a Major Share of Field Crop Acreage

Percent of acreage treated[1]

'1976 data.

The first synthetic, organic insecticides and herbicides were discovered and produced in the early 1900's; this production of synthetic pesticides preceded the subsequent discovery and production of hundreds of synthetic, organic pesticides, starting in the 1940's. Chlorinated hydrocarbons came into commercial production in the 1940's and organic phosphates began to be commercially produced during the 1950's. In the late 1950's, carbamates were developed and included insecticides, herbicides, and fungicides. The 1960's saw a trend toward specific and specialized pesticides which included systemic materials and the trend toward "prescription" types of pesticides. Presently, there are over 900 active pesticide chemicals being formulated into over 40,000 commercial preparations.

Pesticides are used by man as intentional applications to his environment in order to improve environmental quality for himself, his domesticated animals, and his plants. Despite the fears and real problems they create, pesticides clearly are responsible for part of the physical well-being enjoyed by most people in the United States and the western world. They also contribute significantly to the existing standards of living in other nations. In the United States, consumers spend less of their income on food (about 17%) than other people anywhere. The chief reason is more efficient food production, and chemicals make an important contribution in this area. In 1850 each U.S. farmer produced enough food and fiber for himself and three other persons; over 100 years later (1960) he was able to produce enough food an fiber for himself and 24 other people; himself and 45 other people in 1970, and now in the 1980's he is able to produce enough for himself and over 50 other people. World population was estimated at 4.25 billion people in 1980 and is expected to increase to over 5 billion by 1990 and to over 6 billion people by the year 2000. There will be great pressure on the farmers of the world to increase agricultural production in order to feed and clothe this extra population.

At the present time, current world food supply is inadequate to satisfy the hunger of the total population. As much as one-half of the world's population is under-nourished. The situation is worse in under-developed countries where it is estimated that as much as three-fourths of the inhabitants are under-nourished.

Agricultural Losses

In spite of pest control programs, United States agriculture still loses possibly one-third of its potential crop production to various pests. Without modern pest control, including the use of pesticides, this annual loss in the United States would probably double. If that happened, it is possible that 1) farm costs and prices would increase considerably; 2) the average consumer family would spend much more on food; 3) the number of people who work on farms would have to be increased; 4) farm exports would be reduced; 5) a vast increase in intensive cultivated acreage would be required.

Pest Competition

In most parts of the world today, pest control of some kind is essential because crops, livestock, and people live, as always, in a potentially hostile environment. Pests compete for our food supply, and they can be disease carriers as well as nuisances. Man coexists with more than one million kinds of insects and other arthropods, many of which are pests. Fungi cause more than 1500 plant diseases, and there are more than 1,000 species of harmful nematodes. Man must also combat hundreds of weed species in order to grow the crops that are needed to feed our nation. Rodents and other vertebrate pests also can cause problems of major proportion. Many of these pest enemies of mankind have caused damage for centuries.

Some good examples of specific increases in yields resulting from the use of pesticides in the United States are: corn 25%; potatoes 35%; onions 140%; cotton 100%; alfalfa seed 160%; milk production 15%.

Modern farm technology has created artificial environments that can worsen some pest problems and cause others. Large acreages, planted efficiently and economically with a single crop (monoculture) encourage certain insects and plant diseases. Advanced food production technology, therefore, actually increases the need for pest control. Pesticides are used not only to produce more food, but also food that is virtually free of damage from insects, diseases, and weeds. In the United States, pesticides are often used because of public demand, supported by government regulations, for uncontaminated and unblemished food.

Environmental Concerns

In the past, pest problems have often been solved without fully appreciating the treatments and effects on other plants and animals or on the environment. Some of these effects have been unfortunate. Today, scientists almost unanimously agree that the first rule in pest control is to recognize the whole problem. The agricultural environment is a complex web of interactions involving 1) many kinds of pests; 2) relationships between pests and their natural enemies; 3) relationships among all of these and other factors, such as weather, soil, water, plant varieties, cultural practices, wildlife, and man himself.

Pesticides are designed simply to destroy pests. They are applied to an environment that includes pests, crops, people, and other living things, as well as air, soil and water. It is generally accepted that pesticides which are specific to the pest to be controled are very desirable, and some are available. However, these products can be very expensive because of their limited range of applications.

Unquestionably, pesticides will continue to be of enormous benefit to man. They have helped to produce food and protect health. Manmade chemicals have been the front line of defense against destructive insects, weeds, plant diseases, and rodents. Through pest control, man has modified his environment to meet aesthetic and recreational demands. However, in solving some environmental problems, pesticides have created others of undetermined magnitude. The unintended consequences of the long term use of certain pesticides has been injury or death to some life forms. Much of the information on the effects of pesticides comes from the study of birds, fish, and the marine invertebrates such as crabs, shrimps, and scallops. It is clear that different species respond in different ways to the same concentration of a pesticide. Reproduction is inhibited in some and not in others. Eggs of some birds become thin and break while others do not.

Residues of some persistent pesticides apparently are "biologically concentrated". This means that they may become more concentrated in organisms higher up in a food chain. When this happens in an acquatic environment, animals that are at the top of the chain, usually fish-eating birds, may consume enough to suffer reproductive failure or other serious damage. Research has shown that some pesticides decompose completely into harmless substances fairly soon after they are exposed to air, water, sunlight, high temperature, or bacteria. Many others also may do so, but scientific confirmation of that fact is not yet available. When residues remain in or on plants or in soil or water, they usually are in very small amounts (a few parts per million or less). However, even such small amounts of some pesticides, or their breakdown products, which also may be harmful, sometimes persist for a long time.

Pesticides, like automobiles, can create environmental problems, but in today's world it is difficult to get along without them. Those concerned about pesticides and pest control face a dilemma. On the one hand, modern techniques of food production and control of disease carrying insects requires pesticides; on the other hand, many pesticides can be a hazard to living things other than pests, sometimes including people.

Human Concerns

No clear evidence exists on the long term effects on man of the accumulation of pesticides through the food chain, but the problem has been relatively unstudied. Limited studies with human volunteers has shown that persistent pesticides, at the normal levels found in human tissues at the present time, are not associated with any disease. However, further research is required before results are conclusive about present effects, and no information exists about the longer term effects. Meanwhile, the decisions must be made on the basis of extrapolation from results on experimental animals. Extrapolation is always risky, and the judgments on the chronic effects of pesticides on man will be highly controversial.

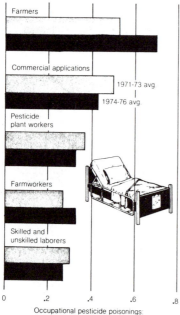

**Pesticide Poisonings:
Farmers Widen Their Lead**

Farmers

Commercial applications

1971-73 avg.
1974-76 avg.

Pesticide
plant workers

Farmworkers

Skilled and
unskilled laborers

0   .2   .4   .6   .8

Occupational pesticide poisonings:
rate per 100,000 hospital admissions

Source: *National Study of Hospitalized Pesticide Poisonings, 1974-76, July 1980, EPA.*

Public concern about the possible dangers of pesticides is manifested in legal actions initiated by conservation groups. Pesticides, like virtually every chemical, may have physiological effects on other organisms living in the environment, including man himself. The majority of the established pesticides have no adverse effect on man, animals, or the environment in general as long as they are used only in the amounts sufficient to control pest organisms. Pest control is never a simple matter of applying a pesticide that removes only the pest species. For one thing, the pest population is seldom completely or permanently eliminated. Almost always there are at least a few survivors to recreate the problem later on. Also, the pesticide often affects other living things besides the target species and may contaminate the environment.

There have been and continue to be unfortunate and generally inexcusable accidents where workers become grossly exposed due to improper and inadequate industrial hygiene or carelessness in handling and use. Children sometimes eat, touch or inhale improperly stored pesticides. Consumers have been inadvertently poisoned by pesticides spilled carelessly in the transportation of pesticides in conjunction with food products. These cases are, however, no indictment of the pesticide itself or the methods employed to establish its efficacy and safety. They are purely due to the irresponsibility of the user.

Pesticides are very rarely used in the form of a pure or technically pure compound, but rather are formulated to make them easy to apply. Formulations may be in the form of dust or granules, which usually contain 5 to 10% of active ingredients, or wettable powders or emulsifiable concentrates, which usually contain 40 to 80% active ingredients. It is important to remember that the formulations which are used as sprays are further diluted with water, oil, or other solvents to concentrations of usually only 1% or less before application. The amount of active ingredient, therefore, which is eventually released to the environment is generally extremely small.

Few responsible people today fail to recognize the need for pesticides and the importance of striving to live with them. Several national scientific committees in recent years have stressed the need for pesticides now and in the foreseeable future. These same committees also recommended more responsible use and further investigation into long term side effects on the environment. It is generally agreed among scientists that there is little, if any, chance that chemical pesticides can be abandoned until such time as alternative control measures are perfected.

TERMINOLOGY

Now that we understand the importance of pesticides and recognize the need for knowledge of these essential chemicals by the concerned citizen and user of pesticides, let's learn the terminology that is necessary to understand what pesticides are all about and to be able to communicate accurately.

The word pesticide is an all-inclusive term which includes a number of individual chemicals designed specifically for the control of certain pests. The illustration showing the "wheel" of pesticides includes 21 spokes which are the generic terms for each specific purpose. All but six of these generic words end in -cide. The word ending or suffix -cide means to kill or killer. Most of the specific pesticide names (insecticides, herbicides, fungicides, etc.) are specific to the pest for which they are intended to control. A few, however, are not as readily recognized by the average person and need study to become familiar with them.

Pesticides As Classified By Their Target Species

acaricide	mites, ticks
algaecide	algae
attractant	insects, birds, other vertebrates
avicide	birds
bactericide	bacteria
defoliant	unwanted plant leaves
desiccant	unwanted plant tops

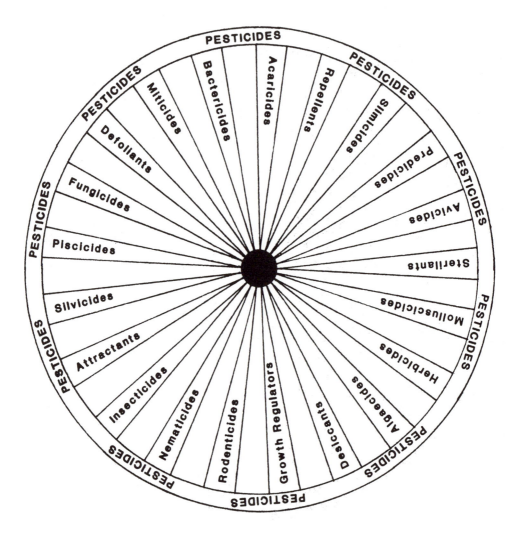

The All-Inclusive Pesticide Wheel

Pesticides As Classified By Their Target Species (cont.)

fungicide	fungi
growth regulator	insect and plant growth
herbicide	weeds
insecticide	insects
miticide	mites
molluscicide	snails, slugs
nematicide	nematodes
piscicide	fish
predacide	vertebrates
repellents	insects, birds, other vertebrates
rodenticide	rodents
silvicide	trees and woody vegetation
slimicide	slime molds
sterilants	insects, vertebrates

The term "pesticide," as defined by federal and state laws, also includes chemical compounds that stimulate or retard growth of plants, known as growth regulators; those that speed the drying of plants, known as desiccants, which are used as an aid in mechanical harvesting of cotton, soybeans, and other crops; and those chemicals that remove leaves, known as defoliants, which aid in the harvesting of potatoes and certain other crops.

The term "pesticide" also applies to compounds used for repelling, attracting, and sterilizing insects. The last two groups do not fit the original definition in that they are not "-cides", but rather they are included only because they fit the legal definition of pesticides.

There are a few other "-cide" terms that you might encounter. The term "Biocide" is sometimes used by environmental groups or others who are opposed to the use of pesticides. A true "Biocide" would be one that kills a wide range of organisms and is toxic to both plants and animals. There are only a few pesticides that would fall in this class and, unless used carlessly or indescriminately, they are used in specific situations with due care for humans and the environment. Therefore, the term "Biocide" is incorrect when discussing materials used for specific purposes in controlling various pests.

Other "-cide" terms include "adulticide", "larvicide", "aphicide" and "ovicide". These are all terms that refer to the use of insecticides. The use of insecticides that are more effective at certain stages of insect growth gives rise to the terms of "adulticide" and "larvicide". An insecticide that is more specific or best for aphid control might be referred to as an "aphicide". There are a few insecticides that are best for destroying insect eggs and these are referred to as "ovicides".

Some pesticides can be used for several purposes. For example, some insecticides can also be nematicides and at least one is an insecticide and also a bird repellent. There are several Restricted Use pesticides that will kill all life forms, and as mentioned above, need to be used with extreme care and caution.

WHO USES PESTICIDES

During the early years of pesticide development, farmers were considered to be the primary users. However, as new chemicals were produced and new methods of formulation were developed and new application techniques discovered, new audiences found uses for pesticides.

Today, pesticides are still a major part of agriculture's production tools, but have also found uses by industry, state and federal governments, municipalities, commercial pesticide applicators, and the public as a whole, including homeowners and back-yard gardeners. The percent of use by agriculture has declined to approximately 50% of the total pesticide production. Purchases of pesticides by those who live in cities and towns is estimated to be approximately 20% of all pesticides sold in the United States. While it is true that agriculture as an industry is still the largest single user of pesticides, we can no longer automatically charge the agricultural segment as being responsible for all of the problems these chemicals can create when used unnecessarily or irresponsibly.

There are approximately 1.5 billion pounds of pesticides used in the United States annually. If this usage were equally distributed to all of the people in the United States, it would amount to almost 7 pounds of pesticides for every man, woman, and child in this country!

2

INSECTS

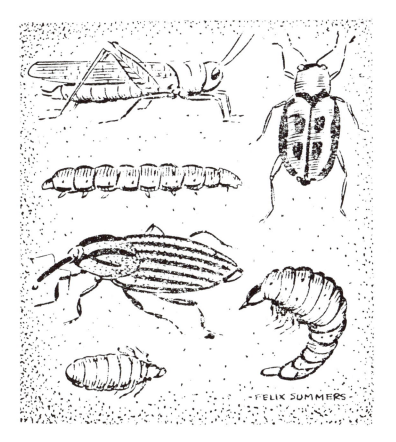

Insects

Man's success in a hostile environment is determined by his ability to adapt or to change his surroundings to his benefit. An area in which man does not exist has no pests; "pest" is a man-made concept and is generally considered to include those organisms which come into conflict with him -- compete with him for his crops and livestock, affect his health or comfort and destroy his property.

Many species of insects are important pests which affect almost all of man's activities.

INSECTS AS PESTS

There are well over one million known species of insects in the world. A very small percentage of these are considered as economically important pests. However, this relatively small number of insect species causes an estimated annual loss of four billion dollars in the United States.

They damage crops by attacking the seed, cutting off young plants, chewing foliage, sucking sap, boring and tunneling in stems and branches and transmitting diseases. After the crop is harvested and stored it is subject to damage by stored-product insects.

What is an Insect?

Insects are spineless; that is to say, they are invertebrates and belong to a group of organisms called the Arthropoda.

These animals are characterized by having a segmented exoskeleton and jointed appendages. Examples of arthropods other than insects are the spiders, ticks and mites; the lobsters, crayfish and crabs; the centipedes and millipedes. These are classes within the larger group, or Arthropoda. Therefore, insects are:

> Kingdom - Animal
> Phylum - Arthropoda
> Class - Insecta or Hexopoda

The class, Insecta, is characterized as follows:

1. The segmented exoskeleton is divided into three body regions, the head, thorax, and abdomen.

2. Three pair of legs on the thorax.

3. May have one or two pair of wings. (Some have no wings.)

4. One pair of antennae

BASIC ANATOMY

It is necessary to know how an insect is "put together" in order to describe it accurately or in order to understand the way in which insects are identified. Proper identification of the pest is basic to control.

It has already been stated that the segmented exoskeleton of an insect is divided into three parts -- the head, thorax and abdomen.

On the head are found the eyes (usually compound eyes), the antennae and the mouth parts. The head is like a hollow capsule formed by the fusion of a number of segments into plates called sclerites. The characteristics of the mouth parts and antennae are used in insect identification.

The thorax is made up of three ring-like segments, the prothorax, meso-thorax, and metathorax. There is a pair of legs on each thoracic segment. When wings are present they arise from the mesothorax and the metathorax. In some insects part of the prothorax extends back dorsally over the other thoracic segments and is called a pronotum.

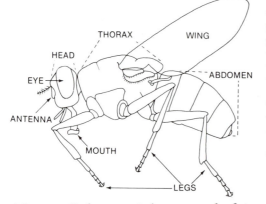

The thoracic segments are joined by flexible membranes which permit movement. Located on each side of the thorax in the membranes are two openings, or spiracles, through which the insect respires.

The characteristics of the legs, wings, and thorax are used a great deal in identification.

The abdomen consists of up to 11 ring-like segments. In most in-sects there are fewer than 11 dis-tinguishable segments in the abdomen. Each segment is composed of two plates or sclerites, one above, or dorsal, and one below, or ventral. The plates and the segments are joined by membranes which permit movement.

There is usually one spiracle on each side of each segment. There are fre-quently appendages on the end of the abdomen.

If we put this all together we have a generalized type of insect. The various modifications in this basic structure found in different insects are used in classification.

Digestive System

Insects have a highly developed digestive system. Starting with the mouth the food passes into the pharynx through the esophagus into a crop. From the crop it moves to a stomach, called the proventriculus, then into a mid-gut, an ileum, a colon, a rectum and the wastes are discharged from the anus.

This is a generalized form and there are many highly specialized types of digestive tracts found in insects.

The Circulatory System

Insects have a open blood system which functions to transport food and wastes. It is not involved in the movement of oxygen as is the blood system of mammals.

There is only one blood vessel, the dorsal aorta and heart, which is a tube-like structure extending through the thorax and abdomen. The blood is pumped forward and out of the front end of the tube. It then flows back through the body cavity and appendages and reenters the tube through slits from where it is again pumped forward.

The Nervous System

There is a nerve cord on the ventral side of the body of the insects instead of along the dorsal or "back", in the case of vetebrate animals. This nervous system starts with a "brain" in the head and a series of ganglia throughout the length of the body connected by the ventral nerve cord. Nerve cells and the branching processes from them reach the muscles, glands and organs.

Respiratory System

Insects have a very unique system to obtain oxygen. It might be compared to a ventilation system using inlets and ducts.

Along the sides of the thorax and abdomen are openings, spiracles, through which air is taken in and CO_2 is discharged.

The spiracles connect to a complex system of tubes called trachae. The trachae divide and branch and finally end in tiny tubes called tracheoles which carry the oxygen directly to the tissues. They also pick up carbon dioxide from the tissues and carry it to the tracheae and then out the spiracles.

Some of the spiracles of some insects are equipped with valves which open or close.

The Mouthparts

An understanding of the mouthparts and feeding habits of insects is important in order to select effective control measures and as an aid in identification. There are two major types of mouthparts -- chewing and sucking.

Chewing

Sucking

Chewing mouthparts are generally composed of a labrum (upper lip), a pair of cutting or crushing or "pinching" mandibles, a pair of maxillae, a labium (lower lip) and a tongue-like hypopharynx. The mandibles and maxillae, or "jaws", work sideways and are used to cut off and chew or grind solid food. A typical example is the type of mouthparts found in a grasshopper or cricket. In some forms of insects, mainly predators, the mandibles are long and sickle shaped. In others, such as honey bees, the hypopharynx, or tongue, is greatly modified.

Insects with chewing mouthparts include the adult Orthoptera (grasshoppers and crickets), Odonata (dragon flies and damsel flies), Neuroptera (lace wing flies), Coleoptera (beetles) and Hymenoptera (bees, ants, wasps) as well as the larvae of several orders.

Damage from Chewing Insects

1. Defoliators
   Those pests which chew portions of leaves or stems, stripping or chewing the foliage of plants. Examples:
       Leaf beetles
       Caterpillars, cutworms, grasshoppers
       Flea beetles

2. Borers
   Those pests with chewing mouthparts which bore into stems, tubers, fruit trees, etc. Examples:
       Corn borer
       White grub - potatoes
       Granary weevil - grains
       Codling moth injury

3. Leaf Miners
   Those pests that bore into and then tunnel in between epidermal layers. Examples:
       Blotch leaf miners
       Serpentine miners

4. Root Feeders
   Those pests that feed on and damage the roots and underground portions of the plant. Examples:
       Seed and root maggots
       Corn root worm
       Wireworms

Sucking mouthparts are those in which the parts described above are highly modified into some form of organ for securing liquid food. They may be piercing-sucking as in the mosquitoes, true bugs, aphids and stable flies; lapping or sponging, as in the house fly; rasping-sucking, as in the thrips; or tube-like, as in the moths and butterflies. Adults of the following orders have sucking mouthparts: Thysanoptera (thrips), Hemiptera (bugs), Homoptera (aphids, scale insects, leafhoppers), Diptera (flies and mosquitoes), Siphonaptera (fleas), Anoplura (sucking lice) and Lepidoptera (moths and butterflies).

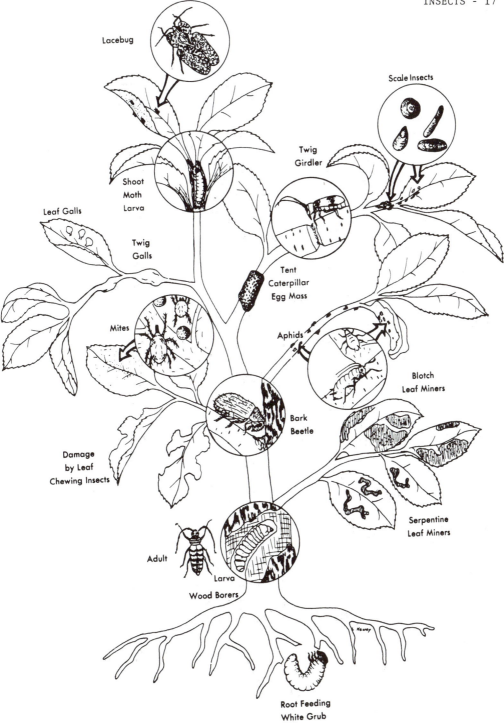

Lacebug

Scale Insects

Twig
Girdler

Shoot
Moth
Larva

Leaf Galls

Twig
Galls

Tent
Caterpillar
Egg Mass

Mites

Aphids

Blotch
Leaf Miners

Bark
Beetle

Damage
by Leaf
Chewing Insects

Serpentine
Leaf Miners

Adult

Larva

Wood Borers

Root Feeding
White Grub

Ways In Which Insects Affect Plants

Damage From Piercing and Sucking Insects

1. Distorting plant growth
   Those pests that cause leaves or fruit or stems to wilt, curl or become distorted. Examples:
     Aphid injury
     Cat-facing of peaches from Lygus bugs
     Cone gall on spruce

2. Causing stippling effect to leaves
   Those pests who may leave many small discolored spots on the leaves which eventually turn yellow. Example:
     Spider mite injury

3. Causing burn on leaves
   Those pests that secrete toxic secretions in the host tissue, causing foliage to appear burned. Examples:
     Leafhopper injury "hopper burn"
     Greenbug injury - sorghum

DEVELOPMENT AND METAMORPHOSIS

Most insects change from the time they hatch from eggs until they are full grown. This change in form is called metamorphosis. It may be a rather gradual change, involving little more than an increase in size, to a very dramatic difference between the young and the old.

There are several ways of characterizing the types of metamorphosis but the most generally used method is to divide them into gradual, or incomplete, and complete.

In gradual types of metamorphosis the insects which hatch from the eggs are called nymphs. As they feed and grow they shed their skins, or molt.

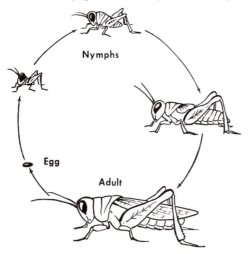

Nymphs

Egg

Adult

In the winged species wings first appear as pad-like buds on the nymphs. Each stage between molts is referred to as an instar. There is no prolonged "resting" period before the adult stage is reached. The common orders of insects which have incomplete or simple metamorphosis are: Odonata (dragonflies), Orthoptera (grass-hoppers, crickets), Isoptera (termites), Mallophaga (chewing lice, Anoplura (sucking lice), Thysanoptera (thrips), Hemiptera (bugs), and Homoptera (aphids, leafhoppers).

Incomplete metamorphosis is also referred to as "simple" metamorphosis as opposed to the more com-plicated complete metamorphosis.

Complete metamorphosis involves a very major change in form between the young and adult. In the winged forms the wings develop internally instead of externally.

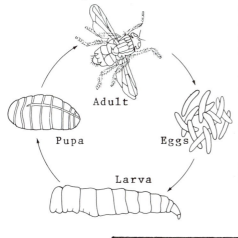

Adult

Pupa   Eggs

Larva

The typical development involves the egg, larva, pupa and adult. The larvae may go through a number of instars and molts as they grow. The pupae may take several forms; they may be exposed, or contained in a capsule like <u>puparium</u> or in a silken cocoon.

Some orders having this type of development are: Neuroptera (lace wings), Coleoptera (beetles), Lepidoptera (moths and butterflies), Diptera (flies and mosquitoes), Siphonaptera (fleas) and Hymenoptera (bees, ants, and wasps).

CLASSIFICATION AND IDENTIFICATION

Proper identification of the pest is basic to its control. There are various aids available to help us do this -- pictures, written descriptions and keys. Keys are the best means of identifying organisms of any kind. There are pictorial, field and "couplet" keys. The following is an example of an analytical couplet key to the major orders of insects. To use such a key the specimen is examined for the characters described after number 1. A choice must be made and this results in a referral to the next couplet, either 2 or 15 in this case. The process is continued until the name of the order is reached.

KEY TO SELECTED ORDERS OF ADULT INSECTS

1. Wings absent 2
    Wings present 15

2. Sedentary insects, legs usually absent; body scale-like or grub-
    like and covered with a waxy secretion. (Females of many Coccidae)
                                             HOMOPTERA
    Not sedentary; with distinct head and jointed legs.................. 3

3. Mouthparts adapted for biting and chewing 4
    Mouthparts not adapted for biting and chewing; piercing-sucking,
    rasping-sucking or siphoning type 9

4. Body covered with scales or mouthparts retracted within the head
    so that only their apices are visible 5
    Body not clothed with scales; mouthparts fully exposed 6

5. Abdomen composed of not more than 6 segments; usually provided
   with a spring near the caudal end; first abdominal segment with
   a forked appendage (sucker) on ventral surface ....
                                                  COLLEMBOLA
   Abdomen composed of 10 or 11 segments; the abdomen may terminate
   in long caudal appendages or immovable forecepts-like appendages
   and the body may be covered with scales .... THYSANURA

6. Small louse-like insects, usually markedly flattened, soft or
   leathery in appearance .... MALLOPHAGA
   Not louse-like in form; not markedly flattened; exoskeleton
   well differentiated 7

7. Adomen sharply constricted at base .... HYMENOPTERA
   Abdomen not sharply constricted at base, broadly joined to the
   thorax 8

8. Abdomen with cerci, body not ant-like in form... ORTHOPTERA
   Abdomen without cerci, body ant-like in form.... ISOPTERA

9. Mouthparts consisting of a proboscis coiled up beneath the
   head; body more or less covered with long hairs or scales
                                            .... LEPIDOPTERA
   Mouthparts not as described above; body not covered with
   long hairs or scales 10

10. Body strongly compressed; legs nearly always long and
    fitted for jumping .... SIPHONAPTERA
    Body not strongly compressed; legs fitted for running 11

11. Mouthparts consisting of a jointed beak (the labium) with
    which are the piercing stylets 12
    Mouthparts consisting of an unjointed flesh or horny beak
    or the beak may be absent 13

12. Beak arising from the anterior end of the head.. HEMIPTERA
    Beak arising from the posterior end of the head, apparently
    between the coxae of the front legs............. HOMOPTERA

13. Tarsi with the apical joints terminating in bladder-like
    enlargements; well defined claws absent; mouthparts forming
    a triangular or cone shaped, unjointed beak .... THYSANOPTERA
    Tarsi not as described above; with well developed claws 14

14. Antennae hidden in pits, not visible in dorsal view; tarsi
    with 2 claws .... DIPTERA
    Antennae exposed, not hidden in pits, visible in dorsal
    view, tarsi with 1 claw .... ANOPLURA

15. Two wings present, hind legs represented by halteres
                                              .. DIPTERA
    Four wings present 16

16. Fore wings and hind legs similar in texture, usually membranous 17
    Fore wings and hind wings dissimilar in texture, fore wings thickened, leathery or horny, hind wings membranous.......... 22

17. Wings entirely or for the most part covered with scales; mouthparts consist of a coiled tube beneath the head, formed for siphoning .... LEPIDOPTERA
    Wings not clothed with scales; mouthpart not as described above.... 18

18. Wings long and narrow with only one or two veins or none; last joint of tarsus bladder-like .... THYSANOPTERA
    Wings not as described above; last joint of tarsus not bladder-like 19

19. Mouthparts enclosed in a jointed beak (labium) and fitted for piercing and sucking; beak arises from the rear of the head apparently between the front coxae .... HOMOPTERA
    Mouthparts not enclosed in a jointed beak; not arising from the rear of the head 20

20. Wings with few longitudinal veins, not net-veined or they may be veinless .... HYMENOPTERA
    Wings with many longitudinal veins and cross-veins, appearing net-veined (12 or more cross-veins) 21

21. Antennae short, setiform or setaceous .... ODONATA
    Antennae longer, not setiform or setaceous ...... NEUROPTERA

22. Mouthparts adapted for chewing and biting 23
    Mouthparts adapted for piercing and sucking 24

23. Front wings horny or leathery, lacking veins (elytra).. COLEOPTERA
    Front wings parchment-like with a net-work of veins (tegmina) .... ORTHOPTERA

24. Fore wings thickened at base, membranous and generally overlapping at the tips (hemelytra); beak-like mouthparts arise from the anterior end of the head .... HEMIPTERA
    Front wings not thickened at base, uniform in texture, not overlapping at tips; beak-like mouthparts arise from the posterior end of the head apparently between the fore coxae .... HOMOPTERA

Pictured on the following pages are insects that belong to the more common orders.

To make it easier to understand and remember the orders of insects, their common names are given along with the order they belong to. Identifiable characteristics of the adult insects are also listed.

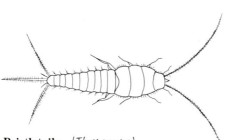

Bristletails (*Thysanura*)

- No wings
- Usually two or three long tails
- Simple metamorphosis
- Often found in houses
- Feed on paper, cloth, and starches
- Are sometimes called "silverfish"

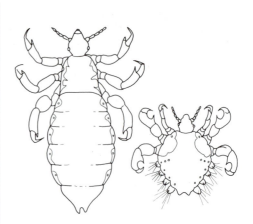

Sucking lice (*Anoplura*)

- No wings
- Piercing-sucking mouth parts
- Simple metamorphosis
- Feed by sucking blood of animals and people

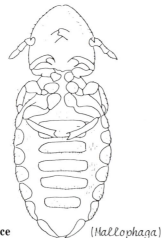

Chewing lice (*Mallophaga*)

- No wings
- Chewing mouth parts
- Broad head
- Simple metamorphosis
- Feed on birds, where they are found on skin and feathers

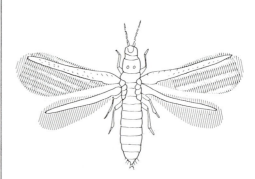

Thrips (*Thysanoptera*)

- May have fringed wings, may lack wings
- Have a combination of sucking and chewing mouth parts
- Simple metamorphosis
- Usually feed in flowers or buds of plants
- Cause misshaped flowers, buds, fruits, and leaves

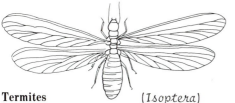

Termites (*Isoptera*)

- Four wings of equal size, or no wings at all
- Chewing mouth parts
- Simple metamorphosis
- Feed on wood and wood products
- Live in complex societies

Grasshoppers, crickets, cockroaches, and katydids (*Orthoptera*)

- Two pair of wings, or no wings at all
- Chewing mouth parts
- Simple metamorphosis
- Pests of agriculture and the home

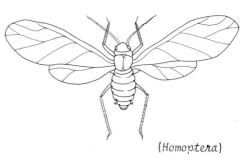

(*Homoptera*)

Aphids, leafhoppers, and scale insects

- May or may not have wings
- Piercing-sucking mouth parts
- Simple metamorphosis
- Suck juices from plants and carry plant diseases

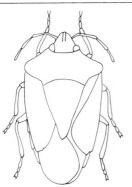

True bugs (*Hemiptera*)
- Two pair of wings, or no wings at all
- Top pair of wings partly leathery, partly transparent
- Piercing-sucking mouth parts
- Simple metamorphosis
- Suck juices from plants and blood from animals and people

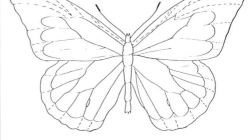

Moths and butterflies (*Lepidoptera*)
- Two pair of wings, usually with scales
- Chewing mouth parts (larvae), sucking mouth parts (adults)
- Complete metamorphosis (larvae called "caterpillars" or "worms")
- Feed on many crops, damaging leaves, stems, tubers, and fruits

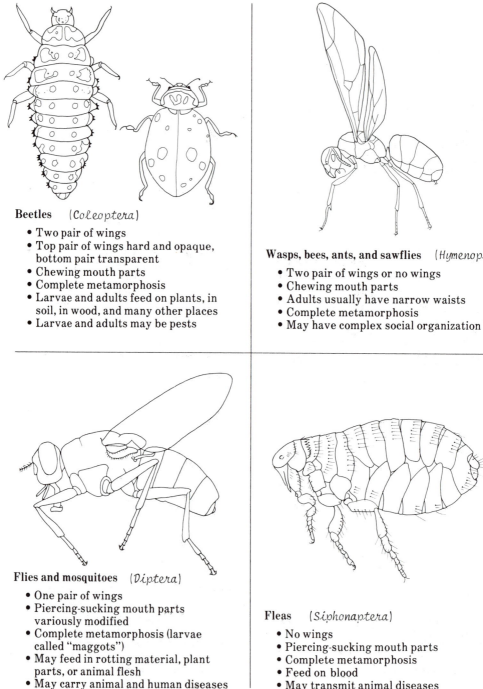

Beetles *(Coleoptera)*

- Two pair of wings
- Top pair of wings hard and opaque, bottom pair transparent
- Chewing mouth parts
- Complete metamorphosis
- Larvae and adults feed on plants, in soil, in wood, and many other places
- Larvae and adults may be pests

Wasps, bees, ants, and sawflies *(Hymenoptera)*

- Two pair of wings or no wings
- Chewing mouth parts
- Adults usually have narrow waists
- Complete metamorphosis
- May have complex social organization

Flies and mosquitoes *(Diptera)*

- One pair of wings
- Piercing-sucking mouth parts variously modified
- Complete metamorphosis (larvae called "maggots")
- May feed in rotting material, plant parts, or animal flesh
- May carry animal and human diseases

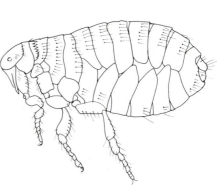

Fleas *(Siphonaptera)*

- No wings
- Piercing-sucking mouth parts
- Complete metamorphosis
- Feed on blood
- May transmit animal diseases

INSECT RELATIVES

Mites, ticks, and spiders are all related to the insects. The spiders are perhaps the best-known members of this class of arthropods -- the Arachnida. Although they bear superficial similarities, the arachanids are quite unlike insects. They have eight jointed legs; two major body regions rather than three; no antennae; simple, small eyes; and they never have wings.

Chemical control measures include the use of miticides and acaricides.

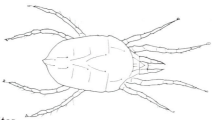

Mites

- Usually very small
- Sucking mouth parts
- Damage plants through feeding; attack animals and people by irritating skin
- May transmit plant and animal diseases

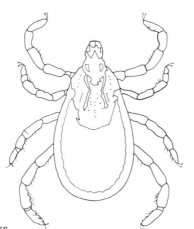

Ticks

- Leathery soft body, without distinct head
- Sucking mouth parts
- Parasitic on animals and people
- May transmit animal and human diseases

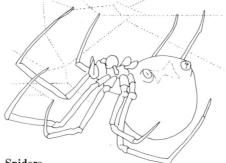

Spiders

- Fang-like sucking mouth parts
- Most are beneficial; a very few are poisonous

SNAILS AND SLUGS

Snails and slugs are included in this section because they frequent many of the same places as insects and entomologists are most often called upon for information about them and for control recommendations.

Snails and slugs are members of the large group of animals called molluscs. They often become pests around the home garden, in lawns, greenhouses, and in ornamental plantings. These animals may damage plants by feeding on the foliage, by their presence in the plant after harvest, and by the slime trails they make as they move from place to place on their muscular "foot".

Snails are soft animals whose bodies are protected by a characteristic coiled shell. Because of the shell, snails are able to survive in fairly dry conditions. They generally reproduce by laying eggs. Control measures include the use of chemicals called molluscicides.

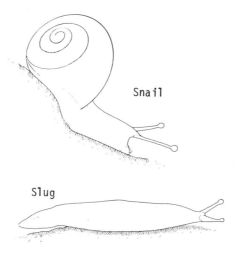

Snail

Slug

WHY INSECT OUTBREAKS?

Insect populations which are successful have two main factors going for them -- adaptability and reproductive capacity -- which permit them to build up to large numbers. Working against this force are a number of hazards or limiting factors in the environment. When the effect of limiting factors is low outbreaks can result. Man has had a direct influence on the environment which may favor insects on the one hand or work against them on the other. For example, the development of large acreages of relatively few crops -- a monoculture -- has resulted in the provision of unlimited food supplies for certain insects.

The reproductive ability of most insects is difficult to imagine. Large numbers of eggs are usually produced and with a few exceptions the life cycles are relatively short. Even with an insect which has an annual life cycle hundred-fold increases may occur in one season.

Insects have an amazing ability to adapt to changing conditions. This even includes those factors which man introduces into the environment so we have the phenomenon of resistance to insecticides.

This all adds up to the fact that economically important numbers of certain insects can develop in a relatively short time and special techniques for their control must be developed.

STEPS TO INSECT CONTROL

An orderly process of decision making must be used intelligently to effectively plan and carry out an insect control operation. The following steps are those which would be followed in handling a new or unfamiliar insect problem. For well known insects some of the steps would be passed over.

1. Detection

    It is necesary to be watchful for those insects likely to be troublesome to a certain crop. It is important to detect infestations early before irreparable damage is done. In most cases this cannot be done through the windshield. It requires careful examination in the field. Areas threatened by new insects moving in from adjacent areas may require special detection methods.

2. Identification

    A pest must be identified accurately before it can be controlled. Just because a certain practice worked on an insect that "looked something like it" is no reason it will be effective on this one. If local authorities are unable to identify the problem, assistance is available at the land grant universities or the United States Department of Agriculture.

3. Biology and Habits

   Knowledge of the seasonal cycles for the specific locality is important in order to pinpoint the most effective time of treatment. The principle is to determine the vulnerable stages in the life cycle against which to direct the control effort to obtain the most effective economical control.

4. Economic Significance

   Control efforts are not undertaken just for the sheer enjoyment of killing insects. There must be a return on the investment. For some insect pests we have sufficient knowledge so that we can offer guidelines and survey methods for determining population levels required to make a given treatment pay. More studies are needed to obtain this information for more pests.

5. Selection of Methods

   After the pest problem has been identified, the life history and seasonal cycle understood and the economic significance has been established the proper method or combination of methods can be selected to do the most effective, practical, economical and safe job of control.

6. Application

   The control methods selected must be applied properly in order to be effective and safe. If the method involves the use of chemicals the application must be done at the proper time, in the best place, at the right rate and using suitable equipment. This means using the proper sprayer or duster or other equipment for the formulation used and the type of coverage or delivery desired.

   This information is specified in the directions for use on pesticide labels and in written recommendations.

7. Evaluation

   It is extremely important to check the field or otherwise evaluate the results of the control operation. This can be done in several ways such as insect counts before and after treatment, comparative damage ratings, yields, etc. In most cases, it is difficult if not impossible to do an adequate evaluation without untreated checks to use as a basis for comparison. The results obtained should be recorded for future reference.

8. Recording

   Records provide the basis for gaining from past experiences. It is especially important to have all chemical applications recorded.

METHODS OF INSECT CONTROL

1. Cultural
   a. crop rotations
   b. tillage methods
   c. resistant or tolerant varieties

2. Mechanical
   a. screens, traps
   b. light and sound

3. Biological
   a. parasites
   b. predators
   c. diseases
   d. male sterile technique
   e. temperature
   f. moisture

4. Legal
   a. inspections and quarantines
   b. plant pest act

5. Chemicals
   a. kill
   b. repel
   c. attract
   d. sterilize

6. Integrated control
   Combination of all possible compatible techniques to manage insect populations at sub-economic levels, usually involving the use of artificial or natural biological methods along with selective chemicals.

The real challenge in pest control today lies in making decisions for each situation and fitting the available techniques into a program which results in the maintenance of sub-economic levels of pest species and which cause the least amount of insult to the environment.

USE OF CHEMICALS IN INSECT CONTROL

The development of effective, economical pesticides has had a profound effect on man's continual battle with insects. In many cases, chemicals have been incorporated as tools in well planned insect control programs without serious hazards to humans or to the environment. However, there has been a tendency for some people to regard these tools as cure-alls and to believe that for every insect problem there is a chemical which will magically solve it. This has resulted at times in some unwise, uneconomical or hazardous applications of chemicals.

It should be the responsibility of every user to learn as much as possible about the insect to be controlled, the chemicals to be used and the potential hazards involved and then to apply pesticides in such a way that hazards are minimized or avoided and so that an effective, economical job of insect control results.

INSECTICIDE CLASSIFICATION

Insecticides may be classified in various ways -- by chemical composition, by formulation or by the way they kill the insects. Before the synthetic organic insecticides were developed the classification was usually made on the basis of whether the chemical killed as a "stomach" poison or as a "contact" poison. Since many of the modern insecticides act in both ways this method of classification is outmoded. Therefore, most classifications are now found in the literature by chemical composition. The following does not list all of the insecticides, but is given as an example of chemical classification.

Chemical Groups

1. Inorganics
    a. Lead arsenate
    b. Calcium arsenate
    c. Paris green
    d. Sodium Fluoaluminate or cryolite
    e. Mercurous chloride or calomel
    f. Boric acid
    g. Sulfur

2. Organics
    a. Botanicals
        pyrethrum
        rotenone
        ryania
        sabodilla
        nicotine

    b. Chlorinated hydrocarbons
        aldrin   endrin
        chlordane   methoxychlor
        toxaphene   etc.
        lindane

    c. Organic phosphates
        Malathion   demeton (SYSTOX)
        parathion   azinphosmethyl (GUTHION)
        diazinon   etc.
        phorate (THIMET)

    d. Carbamates
        carbaryl (SEVIN)
        carbofuran (FURADAN)
        methomyl (LANNATE) (NUDRIN)
        aldicarb (TEMIK)
        propoxur (BAYGON)
        etc.

    e. Biologicals
        Bacillus thuringiencis spores
        milky white disease spores

    f. Miscellaneous
        dinitros
        petroleum oils

3
PLANT
DISEASE
AGENTS

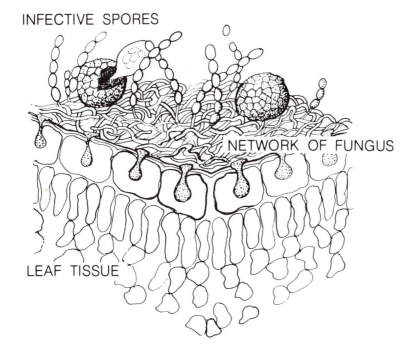

INFECTIVE SPORES

NETWORK OF FUNGUS

LEAF TISSUE

CROSS SECTION OF LEAF WITH POWDERY MILDEW

Plant Disease Agents

The study of plant diseases is a challenging biological science. It pits the energy of a few against more than 50,000 destructive plant diseases. Since man depends on plants for food and other products, plant disease control is essential to feeding the people of the world. By studying microorganisms, host plants, and their relationship to each other, plant pathologists aim at controlling diseases and thereby enhancing the quantity and quality of everyone's food.

ADULT EUROPEAN ELM BARK BEETLE SPREADS FUNGUS SPORES INTO FEEDING SCARS ON ELM TWIGS

FUNGUS SPREADS THROUGH WATER CONDUCTING SYSTEM AND KILLS TREE

EMERGING BEETLE PICKS UP FUNGAL SPORES

BARK BEETLE EGGS LAID IN DEAD OR DYING ELM WOOD

DUTCH ELM DISEASE — DISEASE CYCLE

The ancient Romans worshipped Robigus, a god named for the rust disease of grain. Indeed, people throughout history have feared the fantastic potential plant diseases have for destroying food and other essential crops. The Irish potato famine of 1843-45, which was caused by a fungus, killed a million people and encouraged a mass migration of Irish citizens to America. Within just 25 years, the chestnut blight fungus from Asia essentially destroyed the magnificent stands of American chestnuts that once extended from Maine to Georgia. Likewise, millions of elm trees have been destroyed by the Dutch elm disease fungus that moved across the United States from the East and into the Rocky Mountains. In 1970, the corn blight fungus ravished corn fields throughout our country and threatened the entire corn industry.

Plant diseases are caused by microscopic agents known as fungi, bacteria, nematodes, mycoplasmas, viruses and viroids. These agents and nutrient deficiencies produce symptoms we readily recognize: rotted fruits, blighted and cankered trees, spotted flowers, shrunken grain, and yellowed, dying plants of all kinds.

No plant is immune to disease. House plants are as vulnerable as farm crops. Disease loss occurs during the shippage and storage of fruits, vegetables, flowers, and cereals.

Plant disease agents are everywhere. The earth and the air serve them as both conveyance and habitat. They can move around the world on a crumb of soil. They can move in water and on the wind, or hitchhike on man's vehicles.

Despite modern agricultural practices, plant diseases cause annual losses of billions of dollars. The ever-threatening world food shortage coupled with over population in many areas of the world, emphasizes the need for more efficient crop production, one of plant pathology's prime challenges.

BIOLOGICAL ABNORMALITIES

There is a wide variety of disorders affecting plants. These disorders include disease, injury, and teratosis.

Disease may be pathological, caused by a living microorganism or virus; and disease may also be physiological, caused by the continued deficiency of a nutrient such as lack of available iron in the soil.

Injury is similar to disease in some respects, but is caused by the momentary or transient impact of some agent; for example, injury from machinery or tools. Insects may cause injury; likewise, there may be environmental stress such as hail, wind, and drouth.

Teratosis is a word used for various malformations, usually related to autogenetic factors.

THE PLANT DISEASE CONCEPT

Plant Pathology is a well-established science that is based on a number of principles and concepts:

1. Disease consists of deleterious physiological activity caused by continued irritations by an external primary agent (or agents), resulting in disturbed cellular activity and expressed in characteristic pathological changes called symptoms.

2. Any particular disease is caused by one or more primary irritants or causal factors and several secondary factors of different degrees of importance.

3. The causal factor (or factors) of a disease has its own history and life cycle, which may vary in individual instances depending on its environment, and which is of fundamental importance in the initiation and course of the disease.

4. Disease development passes through three essential stages: innoculation, incubation, and infection. To be meaningful, these stages must be correlated with the activities of the pathogen and interaction with the plant environment.

5. The damage resulting from any disease depends on the time of initiation, the degree of disturbance, the part or portion of plant affected, the previous development of the plant, and the extent of secondary involvement.

6. Control of any disease depends on one or more of the fundamental principles of control: exclusion, eradication, protection, and immunization. These principles also must be based on the concept and knowledge of the disease and therefore related to the stages of disease development and the factors which influence these stages.

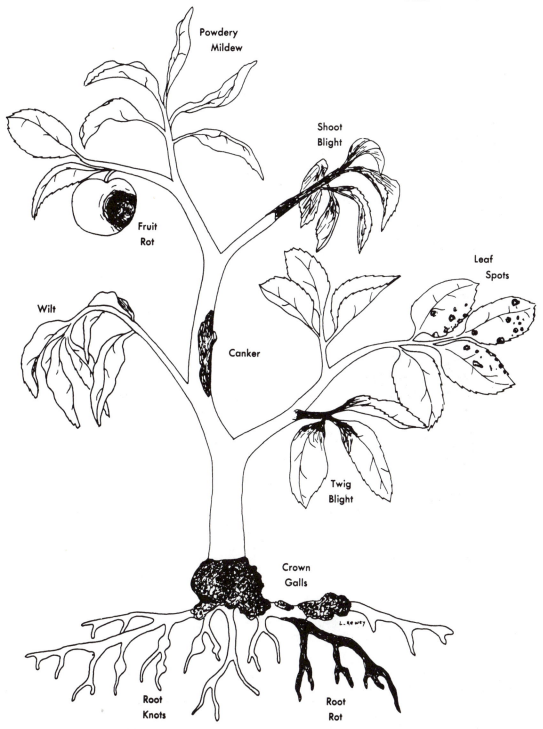

Ways In Which Diseases Affect Plants

PATHOLOGICAL DISEASE

The age-old phenomenon of parasitism occurred along with the evolution of plants from time immemorial. It is the inherent nature of certain microbes to be parasitic. As a result of parasitic activities, a complex and detrimental process of disease occurs. Causes of disease include fungi, bacteria, viruses, mycoplasmas, nematodes, and a few more developed parasites such as mistletoe.

Fungi are plants that lack the green coloring (chlorophyll) found in seed-producing plants and therefore cannot manufacture their own food. There are more than 100,000 different species of fungi of many types and sizes. Not all are harmful, and many are beneficial to man. Most are microscopic, but some such as the mushrooms, are quite large.

Most fungi reproduce by spores, which vary greatly in size and shape. Some fungi produce more than one kind of spore, but a few fungi have no known spore stage.

FIRE BLIGHT
CANKER

**FIRE BLIGHT SHOWING SYMPTOM DEVELOPMENT
FOLLOWING SPREAD OF BACTERIA
FROM BLOSSOM TO LIMB**

Bacteria are very small, one-celled plants that reproduce by simple fission. They divide into two equal halves, each of which becomes a fully developed bacterium. This type of reproduction may lead to rapid buildup of population under ideal conditions. For example, if a bacterium can divide every thirty minutes -- a generation time not especially short for some bacteria -- in 24 hours a single cell could produce 281,474,956,710,656 offspring.

Viruses are so small that they cannot be seen with the ordinary microscope and are generally detected and studied by their effects on selected "indicator" plants. Many viruses that cause plant disease are transmitted from one plant to another by insects, usually aphids or leafhoppers. Viruses also cause very serious problems in plants that are propagated by bulbs, roots, and cuttings because the virus is easily carried along in the propagating material. Some viruses are easily transmitted mechanically by rubbing leaves of healthy plants with juice from diseased plants. A few viruses are transmitted in pollen. Big vein, a lettuce virus, is transmitted by a soil-borne fungus, and a few viruses are transmitted by nematodes.

Mycoplasmas are a relatively new group of cellular microorganisms, some of which were previously considered viruses, particularly in the yellow virus group.

Nematodes are small eel-shaped worms that reproduce by eggs. The number of eggs produced by one female nematode, and the number of generations in a season depends largely on soil temperature. Therefore nematodes are usually more of a problem in warmer areas of the country. Most nematodes feed on the roots and lower stems of plants, but a few attack the leaves and flowers.

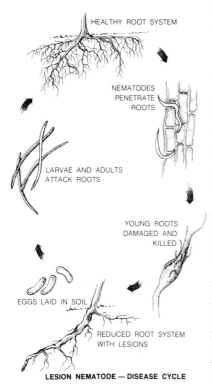

HEALTHY ROOT SYSTEM

NEMATODES
PENETRATE
ROOTS

LARVAE AND ADULTS
ATTACK ROOTS

YOUNG ROOTS
DAMAGED AND
KILLED

EGGS LAID IN SOIL

REDUCED ROOT SYSTEM
WITH LESIONS

LESION NEMATODE — DISEASE CYCLE

Many crops such as strawberries, root crops, bulbs, ornamentals, mint, alfalfa, and potatoes are damaged by nematodes. Stunted or distorted plants may result from invasion by these pests. Symptoms vary from swellings, thickenings, galls, and distortions on above-ground parts to root conditions such as short stubby roots, lesions (dead spots), swellings, galls, and general breakdown. Heavy infestations by the root knot nematode may result in small roots resembling strings of beads. Infested potato tubers may have a bumpy or pebbly, uneven surface, even though the skin is not broken.

Most plant parasitic nematodes range in size from 1/50 to 1/25 of an inch in length. These organisms may occur almost anywhere life exists.

All plant parasitic nematodes possess a spear or stylet, which is usually hollow, by which they puncture plant cells and feed on the cell's contents. Nematodes may develop and feed within plant tissue (endoparasites) or outside (ectoparasites). A complete life cycle involves the egg, four larval stages, and the adult. The larvae usually resemble the adults, except in size. The females of some, such as root knot and cyst nematodes become fixed in the plant tissue and the body becomes swollen and rounded. The root knot nematode deposits its eggs in a mass outside of the body while the cyst nematode retains part of its eggs within the body where they resist unfavorable, environmental conditions and may survive for many years.

HOW PLANT DISEASES DEVELOP

Development of any parasitic disease is critically dependent on the life cycle of the pathogen. The life cycles of all disease organisms are greatly influenced by environmental conditions affecting both the host and the pathogen. The temperature and moisture are probably the most important factors that affect the severity of plant diseases. They not only influence the activities of the disease organism but also affect the ease with which a plant becomes diseased and the way the disease develops.

The life cycle (or life history) of a pathogen begins with the arrival of some portion (fungus spore, nematode egg, bacterial cell, virus particle) at a part of the plant where infection can occur. This step is called innoculation. If environmental conditions are favorable this infectious material (called inoculum) will begin to develop, i.e., spores will germinate, and

bacterial cells begin to multiply. This stage is called <u>incubation</u>.
If the inoculum is on a plant part which it can enter, such as natural
openings or wounds, or can penetrate the plant's surface directly, the
pathogen enters the host plant to begin the stage called <u>infection</u>.
This is the real beginning of disease, but the plant is not yet diseased.
Only when the plant responds to the invasion of the pathogen in some way,
i.e., cells die or multiply abnormally, has disease developed.

The plant generally expresses its response to invasion of a pathogen and
development of disease in the form of <u>symptoms</u>. These are outward expressions of the plant disease and consist of three general types.

1. Over development of tissues -- galls (crown gall, club root,
    western gall rust); witches broom (dwarf mistletoe, big vein
    of lettuce, alfalfa virus); swellings (white pine blister rust);
    leaf curls (peach leaf curl, leaf blister).

2. Under development of tissues -- stunting (many virus diseases);
    chlorosis or lack of chlorophyll (iron deficiency); incomplete
    development of organs.

3. Necrosis (death) of tissue -- blights (fire blight); leaf spots,
    wilting, (Verticillium wilt); decays (soft rot); cankers (anthrac-
    nose, bacterial canker).

In order for a disease to develop, three ingredients must be present:
(1) a susceptible host plant (the suscept).
(2) a disease-producing agent (the pathogen).
(3) an environment favorable to disease development.

Eliminate any of these three ingredients and disease cannot develop. Since
each of these ingredients is variable, the interrelationships are variable
and complex. Any of the three ingredients can influence one of the other
two independent of the third, yet ultimately have an impact on disease
development.

To understand the interrelationships of the three ingredients, look at the
diagrams in Figure 1. There are three circles, each representing one of
the three ingredients. These circles are free to move in any direction.
Only when the three overlap will disease develop. If all conditions were
ideal, the circles will be almost one on top of the other. This would
represent a very severe disease situation or epidemic. If conditions are
average or moderate, only a small portion of each circle would overlap the
others and the disease situation would be mild. This would be an endemic
disease. If one or more of the ingredients is not present under otherwise
favorable circumstances, the circles cannot overlap and no disease would
occur.

HOW DISEASES ARE IDENTIFIED

<u>Symptoms</u> by themselves may not allow an accurate diagnosis of a plant
disease because several distinctly different causal agents may produce

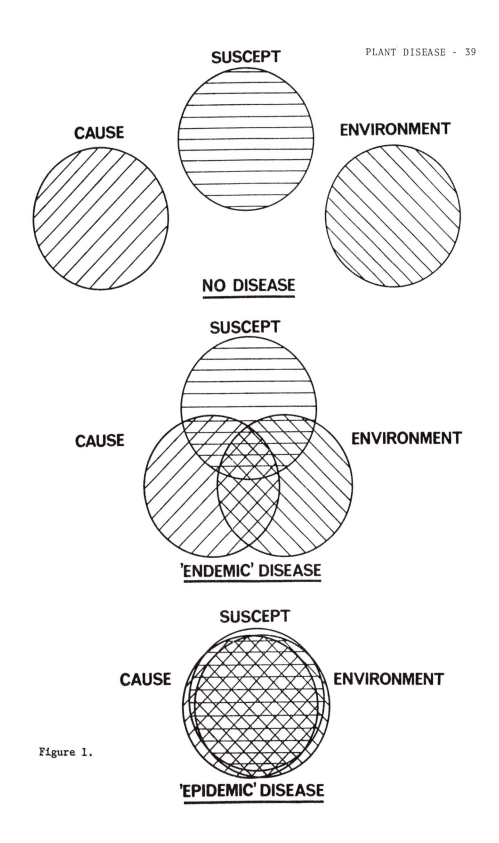

Figure 1.

identical symptoms. However, symptoms used with other evidence, plus experience, can often produce a satisfactory diagnosis. The name of a disease is often derived from the symptoms produced.

Signs are more reliable indicators of the cause of a disease, but their detection and identification require more specialized training and techniques than does the observation of symptoms. Signs are structures produced by the causal organism. They may be fungus spores or other structures, nematodes or their eggs, bacterial ooze or similar material. Usually signs are too small to be seen by the unaided eye and lenses or microscopes of varying types must be used.

The extremely large number of plant diseases (about 50,000 in the U.S.) makes it impossible for any one person to be familiar with very many of them. However, there are certain facts that help in the identification of plant diseases. In a diagnosis of a suspected plant disease there is a logical and convenient series of steps to follow:

a. The first piece of information to obtain is the identity of the plant affected. If possible this should include the scientific name as well as the common name because the same common name may have been used for several distinctive plant species. Even the variety of the diseased plant should be determined whenever possible. Since great variation in susceptibility to plant diseases may exist between different cultivars.

b. Careful examination of the diseased area, whether a bench in a greenhouse or a large field, is a logical second step in determining the cause of a plant problem. Note how the diseased plants are distributed over the affected area. Are they uniformly distributed or localized in certain spots? Definite patterns of distribution such as the edges or the center of a greenhouse bench or along roadways, fences, corners, or low spots in a field, frequently indicate climatic soil factors or toxic chemicals but do not necessarily exclude parasitic causes.

Is only one type of plant affected or are many unrelated plants involved? If more than one kind of plant is affected, parasites are probably not involved. Rather, one should suspect climate or chemicals.

How did the disease develop in the affected area? Did it appear overnight? If so, suspect a climatic factor such as fog, or the application of a toxic chemical. However, if the condition started at one point and spread slowly in extent and severity on a single species, a parasitic disease is probably responsible.

Obtain a record of environmental conditions preceding the appearance of the disease. Check for short periods of below freezing temperature or prolonged drouth periods. Did hail and/or lightning occur? Air pollutants might have been involved. If so, information on air inversions and a prevailing wind will be helpful in explaining the pattern of damage.

Has there been any treatment that could have resulted in plant injury? Soil sterilant type herbicides can burn foliage on trees some distance away because their expensive root systems have intercepted the downward movement of the herbicide. Excessive amounts of fertilizers can cause similar damage to plants.

c. The typical appearance of a diseased plant should be described. The symptom picture should not be based entirely on the early stages of the disease nor on a plant that has deteriorated to a point that secondary organisms have obscured the primary cause of the problem. Ideally, the symptom picture should be a composite based on the progression of symptoms from earliest to latest.

The "typical" diseased plant should always be compared with a healthy or normal plant since normal plant parts are sometimes mistakenly assumed to be evidence of a disease. Examples are the brown-spore producing bodies on the lower surface of leaves of ferns. These are the normal propagative organs of ferns. Also in this category are the small, brown, club-like tips that develop on arborvitae foliage in early spring. These are the male flowers, not deformed shoots. Small galls on the roots of legumes such as beans and peas are most likely nitrogen-fixing nodules essential to normal development, and are not indicative of a typical disease.

Premature dropping of needles by conifers frequently causes alarm. Conifers normally retain their needles for 3-6 years and lose the oldest gradually during each growing season. This normal needle drop is not noticed. However, prolonged drouth or other factors may cause the tree as a whole to take on a yellow color for a short period of time and may accelerate needle loss. If the factors involved are not understood, this often causes alarm. The needles that drop or turn yellow are usually the oldest needles on the tree and the dropping is a defense mechanism which results in reduced water loss from the foliage as a whole.

A portion or parts of a plant affected should be noted. Are they the roots, leaves, stems, flowers, or fruit -- or is the entire plant involved? The root system, which may constitute half of the plant, is often overlooked because it is underground and not visible.

d. What are the primary symptoms of the disorder under study? Symptoms are expressions of the affected plant that indicate something abnormal and are grouped into three general classes:

(1) Under development of tissues or organs.
(2) Over development of tissues or organs.
(3) Necrosis or death of plant parts.

DIAGNOSIS OF A DISEASE UP TO THIS POINT REQUIRES NO SPECIAL EQUIPMENT OTHER THAN A HAND LENS AND KNIFE.

e. Signs of disease are structures produced by the causal agent of
    that disease. Since signs are more specific than symptoms, they
    are much more useful in the accurate diagnosis of a disease. For
    example, distortion of leaves and shoots usually points to damage
    from a hormone type herbicide such as 2,4-D, but similar symptoms
    are sometimes produced by frost, viruses, or some fungi. Discovery
    and identification of signs of disease may require special equip-
    ment and knowledge, but many signs can be found with only a hand
    lens and a knife. The presence of spore-producing bodies on bark
    cankers; by mycelial mats of the root rot fungus, and Armillaria
    mellea; nematodes in alfafla roots; bacterial ooze from fire
    blighted pears; masses of rust spores; and the gray and white
    mycelia of powdery mildew are all signs of different diseases.

f. Sometimes neither symptoms or signs are specific or characteris-
    tic enough to pin down the cause of a disease. Then additional
    specialized techniques are required to isolate and identify the
    causal agent.

    Isolation of bacteria and fungi usually requires that small pieces
    of disease tissue be placed on a nutrient medium. The organisms
    growing out of the tissue are then isolated into a culture. However,
    many disease-producing organisms, especially obligate parasites
    cannot grow on artificial media, but can grow only on living tissue.
    Obligate parasites require still other special methods for their
    isolation. Obligate parasites include rust, powdery mildew, and
    viruses.

    Invasion of diseased tissue by saprophytic organisms often makes
    isolation of the primary cause of disease difficult if not impos-
    sible. Because of their ability to grow rapidly and utilize the
    artificial substrate more efficiently than can many disease
    organisms, these saprophytes may rapidly overgrow and crowd out
    the primary pathogen on synthetic media. Experienced plant patho-
    logists overcome this difficulty by isolating from the margins of
    the diseased tissue and by employing selective media especially
    developed to favor growth of the pathogen.

    Assuming that isolation has been successful and you are reasonably
    certain you have isolated the primary cause of the disease, there
    is still a problem of identifying the organism. There are 20,000
    to 40,000 species of bacteria and fungi, as well as many viruses,
    nematodes, and even higher plants that cause disease. The charac-
    teristics upon which the identification is based are often complex
    and not easy to determine. Frequently only a specialist who deals
    with a small group of organisms can correctly identify the disease
    organisms in question. For example, identification of a parasitic
    nematode is most difficult for anyone except a trained nematologist.

    Viruses are especially difficult to identify because their submicro-
    scopic size makes them too small to be seen except by very special
    and costly equipment. Their identification usually is based on the
    reactions (i.e., symptoms) of selected hosts called indicator plants
    and on their physical and chemical properties. The determination
    of all characters require very special knowledge and techniques.

Our primary interest is usually to control a given disease. Only after positively identifying the disease can the best control measures be employed. Therefore, when all of the previously mentioned information has been successfully collected, we should then consult the available literature to determine what is already known about this disease. This information may be found in books, technical journals, commodity newsletters, experiment station records, or correspondence with other experts on plant diseases. County extension agents and specialists in plant pathology should be consulted.

HOW PLANT DISEASES ARE CONTROLLED

The ultimate concern about a plant disease is to reduce or eliminate the economic or esthetic loss it causes. This is called the control of a a disease. Plant disease control involves one or more of four basic principles. These are called exclusion, eradication, protection, and resistance.

Exclusion involves measures to prevent a disease organism from becoming introduced into and established in an area where it does not occur. Plant quarantines are one means of exclusion.

Eradication is the elimination of a pathogen from an area, usually when it has limited or restricted distribution.

Protection consists of the placement of a protective barrier, usually a chemical between the plant and the pathogen.

Resistance involves the use of plants that are not susceptible to a disease. Immunity is the ultimate degree of resistance and is usually not obtained in genetic programs aimed at developing resistance in a given plant. The level of resistance may vary considerably depending on a large number of factors, such as the age of the host plant, aggressiveness of the pathogen, relative favorability of the environment, etc. Very often, a plant variety or selection that is resistant to disease lacks desirable qualities wanted for commercial purposes.

Many diseases can be controlled by cultural practices alone, which includes exclusion, eradication, and resistance. Such practices include:

    a. Selection of resistant or tolerant varieties of plants.

    b. Proper establishment of plants.

    c. Rotating planting locations.

    d. Maintenance operations, such as raking and destroying fallen leaves, thinning, pruning, and regulating fertilizer and water.

Chemical control of plant diseases involves the principle of protection whereby a protective chemical barrier is placed between the plant and the pathogen.

Chemicals used to control plant diseases are generally called fungicides, but correctly these include only those chemicals that kill or retard the development of fungus pathogens. Those that control bacteria are bactericides and those controlling nematodes are called nematicides.

In general, fungi are much more difficult to control with chemicals than insects or weeds. A fungal pathogen is a plant which is living in or on another plant, its host. Thus, one is trying to kill one plant, without injuring a second plant. Finding chemicals with such selectivity is a time consuming task, which is not always successful. In addition, most of the fungal pathogens have asexual repeating cycles only a few days long, thus the crop plant may be subjected to as many as 10-25 generations of a pathogen during the growing season. This requires repeated application of the fungicide in order to control the pathogens. In addition, many plant pathogens are either below ground or within the plant's interior tissues and hence are not reached by most common fungicides.

In general, application of fungicides differs markedly from the application of herbicides or insecticides. An herbicide placed anywhere on the plant foliage will often kill the plant. Likewise, an insecticide placed at random on the plant is liable to kill the insect because most insects tend to move around from leaf to leaf or branch to branch. However, with a fungicide, only that portion of a plant which is covered by the fungicide is protected. Thus, it is essential that a fungicide give complete coverage of all tissues. A dust may be quite suitable as an insecticide, but because most dusts are deposited only on the upper surfaces of the leaf, they are relatively ineffective as fungicides. Sprays should be applied to both upper and lower surfaces for maximum protection.

Protection of plants with chemicals involves killing the pathogen (or limiting its growth) before it invades the suscept; this may be accomplished by two different means:

(1) Application of the fungicide to the pathogen in place. Such a fungicide would be termed a contact fungicide.
(2) Applicatiion of the fungicide before the pathogen is in place. Such a fungicide would be termed a residual fungicide.

The use of contact fungicides to eliminate potential sources of inoculum or to eradicate inoculum after it has been deposited on the target site is risky and hence rare. In most instances, residual protectant fungicides are used.

A good residual protectant fungicide is one which:

(1) Remains active for a relatively long period of time. Quite obviously, if a fungicide must be applied everyday to maintain the necessary level of protection around the clock, it would be next to impossible to apply and too expensive if it were possible.
(2) Has good adhesive properties. Since the dissemination of most pathogens is favored by rainy weather, the fungicides must resist the erosive action of water.
(3) Has good spreading properties. Since it is necessary to completely protect leaf and stem surfaces, the fungicide should spread evenly over

the surface of the leaf. This is usually accomplished by adding a
wetting agent to reduce surface tension. If too much wetting agent
is added to a formulation, "run-off" may occur which results in lower
concentrations of fungicide than required. Wetting agents also facilitate penetration of fungicides into leaves or stems and if too much
is used, phytotoxicity may occur.

(4) Is stable against photodeactivation.

(5) Is toxic to plant pathogenic microorganisms, but non-toxic to the plant
and non-target organisms.

(6) Is active against a wide range of pathogenic microorganisms. Actually,
most of the presently available fungicides are rather specific. Thus,
a particular fungicide must be selected for control of a particular
disease. Frequently, spray programs are devised involving the use of
more than one fungicide, either as "combination" sprays or in alternate
applications. Formulations of two or more fungicides in a single
package are marketed for use in the home and garden, but the commercial
producer must prepare his own combination sprays as tank mixes.

(7) Is compatible with other pesticides, such as insecticides and other
fungicides.

(8) Is relatively easy to apply and does not present an undue hazard to the
applicator or to the environment.

(9) Is non-corrosive to the application equipment.

TYPES OF PROTECTANT FUNGICIDES

Fungicides can be classified either as organic or inorganic; they can be
further broken down by the heavy metal element(s) involved in inorganic
fungicides or by the basic class of chemical compounds in organic fungicides.

Inorganic fungicides -- some of the inorganic pesticides date back into
man's pre-history. The use of these compounds has continued until relatively recently. Many of the heavy metal compounds have now been banned
for use because of their persistence in or on the treated plants themselves,
leading to pesticide poisoning of man and animals, or because of the buildup of toxic residues in the soil.

(1) Sulfur -- probably the oldest pesticides known to man are various
sulfur compounds. These were used either as dusts or mixed with fats
and used as ointments to treat various diseases of man. It was probably
sulfur dust that was used by the early Greeks and Romans to control
certain insects and diseases on crops. Other formulations have been
developed since those times.

Elemental sulfur kills some insects, bacteria and fungi by direct
contact. Sulfur is still used to control powdery mildew because the
mycelium of these pathogens is superficial on the leaf surface.
Sulfur acts by interfering with electron transport in the cytochrome
system.

(2) Copper -- The use of copper goes back to the early 1800's when it was
discovered that copper inhibited germination of the spores of covered

smut of wheat. In 1887, Millardet discovered Bordeaux mixture as a control for downy mildew of grape. It was the first successful and widely used fungicide. It is still chemically undefined and no one is really sure of the active ingredients. It is formed by a mixture of copper sulfate and hydrated lime.

In general, copper compounds are relatively insoluble in water and are not easily washed from the leaves by rain. A very small portion of the copper goes into solution. This is absorbed by living cells and then more copper goes into solution to replace that which has been removed. The living cells continue to absorb the toxic ions and eventually accumulate enough so that they are poisoned. Because of this factor, many copper compounds are fairly phytotoxic.

(3) Mercury, Cadmium and Other Heavy Metals -- Some of the most toxic fungicides are those containing inorganic mercury, cadmium or certain other heavy metals. However, these compounds are also very toxic to other forms of life, particularly to warm-blooded animals. Thus, mercury and other heavy metal residues are not permitted in food or feeds and most of these compounds have been banned from use.

Organic fungicides -- Synthesized, man-made compounds.

(1) Dithiocarbamates -- A whole group of compounds belong to the dithio-carbamates, most of which were developed in the 1930's and 1940's and are some of the oldest organic fungicides known. These include Polyram®, Manzate®, Dithane® and Fermate®.

(2) Dicarboximides -- The first of this group of fungicides, captan, appeared in 1949. Two others, folpet and captafol, appeared in the early 1960's. They are widely used as protective sprays or dusts for fruits, vege-tables, ornamentals and turf, and as seed treatments. They probably act by an inhibition of synthesis of proteins and enzymes containing sulf-hydryl groups.

(3) Oxathiins -- Two of these compounds, carboxin and oxycarboxin, were in-troduced in 1966 as systemic fungicides. They are selectively toxic to some of the smut and rust fungi and to Rhizoctonia. They apparently act by inhibition of succinic dehydrogenase, a respiratory enzyme im-portant in the mitochondrial systems.

(4) Benzimidazoles -- These compounds were first introduced in the late 1960's and have been used as systemic fungicides against a great many different types of diseases. One of the most widely used of these is benomyl, which can be used to control a great many fruit, vegetable, ornamental, and turf diseases. Two others are thiabendazole and thio-phenate. These compounds have been used as foliar fungicides, seed treatments, soil drenches and as dips for fruit or roots. They appar-ently act by interfering with the synthesis of DNA, affecting spore germination and growth.

Systemic fungicides -- Are absorbed by the plant through the leaf, stem, or root surface and translocated varying distances within the plant. This distance may be as small as from one leaf surface to the other or as far as from the roots to the shoot apex. The advantages of systemic fungicides are:

a. The plant can be continuously protected throughout the growing season without repeated applications of fungicides.

b. The systemic may be taken up by the roots and transported to newly formed tissues.

c. The systemic is not subjected to weathering as are fungicides applied to the foliage.

d. Unsightly residues on flowers and foliage can be avoided.

e. The systemic may provide a means of controlling and eradicating vascular wilt diseases as well as other internal disorders of plants.

f. Since the toxicant is in the plant, there is minimal toxic effect to people working in the greenhouse or with the crop during the growing season.

The disadvantages of systemics are:

a. Resistance to many of these compounds is becoming quite common. This is because many of these more selective organic fungicides have only a single mode of action, thus disrupting usually only one process in metabolism.

b. Most systemic fungicides are actually fungistatic, not fungicidal. Thus, an organism frequently can recover as the chemical dissipates. Also, although the growth of an organism may be inhibited, the organism may continue to reproduce.

c. The systemic fungicide exerts a selection pressure, and any resistant genotypes will be selectively propagated, soon establishing a resistant or tolerant population.

(5) Dinitrophenols -- Dinocap (Karathane) was developed in the late 1930's not only as an acaracide, but also for control of powdery mildews. Because it acts in a vapor phase, it is effective against the powdery mildews, whose spores are often able to germinate in the absence of water. They act by uncoupling oxidative phosphorlation which upsets the energy systems within the cells.

(6) Substituted aromatics -- This group contains several different fungicides which basically have a simple benzene ring, with various attached radicals. Pentachlorophenol has been used by itself or as a salt of sodium, copper or zinc. Great care must be used in using this compound in greenhouse benches or flower pots, because it is phytotoxic to a

great many plants. Pentachloronitrobenzene (PCNB) was also introduced in the 1930's as a seed treatment and as a foliage fungicide in some cases. It has also been used as a soil treatment to control damping off fungi. Hexachlorobenzene was introduced in the mid 1960's for use in controlling diseases of turf grasses and cotton seedlings.

(7) Quinones -- Dichlone is a quinone which has been used on various food and vegetable crops as a foliar fungicide and also in control of blue-green algae in ponds. It apparently acts by affecting the sulfhydral groups in enzymes, inhibiting them and affecting oxidative phosphorlation.

(8) Aliphatic nitrogenous compounds -- Dodine is an example of this class of fungicide, introduced in the mid 1950's for control of various fungal leaf spot diseases. It is quite specific. The mode of action is unknown in total, but it acts by affecting membrane permeability, allowing a leakage of metabolites from affected cells, and also by inhibiting synthesis of RNA.

(9) Antibiotics -- There are many different antibiotics which are used to control bacterial diseases of humans and other animals. Antibiotics are metabolites produced by microorganisms, particularly the actinomycetes. Three compounds are fairly commonly used; streptomycin is used to control bacterial diseases on fruits and vegetables. It is commonly used to control fire blight of apples and pears. The mode of action is unknown, but it presumably interferes with protein synthesis and with the synthesis of organic acids. Another compound is tetracycline, which is used as a chemotheraputant against micoplasma diseases. A third compound is cyclohexamide, also called actidione, which has been used to control various foliage diseases.

4
VERTEBRATE
PESTS

Vertebrate Pests

Man holds a unique and tenuous position in the complex world of existence. He is perhaps the most formidable creature on earth. He is, at the same time, one of the most vulnerable. Man can modify and adapt to a myriad of climatic conditions and food niches. When men pick a place for themselves in each new system, it is at a very great expense to the creatures already present. In a system which produces a finite amount of useable, digestible energy, creatures that take their cut, survive. Man, in his quest for survival, has done well in this power struggle because he has reasoned methods and tools for eliminating or suppressing his competition. During man's 2,000,000 year existence, he has written or drawn his thoughts for about 10,000 years. From the beginning of recorded thought, man's contests with wildlife for food, shelter and protection have held a place of honor in song, legend and ritual. Those contests remain with us today. If you doubt it, read the national and the state agricultural presses' regular features on the coyote, prairie dog, and starling problems.

In a period of affluence and food abundance the competitive feeling with wildlife is suppressed by elements of society. There is felt to be an abundance for all and that man can tolerate more competition with wildlife than was the case 35 or 350 years ago. The result has been governmental management regulations to reduce the extent and effectiveness of wildlife pest control. This has led to severe financial losses on the part of segments of agricultural industry, particularly livestock producers. It has also led to severe dilemmas among wildlife management agencies trying to satisfy widely divergent views and needs.

The current theme of wildlife pest control is to center control on managing out the opportunity for a wildlife species to cause a problem, chiefly by "non-lethal" means. The minimum action that can be taken is promoted over general higher impact action. Prophylactic reduction of wildlife populations to lower the probability of depredations has been largely discarded because the cost-effectiveness of such programs have been discounted. The effects of prophylactic population suppression have been considered too severe for environmental harmony.

Wildlife generally encompasses fish, amphibians, reptiles, birds and mammals. Wildlife and man share backbones, sexual reproduction, red blood and have to hustle food by eating plants, lower forms of life, or each other. Wildlife is generally larger than invertebrate forms of life and have a more visible impact on the ecosystems in which they live, even though their overall effects are generally less than those by invertebrate life.

$$\boxed{\text{SOME VERTEBRATE PEST ANIMALS}}$$

Small Animals

Bats -- Bats are unique in the animal kingdom in that they are the only true flying mammals. A high degree of structuralization makes it possible for bats to fly. Bats' wings are thin membranes of skin stretched from fore to hind legs, and from hind legs to tail. This skin is barely or thinly furred.

Most bats in the United States are members of the evening bat family. By choice, none are active during the brighter hours of the day; they prefer late afternoon, evening, and early morning for their feeding flights. When not in flight, they rest in the dark seclusion of natural places such as caves, hollow trees, and rock crevasses. They may also occupy vacant buildings, church steeples, attics, spaces between walls, and belfrys. Bats can enter places of refuge through very small openings--some observers squeeze through a crack no

say a small bat can wider than 3/8" thick.

Most bats eat insects which are captured in flight. During the winter when food is in short supply, bats must either hibernate, or migrate to warmer climates.

Because of their insectiverous habits, bats are beneficial. They occupy a very special niche in the animal kingdom because they can fly. They do not compete with other mammals for food and shelter. For these reasons, even though bats are unprotected by law, they should not be needlessly destroyed.

Bats frequently congregate in significant numbers at favored roosting sites. If these sites are in buildings, the accumulation of droppings and the odor of bat urine are objectionable. Their squeaks and the noise they make as they enter or leave the roost may also prove bothersome to the buildings' occupants. The incidence of bats' transmitting disease to man is not high, but if such a situation occurred, it would suggest a need for control. As a precaution against exposure to disease, do not handle live bats. Bat bites are dangerous. In case of bat bites, prompt medical attention should be obtained and if possible, the bat should be captured without damaging the head. The head should be placed in a jar or a plastic bag, kept under refrigeration, and submitted to health authorities for rabies tests.

Chipmunks, Deer Mice, Ground Squirrels, Moles, Pocket Gophers, and Shrews --

The small mammals in this group are known to consume a large number of seeds during an evenings feeding. Deer mice and shrews, ground squirrels and chipmunks may consume large numbers of pine or fir seeds. Such seed-eating activities are not readily apparent to the casual observer. These rodents can cause serious problems in attempts at reforestation by means of re-seeding.

Pocket gophers develop extensive burrow systems among roots for feeding purposes. The pocket gopher can be a problem on alfalfa fields, pastures, and lawns. They may seriously damage a crop of alfalfa by feeding on the leaves, stems and roots. Plants may be killed when their roots are cut or when they are covered by dirt from the mound. Gopher mounds in hay fields cause breakdowns and wearing of sickle mowers. The scattered flat fan-shaped mounds created by the gophers may also provide good seed beds for noxious weeds.

Moles and shrews in their natural environment cause little damage. They are seldom noticed until their tunneling activities become apparent in

lawns, gardens, golf courses, pastures, or other grass and turf areas. These are the rare times that moles and shrews may require control. Otherwise, these mammals are beneficial due to their feeding on insects and other soil organisms. They feed on insects, both mature and larvae, snails, spiders, small vertebrates, earthworms, and only a small amount of vegetation.

SHORT-TAILED SHREW

Marmots (Woodchucks, Rockchucks)-- These mammals can cause damage to many crops. Damage to legumes and truck gardens is often severe. The majority of the damage is caused by the animals consuming the plants. Earthen

mounds from their burrows and the burrows themselves may also damage haying and other farm equipment. Their burrows may also cause a loss of irrigation water when dug along water-conveying ditches.

Marmot damage is usually evidenced by areas where plant production has been terminated or reduced by grazing by these rodents. Supplemental signs include droppings, burrows, and trails leading to and from the damaged area to dens or loafing areas. Their burrows and

mounds may also be hazardous to horses and riders.

Prairie Dogs -- Prairie dogs can cause damage to rangeland by feeding on and cutting vegetation within their "towns". Studies have shown that

prairie dog grazing changes the composition of the vegetation toward species that are more tolerant of their grazing, but not all of these changes are detrimental. In general, prairie dogs cause a reduction in grasses such as blue grama, and an increase in buffalo grass. In areas of tall and mixed-grass prairies, overgrazing often leads to infestations of prairie dogs. Prairie dog towns are easily visible, particularly in early spring when the green grass makes the light-colored mounds easier to spot.

<u>Rabbits</u> -- Rabbits are responsible for damage to young trees, truck crops, grain fields, gardens, and ornamentals. They can also strip the bark from established orchard trees.

All rabbits, including both cottontail and jackrabbits produce similar clipping injuries to tree seedlings. Repeated clippings can suppress the height or deform seedlings, seed grains, and other crops. Rabbits may also girdle trees by gnawing off the bark around the base.

Jackrabbits may cause damage to cereal grains, alfalfa, and even range plants under high population densities. Jackrabbits often cut distinctive paths through fields, or can be observed doing damage.

<u>Rats and Mice</u> -- Rats feed on garbage, meat, fish, cereal, grain and fruits. They require about an ounce of food and an ounce of water daily. The normal roaming range is about 100-150 feet. Domestic rodents will move greater distances when their shelter is destroyed or disturbed.

Identifying Rats and Mice

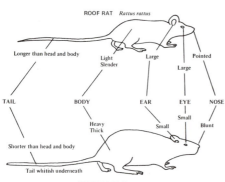

Human health problems associated with rats involve direct contamination of human food from excrement and from the rodent filth. Indirectly rat parasites such as fleas can transmit disease to humans.

Rats and mice are carriers of a number of important diseases including plague, murine typhus, leptospirosis, and salmonellosis.

The rat and its cousin, the mouse, have been important pests to man since ancient times. They are especially important as disease carriers. They also cause enormous destruction and loss of food and property. Their control is not easy due to their ability to adapt to changes and their capacity to reproduce.

Rodents such as rats and mice consume, contaminate, and cause extensive damage to food in agricultural crops. For every two dollars worth of food they eat, they cause twenty dollars worth of damage.

Rat behavior is influenced by thirst, hunger, sex, maternal instinct, and curiosity. Rats cannot go without water for more than 48 hours or without food for more than 4 days. Thirsty or hungry rats become desperate and are therefore easier to control because they are less wary. Judicious use of traps, poisons, and other control measures thus become doubly effective. Properly utilized environmental and physical controls along with rodenticides will prevent rapid population buildups and reinfestations of treated areas.

Carnivores

Bears -- Bears are powerful animals with omnivorous feeding habits. They occasionally can create problems by killing domestic livestock. Torn and mutilated carcasses, often with many broken bones, characterize

BLACK BEAR

the victims of bear attacks. Often, the udders of female victims are removed and eaten. While feeding, bears will often peel back the hide similar to an animal that has been skinned. They move the carcass to a more secluded spot to eat if the kill was made in the open.

Bears may also inflict severe damage to apiaries in attempts to obtain the honey contained in them. Bears can also cause damage to trees. A feature of bear-girdling is an array of long, vertical grooves on exposed sapwood and large strips of bark at the base of the tree. Barking of trees by bears differs markedly from that caused by rodents, which eat the bark and leave horizontal or diagonal tooth marks.

Bobcats and Mountain Lions -- These animals occasionally create problems by killing domestic livestock, especially sheep and lambs. Bobcats usually attack sheep by leaping on their backs and biting at the top of the neck or the throat. If the carcass is skinned, hemorraging and numerous

small punctures through the skin will be noted on the shoulders and possibly the hips. These wounds are inflicted by the claws and are diagnostic of a bobcat kill. Animals that have been fed upon by bobcats usually have the skin peeled back neatly around the area that was fed upon.

Mountain lions kill sheep with a bite in the neck inflicted from above. Removal of the hide from the sheep will reveal large holes made by the cougars canine teeth. Mountain lions usually disembowl their prey, but feed first on the ribs or loin. They frequently drag their prey from the kill site to a more remote area and have a tendency to cover a partially eaten carcass with leaves or loose soil.

Coyotes and Domestic Dogs -- Coyotes more commonly take lambs than mature sheep and characteristically kill with a bite in the throat. The throat wound, if visible, usually consists of pairs of punctures by the coyote's large canine teeth. Death is usually caused by suffocation, caused by damage or compression of the trachea. Blood on the throat wool is  usually indicative of predation, but external bleeding is not always apparent. In that case, the hide should be skinned from the neck, throat, and head of the carcass. If the animal has been killed by a coyote, this will reveal subcutaneous hemorrage (bruises and clots) fang holes in the hide, and tissue damage. The hemorrage occurs only if the animal was bitten while still alive.

Although coyotes typically attack sheep in the neck, it is not a hard and fast rule. In late winter, when the winter wool is long and thick on the neck, coyotes may attack sheep at the more exposed hindquarters.

Domestic dogs can be a serious problem to livestock, particularly in areas near towns or cities. Dogs can occasionally cause severe losses and injuries to penned sheep or turkeys without touching an animal. Harrassment from outside the pen or simply running the enclosed animals can cause them to stampede and pile up.

It is sometimes difficult to distinguish between a coyote kill and a dog kill. In general, a dog will mutilate an animal much more severely than will a coyote. Dogs typically attack the hindquarters and front shoulders, but little flesh is actually consumed. The ears of mature sheep are often badly torn by attacking dogs. Sheep-killing dogs often work in pairs or larger groups and can inflict a considerable amount of damage in a short period of time. A few dogs are very efficient killers and will kill sheep by attacking the throat or neck. These kills are indistinguishable from kills made by coyotes without looking for further evidence. Dogs often kill or injure large numbers of animals in a single episode, while coyotes normally will kill only one or two in a night. Any time more than three sheep are killed in a single episode, it would be wise to investigate the possibility of dogs.

Killing behavior and feeding patterns are useful in making a first general determination of cause of death. Often times, however, it is necessary to look for more evidence before you can be certain what did the killing, particularly in situations where either dogs or coyotes could be involved. Additional evidence such as tracks, droppings, hair on fences, or vetetation, or past experience are useful in making a more positive determination.

Foxes -- Grey, red, and kit foxes are sometimes a problem with small domestic animals such as lambs, pigs and turkeys. Foxes will feed upon larger animals as carrion, but their largest prey are young lambs and pigs, adult jackrabbits, and turkeys, although mice and other rabbits are their major food.

The prey may be carried some distance from the kill location, often to a den. The remains are often cached or partly buried in a hole scratched in the soil. Foxes often urinate on uneaten remains and that odor is an identifying sign. A fox can climb over a fairly high chickenwire fence while removing poultry.

RED FOX

The breast and legs of birds killed by foxes are eaten first and the other appendages are scattered about. Foxes often kill more young birds than they can eat when they find a nest. The young birds are left where killed, with tooth marks under the wing on each side of the body, and the head is often missing.

Eggs are usually opened enough to be licked out, and the shells are left beside the nest and are rarely removed to the den. Fox dens are noted for containing the remains of the prey, particularly the wings of birds.

Raccoons -- There is little evidence that raccoons prey heavily on mammals. They may, however, occasionally kill poultry. They also prey on birds and their eggs. The heads of adult birds are usually bitten off and left some distance from the body. The breast and crop may be torn and chewed, the entrails are sometimes eaten, and there may bits of flesh near water. In a poultry house, the heads of many birds may be taken in one night. One or more of the eggs may be removed from poultry or game-birds' nests and may be eaten way from the nest. The shells are heavily cracked with the line of fracture being along the long axis of the egg. There is often some disturbance of nest materials. In addition to eggs, raccoons sometimes dig a small hole in watermelon and rake out the contents.

Skunks -- Skunks, although often accused of killing poultry and small game species, kill few adult birds compared to the number of nests they rob. Most rabbits, chickens, and pheasants are eaten as carrion, even though they are dragged to the skunk's den. When skunks do kill poultry, they generally will kill only one or two birds and maul them considerably. Spotted skunks are quite effective in controlling rats and mice in grain and corn buildings. They kill these rodents by biting and chewing the head and foreparts and the carcasses are not eaten.

STRIPED SKUNK

Striped and spotted skunks are notorious egg robbers and the signs of their work are quite obvious. Eggs are usually opened up at one end and the edges are crushed as the skunk punches its nose into the hole to lick out the contents. The eggs may

appear as hatched, except for the edges. They are sometimes removed
from the nest, though rarely more than three feet. Canine tooth marks
on the egg will be at least one-half inch apart. When in a more advanced stage of incubation, the eggs are likely to be chewed into small
pieces.

Skunks may sometimes damage lawns or golf courses by digging numerous
small holes while looking for beetle larvae and other insects.

Birds

Crows, Ravens, Magpies, and Gulls -- Crows, ravens, magpies and
gulls are well known robbers of other birds' nests. Crows usually remove the egg from the nest before breaking
a hole in it. The raven and magpie break
a hole up to an inch in diameter. The
raven leaves a clean edge along the break,
never crushed, whereas the magpie often
leaves dented, broken edges. Ravens eat
young birds as do gulls.

Crow

In certain areas, crows, ravens, and magpies may occasionally kill newborn lambs,
by pecking at their eyes, navel, and anal
regions. At times, they may also damage
cattle by picking on fresh brands or sores
where they may cause infection of the area, loss of weight, or even
severe wounds.

Eagles and Hawks -- Eagles may occasionally kill small lambs in the western United States. The carcass will show puncture wounds from talons on
the shoulders or lower back. These wounds should
be looked for by skinning the carcass before making
a positive determination of an eagle kill. Feeding
takes place on the uppermost chest area and the
meat between the ribs is characteristically picked
clean by eagles. Eagles will readily feed on carrion,
particularly when other food supplies are scarce.
For this reason, a positive determination of an eagle
kill requires the finding of talon wounds.

BALD
EAGLE

Hawks are sometimes accused of causing small domestic animal kills, but
these are very rare. More often, hawks are seen feeding on small animals
such as lambs which have died of natural causes. One should remember
that eagles, hawks, and falcons, and owls, due to their value in eating
insects and rodents, are protected by law.

Blackbirds, Cow birds, English or House Sparrows, Grackles, Pigeons and
Starlings -- Many of man's conflicts with birds, whether crop damage or
nuisance problems, really represent the inevitable consequences of man
attempting to co-exist with wildlife in an environment modified to serve
man's interests.

Blackbirds, such as the red-winged blackbird and Brewers blackbird, are often seen together with the common grackle and the brown-headed cowbird as well as starlings because these birds all have similar habits and are often found together in mixed flocks. Although these birds do feed in ripening grain fields at times, only 5-10% of their total annual diet may consist of grain. They do feed extensively on insects, including grubs, caterpillars such as army worms, cutworms, and corn earworms, as well as beetles and other insects.

STARLING

The most serious problems associated with these birds is their gregariousness and formation into rather large flocks which causes competition with song-birds and other more desirable species. Urban roosts pose a number or problems to residents of cities, suburbs or towns. Starlings are the most common problem, although other species may be present. The filth caused by their droppings and nests is highly undesirable as well as the fact that large roosting concentrations of birds can lead to a potential public health problem. The droppings also form a medium for the growth of bacteria and fungi. In addition, birds may act directly as carriers or vectors for some disease. Histoplasmosis is a respiratory disease in humans caused by inhaling spores from the fungus Histoplasma capsulatum. Birds do not spread the disease directly -- the spores are spread by the wind and the disease is contracted by inhaling them -- but the birds' droppings enrich the soil and promote growth of the fungus.

Salmonellosis, a form of food poisoning, is a common disease. It is an acute gastroenteritis produced by members of the salmonella group of bacteria pathogenic to man and other animals. The organism can be spread in many ways, one being through food contaminated with bird feces or with salmonella organisms carried on the feet of birds.

Pigeons, similar to those now living in a semi-wild state in towns and cities have been closely associated with man since before recorded history. Pigeons utilize man-made structures such as barns, city buildings, bridges, and overpasses almost exclusively for their roosting and nesting sites. Excessive numbers of pigeons can cause property damage and may constitute a health hazard.

The pigeon is the wild bird species most commonly associated with the transmission of ornithosis (Psittacosis) to humans. Ornithosis is caused by a virus-like organism and is usually an insidious disease with primarily pneumonic involvement, but it can be a rapidly fatal infection. Birds have become adapted to the disease and show no symptoms, but act as "healthy carriers", shedding the organism in their feces which later become airborne in dust. The disease may also be contracted from parakeets or farm poultry.

Reptiles

Snakes -- Snakes are secretive and usually prefer to move away when disturbed. During their active period, snakes feed on a variety of animal life, such as frogs, toads, salamanders, insects, worms, small rodents, and birds.

The majority of snakes are harmless. However, poisonous snakes do occur in the United States and it is important to know the difference between

Harmless Snake

Poisonous Snake

a harmless and a poisonous one. Poisonous snakes such as copperheads, rattlesnakes, and cottonmouth moccasins, belong to the pit viper family. The name comes from the "pit" or opening, in the side of the head between the eye and the nostril. The pit is absent in non-poisonous snakes. The poisonous snakes mentioned above have a vertically eliptical eye pupil, and non-poisonous species have a round eye pupil. Another distinguishing characteristic is the pattern of scales on the underside of the tail. Poisonous snakes have undivided scales while harmless species have divided scales on the underside of the tail. Because observation of these distinguishing characteristics requires rather close examination and handling of the snake, they have limited usefulness as field identification markings, but they are useful in correctly identifying a dead snake suspected of being poisonous.

The incidence of snake bites is very low in the United States, yet sensible precautions should be observed. Do not handle snakes, alive or dead, without being thoroughly familiar with harmless or poisonous ones. There is no reason for concern about the occasional non-poisonous snake found in the field, forest, or home garden. Even an occasional poisonous snake should not cause panic. However, when snakes invade homes or become common in urban areas, or when poisonous species are frequently observed, reductional measures may be required.

The most effective and lasting method to get rid of snakes is to make the area unattractive to them. Remove cover or other shelter such as board piles, debris or trash piles -- all of which are good protective cover. Keep vegetation mowed around areas and be sure that there are no rodents available as they are important prey of many snakes. All cracks and crevasses that snakes may be able to enter in houses and other buildings should be plugged and snake-proofed so that they cannot be entered. There are no pesticides registered to repel or kill snakes.

Other Animals -- Other animals which can, on certain occasion, cause problems include badgers, weasels, mink, oppossums, deer, porcupines, beavers, muskrats, and a few other bird species. Tracks, droppings, tooth marks, dens, feathers, burrows, and trails cut through vegetation by small rodents, will help determine if the damage was caused by wildlife and, if so, by what species.

GENERAL WILDLIFE DAMAGE CONTROL MEASURES

A. All wildlife species have both positive and negative social and economic values. When a wildlife species or congregation causes serious economic losses, becomes a public health threat or an unbearable nuisance, control of the problem is the legitimate right or obligation of the individual or agency in the position of responsibility.

B. Wildlife damage control techniques should be applied in proportion to the problem.

C. Damage control techniques include the following general categories, listed in general order of public acceptance.

1. Tolerance of losses
2. Mechanical exclusion or protection
3. Repellent devices and sounds
4. Live trapping and transfer
5. Habitat manipulation - flooding, clearing, burning
6. Biological control - disease introduction, encouraging predators
7. Mechanical lethal control
8. Chemical control, including toxicants and repellents
9. Divine intervention

D. Control methods should be selected on the basis of cost-effectiveness, and after careful weighing of available alternatives.

E. The most feasible control methods are those which result in the minimum effect on the problem species, maximum reduction of losses and the least negative impact on related natural systems.

F. Control methods should be as efficient, safe, economical, selective, and humane as possible under the circumstances presented.

G. Control methods should be selected in conformity with local, state and federal law.

H. In situations where endangered species may be affected, special care must be taken before lethal control measures are applied.

I. Control processes should be applied in a low key, responsible manner, without vindictiveness or public display.

DAMAGE CONTROL TECHNIQUES

The Second Law of Thermodynamics states: For every action in the universe there is an equal and opposite reaction. This concept applies equally to wildlife damage control.

A. Tolerance of Losses

Human tolerance of wildlife damage is generally in direct proportion to:

1. The intensity and cost of the damage and,
2. The directness of the impact on the human who suffers as a result of that damage. Damage by wildlife is easy to tolerate when it happens to a neighbor.

Some people are willing to tolerate very serious depradations, others will tolerate very little depending on their backgrounds and personal beliefs.

Considerations:

1. Toleration of the damage allows the pest to continue to exist and perhaps multiply to further compound damage.
2. The pest may infest neighboring situations and become a hardship for others because of the lack of control.
3. The pest may reach carrying capacity and the population decline to a more suitable tolerance level without action.
4. The pest may over-shoot the carrying capacity, contract an epizootic and die back to tolerable levels.
5. Pest species may produce an effect on other creatures which reduces over-all damage. The pest may pay his way by its action on mice and rats, insects or weeds.
6. The pest may provide recreation, food or income values outweighing its negative values.
7. If it costs more to control a species than that species can be expected to destroy, the logical approach is to put up with the damage.

B. Mechanical Exclusion or Protection

The most effective method of wildlife damage control is to fence out or exclude depredating species. The construction or modification of buildings, fences and machinery to exclude wildlife is often the cheapest method over the long run.

Considerations:

1. Some crops and livestock production require large areas making exclusion fencing prohibitively expensive and impractical.
2. Fencing may pen the depredating species in and compound damage.
3. Fencing may severely impact non-target species, ie. fencing to exclude coyotes may result in a serious disruption of deer, antelope, and elk migration lanes. In the case of antelope, it can result in large reduction of carrying capacities by fencing out water and necessary plant groups and space.
4. Fencing may create added mortality as a result of accidents and restricted escape routes from predators.

5. Fencing an open range and forested areas are difficult and costly to maintain. Wildlife often figure out means of defeating the fence.

6. Temporary netting and screening work well for bird and bat control but again may be too costly in time and labor for practical use and do not add to the aesthetics of the building.

7. Exclusion from food or protection often ultimately results in lower carrying capacity for the pest species involved. Exclusion may result in hardship or death of the pests involved.

C. Repellent Devices and Sounds

Repellents generally work on the principle of scaring or warding off pests without consequence to the health of pests.

Sound:

1. Zon Guns, firecrackers, firearms, music, distress calls, ultrasonic frequencies.

Sound repellents work through the sense of hearing and animal communication. They are generally irritating to the animals, or as a result of past experience associating danger with the noise, the pest is frightened off or cannot tolerate the sound and leaves.

Considerations:

a. Advantages - The pest is not killed and can go about its beneficial role with little impact on non-target critters and surrounding environment.

b. Disadvantages - The offending pests may be driven over to the neighbors to create a problem there.

c. The pest may get used to the noise and continue the damage.

d. Cost and time involvement may be high with questionable results.

e. When the pest becomes accustomed to the noise associated with food, the noise may attract the pest.

2. Visual repellents - Scarecrows, silhouette images and similar devices, lighting, flashing lights.

Scarecrows, silhouette images and similar devices have been used for centuries to scare off pests. Lighting may be effective for nocturnal species, especially when accompanied with noise.

Considerations:

a. The effectiveness of repellents depends on the pest involved, the duration of time over which damage occurs, the intensity of the pests' past experience, and the delivery method of the repellent stimulus.

b. Repellents all vary considerably in effectiveness.

c. The longer they are used, the less effective they are.

d. Several repellent devices delivered on alternate days, separately, then in concert are more effective than one method used continuously.

e. Repellents in many cases simply transfer problems from one place to another, defer losses to a later date or condition the pest to tolerate higher noise and light levels. As one sheep producer put it, "Sure, I put up high density light in my feed lot and I also put on hard rock music from dark 'til dawn. I had the most contented coyotes in the state. They picked out the choicest lamb by the light and sat down to eat it to music."

D. Live Trap and Transfer

Live trapping and transfer is a preferred method for endangered species, valuable game animals, and in areas of delicate public relations, it involves catching the pests, removing them to a sufficient distance to insure against their return and releasing them alive.

Advantages: Valuable animals are placed out of the depredating situation without destruction of the animal. Black bear, mountain lion, eagles and other protected species are often handled this way. It is popular with the public.

Disadvantages: It is expensive in time and materials cost. It often transports the problem to a new area. Migration back to the complaint site is common with some species. Survival of transported animals is often very low as a result of interspecific - territorial strife that results with the animals already at the introduction area. The net effect of the transplant is often death of the animal at a tremendous cost in time and money.

E. Habitat Manipulation

Habitat manipulation is a widely accepted method of control. It entails destruction of food, shelter, protection, or other habitat requirements of the pest species. Burning of fence rows, irrigation ditch banks, cleanup of old wood piles, machinery lots and destruction of trees are effective in removing nest and roosting areas of pest birds and mammals. Destruction of winter food sources by cultivation and burning is also used successfully to reduce depredations by some wildlife species. Flooding areas to drown out rodents (gophers, ground squirrels) is an example of habitat manipulation.

Advantages: Once these methods are established and the initial costs are absorbed it is fairly easy to maintain depredation free situations from many pest species. It is acceptable to the ignorant protectionist since wildlife is not directly killed. It is often extremely effective in a negative way.

Disadvantages: It sterilizes the area for both pest and beneficial species. The end result is very low levels of wildlife for recreational and aesthetic enjoyment. Of all methods listed this one is perhaps the most damaging to wildlife and should be recommended as a last resort!

F. Biological Control

There is a story about an old range rider that used to pick up black plague stricken prairie dogs, put them in in a gunny sack and release them into healthy prairie dog colonies. The black plague reduced the prairie dogs very effectively. The rider somehow survived it. This is an example of biological control.

Biological control has been highly touted as the way of the future. It, however, has some serious problems incident to it. Environmentalists have applauded this as a general alternative to chemical control.

Biological control agents can be categorized into infectious disease agents, parasites or predators. Myxomatosis virus disease was introduced into Australia to control rabbits. It worked. However, each species of pest requires different disease organisms to affect it. Infectious diseases with the capacity to greatly reduce some pests have the capacity to reduce humans as well. Bubonic plague, rabies, and tularemia might be developed for use on mammal pests but the public health implications are too negative to allow it.

Generally, parasitic species are host specific and do not offer the potential for application and effective treatment in pest control situations under the most optimistic assessments.

Predators are commonly given general credit for controlling rodent populations. The literature of wildlife does not support those claims, the reverse is more accurate. Depending on the pest involved, sometimes predators can be of assistance, ie. (bull snake and Norway rats) but very few creditable examples are available. The livestock producer losing lambs or calves to coyotes is not generally impressed with the amount of grass a coyote saves for his cows by eating mice and rabbits.

The space - territorial requirements of predators is usually much larger than the area over which those predators could feasibly control or erradicate rodent pest populations.

Although, biological control is a popular idea it has not been shown to be of much practical help in solving wildlife pest management problems.

A regime of excellent range and wildlife management may contribute to lower livestock losses due to large predators by encouraging a surplus of rabbits, mice and other alternate prey. It also encourages resistant, alert and healthy livestock to better ward off predator attack. Many losses of livestock are random events, impossible to predict or protect against. They may occur again the next day, or not for another five years.

G. Mechanical Lethal Control

Lethal mechanical control methods (traps, snares, and shooting from airplanes, snowmobiles, with the aid of calls and spotlights) have come under considerable attack recently by preservationist groups. Each tool has limitations in effectiveness and to some people the tools violate the "fair chase" hunting ethic. Each tool has a necessary place in control. Because, contrary to popular belief, the direct killing of a depredating individual animal or animals has the least environmental impact of all the methods discussed to this point except tolerance and perhaps live trapping. In the overall management of pest species, the killing of a few individuals which are a problem does little more than duplicate the natural mortality forces working on that species. There is no destruction of habitat, no transfer and continuance of problems or other adverse effects. Widely used mechanical lethal control means are:

1. Steel leghold traps, snares and killer traps. Used properly these are effective, humane and efficient. They have no secondary effects and used properly are very selective. Traps are inexpensive and can be applied and withdrawn at will to meet temporary problems.
2. Shooting - calling, aerial shooting, shooting from a vehicle, catching with hounds. These methods are selective, of moderate efficiency, of varying expense from very inexpensive as in calling, or very expensive as in shooting from helicopters. These methods have been often condemned by some groups as being unfair, cruel, and inhumane.

Advantages: These methods are generally very selective to individual pests or pest species causing depredations. They are of varying expense but response time is quick and termination of losses may be immediate.

Disadvantages: These methods are susceptible to weather conditions and are based on human involvement. Control with air or land vehicles is dangerous and expensive. Some pests have habits which do not leave them vulnerable to these types of control methods. Cost in human effort is often high. In situations where pest density levels require reduction, mechanical methods have not been shown to be adequate to achieve population reduction.

H. Chemical Control

Chemical control of wildlife can be broken into several major categories as follows:

1. Lethal toxicants applied orally, ie. strychnine, warfarin, 1080.
2. Lethal toxicant contact poisons, ie. endrin.
3. Fumigants, ie. gasoline, phostoxin, chloropicrin, sulphur dioxide.
4. Chemosterilants, ie. diethyl stilbestrol
5. Aversive conditioning agents, ie. lithium chloride.
6. Saponificants, ie. soaps.
7. Repellents, ie. mothballs, Arason, bone tar oil, lion dung, and cinnamonaldehyde.

Since the amended FIFRA and EPA assumption of its enforcement, chemical use on vertebrate pests has diminshed considerably. Options on chemicals and application alternatives are extremely narrow. According to the EPA

Compendium, there are no chemicals registered for use on amphibians and reptiles. Several of the commonly recommended repellent and fumigant chemicals are not formally registered for such use, ie., mothballs cannot be legally recommended for repelling skunks but can be for bats. Phostoxin is not registered for vertebrate use. Several of the old faithful toxicants, strychnine and 1080 have had the registrations cancelled for use on pradatory wildlife leaving only sodium cyanide used in the M-44 device abailable for controlling canine damage. Predator toxicants on hand prior to the 1972 cancellation can be used up on private property if state regulations are followed. There are no orally delivered toxicants registered for non-canine predators.

EPA compendium of Registered Pesticides, Volume IV, Rodenticides and Mammals, Bird and Fish Toxicants lists chemicals registered for use on wildlife. Many college libraries carry this volume.

Use of chemical toxicants to control wildlife is not popular and has been the subject of heated debate, public pressure and a great deal of inaccurate description and legend. Like any tool man uses, chemicals can and have been abused. Judiciously and carefully used, chemicals are safe, selective, economical and often the most feasible of alternatives.

 1. Lethal toxicants applied orally, ie., strychnine, warfarin and 1080.

These chemicals are ingested into the digestive tract through baits and lead to the death of the animal. They have a wide range of effects, residual half lives, modes of action and effectiveness. Each chemical carries unique qualities of taste, smell, acceptance, physiological reaction, and delivery modes. Toxicity of each chemical varies with the physiology of each species, ie., warfarin is deadly to rats and mice, is relatively harmless to chickens and predators. Strychnine is lethal to anything that eats it in sufficient dosages. It does, however, break down quickly in the soil or dead animals, rendering it harmless.

The baiting system used, placement of baits, dosage levels used, time of placement, all affect the selectivity of the toxicant to the target animal.

 Advantages:

 a. A great degree of selectivity to target animals can be attained.
 b. A range of chemicals can be used for the conditions presented.
 c. Often very effective in terms of time and cost investment.
 d. Often the most humane method available.
 e. Offer a great deal of flexibility in terms of bait systems, placement, and exposure time for handling the wide variety of terrain and weather conditions encountered in wildlife damage control.
 f. Usually quite efficient in solving problems quickly and economically.

Disadvantages:

a. Risk to humans, pets and livestock varies with each chemical and the skill of the applicator.
b. Emotionally and politically a hot issue.
c. Some risk of secondary poisoning to non-target species with some chemicals and delivery systems.

2. Lethal Toxicant - contact poisons, ie., endrin.

These chemicals are placed such that the chemical is absorbed through the skin by contact. Much of what was covered in the previous section applies here. Very few chemicals are registered for this application method, most of these are for birds and bats. Advantages and disadvantages are the same as for orally applied toxicants.

3. Fumigants, ie., sulphur dioxide, calcium cyanide, etc.

These chemicals release toxic gases which are inhaled and result in death of the pest. They are widely used and are effective for some building and burrow dwelling wildlife.

Advantages:

a. Extremely effective and usually economical. When used correctly they are quite selective.
b. Little risk of secondary hazard to people or non-target species, if properly applied.
c. Usually results in immediate termination of damage.
d. Results in quick, humane death of the pest.

Disadvantages:

a. Requires trained people to handle them. Fumigants are hazardous and highly toxic to people unless proper precautions are taken.
b. Not publicly popular.

4. Chemosterilants, ie., diethyl stilbestrol.

Chemosterilants have been researched quite extensively during the past 20 years. There are none registered for use on mammalian pests. Speculation as to their promise has been abundant but is without a great deal of substance. The basic problems are two:

a. How can chemosterilants be delivered to 2.5 million coyotes over the western U.S. with enough efficiency to make a difference? Some coyotes will not eat dead bait materials, if they did 1080 would have controlled them with more efficiency.

b. No chemical is available that has all of the needed characteristics, ie., tasteless, odorless, residual effect for 2+ months,

100% sterility for the life of the animal, no effect on non-target birds and animals, and biodegradable, among other requirements.

Frankly, science is not very close to releasing a practical chemosterilant for most wildlife pests.

   5. Aversive conditioning agents, ie., lithium chloride, Bitrex, cinnamonaldehyde.

Basically, these are chemicals that when ingested or tasted make the pest sick or give off a bad taste. The pest associates its illness with the ingestion of the bait and supposedly will not eat that material again. Lithium chloride was given a lot of publicity recently as showing a great deal of promise for use in coyote damage control but subsequent research has failed to show an efficient delivery system or successful field application. Basically there are many problems involved with the idea.

   Advantages:

   a. Quite specific to the pest animal.
   b. Non lethal to the pest animal

   Disadvantages:

   a. A delivery system must be developed to get the material to the coyote and the coyote must eat it.
   b. The coyote must make the association between illness, the bait, and the live animal that needs protection (the sheep).
   c. The coyote has to have a memory so it will avoid eating the sheep again.
   d. The chemical has to fulfill all other requirements listed under the discussion of chemosterilants except sterility, plus others to get EPA registration for that use.

Aversive conditioning chemicals are a long way from a field ready tool.

   6. Saponificants, ie., soaps.

These are used primarily in controlling large concentrations of bird pests; ie., starlings and blackbirds. The chemicals are sprayed on in water solution during cold weather. The soap breaks down the feather oils and wets the bird so that its body heat dissipates. The bird dies of hypothermia.

   Advantages:

   a. It can be quite selective, effective and economically feasible.
   b. Very little secondary or negative side effect.

Disadvantages:

a. Some people like millions of starlings and blackbirds, particularly when they are deficating on someone else.
b. The effectiveness of this method varies with weather conditions, tree cover and densities of the roosting birds.
c. It requires planning and high pressure spraying equipment, or aerial spraying equipment.
d. Some states and communities require permits before this method is used.

7. Repellents, ie., Arason, bone tar oil, mothballs, etc.

These chemicals by their action scare off or irritate the pest, resulting in less economic loss.

Advantages:

a. The pest is not affected, so it can continue its positive roles.
b. Often inexpensively applied and effective, depending on the pest.

Disadvantages:

a. Materials may adversely affect the domestic livestock and handlers.
b. May just delay the problems or drive them over to the neighbors.
c. Odor may become acceptable or even an attractant to the pest after prolonged use.
d. Most repellent chemicals are not readily available.

I. Divine Intervention - Miracles

This control method results from serious petitions or extraterrestrial forces. Historical precedents are abundant. The gulls saving the Morman's grain crop by eating Morman crickets in an hour of great need; Moses' petitions for plagues to control Egyptian depradations on the Jews; Sampson's ability to break lion's jaws with his bare hands; David's uncanny ability with a slingshot in slaying depredating Philistines. These alternatives have been highly recommended by environmental agencies and organizations, in lieu of methods previously discussed and particularly where chemical control would be otherwise indicated. Animal protection societies, wildlife groups and humane organizations also support this concept but question the humaneness of some of the application techniques, especially Sampson's methods!

Occasionally an animal species is affected by some sort of disease which helps to reduce the total population. Unfortunately, diseases frequently occur in species whose numbers are small (this is probably why they are small) and often the species is not one that is of concern as far as damage or other problems. An example of this is lung disease in Big Horn mountain sheep and sometimes elk.

WHAT DOES THE LAW SAY?

Laws affecting wildlife damage control are extensive, specific, and generally allow for quick relief for most serious problems. Persons and agencies dealing with wildlife damage control are advised to read state and federal laws carefully. Legal control measures for some species are very limited.

Wildlife is considered public held property, the use of which is regulated by the federal and state governments. The Migratory Bird Treaty protects all migratory birds, with authority for regulation of hunting seasons and management given generally to the USDI, U.S. Fish and Wildlife Service and state game and fish agencies. Certain pest species are listed as exceptions, ie., English sparrow, European starling, and domestic pigeon. State laws generally are applied to these. Blackbirds, magpies, crows and great horned owls can, under certain conditions, be destroyed without federal permit by mechanical and chemical means. Eagles are protected under the Migratory Bird Treaty and the Bald Eagle Act. The Endangered Species Protection Act protects threatened wildlife. The Lacey Act prohibits interstate shipment of wildlife taken in violation of state laws. These laws and treaties were enacted and are enforced from federal and state levels.

Very few chemicals are registered for use on wildlife. The few labels that are cleared are specific to a narrow range of target species and circumstances. Chemicals (lethal or repellent in nature) cannot be legally used on species other than those listed on the label. The 1972 amendment of FIFRA and EPA's restrictive interpretations of FIFRA have resulted in severe restrictions on lethal and non-lethal chemical control of wildlife. Registration requirements are extensive and expensive. Since chemical use on vertebrates is periodic and minor in terms of amounts of chemicals and monetary return to the chemical industry, it is not feasible for the industry to register many effective vertebrate control products. This leaves the responsibility for registration of wildlife control chemicals to state or federal agencies who have shown little enthusiasm for the task.

Because of the bio-politics and public pressures involved many of the historically effective chemicals like strychnine and 1080 are not appreciated by the federal agencies which are the only groups with the money and capability for collecting the data required for re-registration of the products.

SUMMARY

Wildlife Damage Control is presently in an adjustment period from one of wide flexibility and personal freedom to one of narrowed alternatives and heavy state and federal agency regulation.

Select the control methods and tools carefully with prior consultation with responsible agencies. Anticipate problems before they occur and manage around them. Most problems with wildlife can be simply and inexpensively handled with the proper information and a little imagination and good sense.

5
WEEDS

CANADA THISTLE

RUSSIAN KNAPWEED

LEAFY SPURGE

FIELD BINDWEED

Weeds

Man is more responsible for the spread of weeds than any other single factor. Most of the plants that are commonly thought of as weeds have existed in parts of the world for many years, but were relatively insignificant before man started growing plants for food. Thus weeds have evolved along with crops and are objectionable because of their ability to compete with the plants man tries to grow for food or fiber. The original habitat of many weeds is unknown. Many weeds are closely related to cultivated crops or to ornamental plants and it is this close relationship that makes some weeds more difficult to control than others.

A weed can be defined in many ways. Some of the more common definitions are: a plant species growing where it is not desired; a plant out of place; a plant that is more detrimental than beneficial. Kentucky Bluegrass that spreads from a lawn into a flower bed is very much a weed; likewise, volunteer corn in a sugar beet or bean field is as much a weed as lambsquarters or red-root pigweed. Thus, a plant is a weed only in terms of human, or even more specific, individual, definition. A plant that is a weed to one person may be a desirable plant to another. Any plant can be a weed in a given circumstance.

ANY PLANT CAN BE CONSIDERED A WEED IF IT GROWS WHERE YOU DON'T WANT IT TO GROW.

CHARACTERISTICS OF WEEDS

Plants that are commonly referred to as weeds have certain characteristics that give them the ability to spread and exist where most cultivated plants would soon die out. Some of these characteristics are:

1. Most weeds produce an abundance of seed.
2. Many have unique ways of dispersing and spreading their seed.
3. Many weed seeds can remain dormant in soil for long periods of time.
4. Most weed-like plants have the ability to grow under adverse conditions.
5. Weeds can usually compete for soil moisture, nutrients, and sunlight better than crop plants.

Modern agricultural practices favor invasion by weeds. Plant communities are a complex thing and under any particular set of environmental conditions (climate, temperature, rainfall, soil, etc.) there is a natural progression to a "climax" vegetation. Man changes this natural progression by the growing of crops. Man attempts to grow crops in pure stands or as individual plants. Under natural conditions, single plant species will usually not monopolize an area because any single species will have too short a growth cycle to cover the ground for the entire single growing season, or cannot use all the available sunshine because of leaf area development, or will not use all the available water or nutrients because of its root system. In nature, other plant species (weeds) move in to use these wasted resources. Man's ability to manage vegetation to meet his needs for food, livestock feed, and fiber, is one of the most important factors to his survival. To produce crops efficiently, or to grow plants for ornamental purposes, it is necessary to minimize the competitive effect of weeds. This must be accomplished selectively, that is, a weed control method or combination of methods that will minimize the yield-reducing effect of weeds, while in turn, assisting the crop.

WAYS IN WHICH WEEDS INJURE PLANTS OR AFFECT LAND USE

Weeds affect our lives in many ways. They not only cost us money, but they can cause untold misery and grief to hay fever sufferers as well as being poisonous to humans, livestock and wildlife. Some of the ways weeds affect us are:

a. Reduce yields -- Competition for moisture, light, and nutrients.
b. Reduce crop quality -- Weed seeds, dockage, weeds in hay, straw, etc.
c. Increase production costs -- Additional tillage of farm crops and cultivation of nursery crops.
d. Increase labor and equipment costs -- machinery wear and tear, etc.
e. Insect and disease carriers or hosts -- e.g., wheat stem rust, corn borer, pine needle rusts, and numerous viruses.
f. Poisonous or irritating to animals or people -- Cocklebur seedlings, poison ivy, hay fever, etc.
g. Increase upkeep of home lawns and gardens.
h. Create problems in recreation areas such as golf courses, parks, and fishing and boating areas.
i. Increase upkeep and maintenance along highways, railroads, and irrigation ditches.
j. Land values may be reduced, especially by the presence of perennial weeds.
k. Cropping system choice may be limited. Some crops will not compete effectively against heavy weed growth.
l. Reforestation costs may be increased due to slower rate of growth due to weedy competition.

WEED CLASSIFICATION

Green plants are basic for life and are indispensable in man's environment. They are a complex life form that utilizes energy from the sun, combined with minerals, water and carbon dioxide to provide food for man and wildlife, to beautify the landscape, and to reduce soil erosion. Plants can be classified in several ways; one simple classification is according to life cycles. Weeds may be classified into four major groups: Summer annuals, winter annuals, biennials, and perennials. Examples are shown on the following pages.

Summer annuals live one year or less. They grow every spring or summer from seed; they produce seed, mature, and die in one growing season. Seed from most summer annual weeds germinates during a two-month period in the spring. Then the seed lies dormant until the next spring. There are usually a few seeds of every species that germinate during the summer or fall, but seldom in numbers equal to the spring flush. Of the millions of weed seeds lying beneath every square foot of land, probably not more than 5% will germinate in any growing season. Summer annual weeds are the greatest problem in spring when annual crops are planted. The new seedlings of field crops are competing directly with weeds because their seeds germinate most readily in disturbed soil. Examples of summer annuals are: russian thistle, redroot pigweed, lambs-quarters, puncture vine, wild oats, and crabgrass.

Winter annuals germinate in the fall or early winter and overwinter in a vegetative form (without flowering). In the spring, they flower, mature, seed and then die in late spring or early summer. Seed from most winter annual weeds is dormant in the spring, but germinates in the late summer or fall. Some species such as common chickweeds, can germinate under snow cover. These weeds start growth at the first sign of spring and many species bloom and produce ripe seed by mid-May or June. This means that some winter annuals can re-seed themselves before late planted crops. Examples of winter annuals are: tansy mustard, blue mustard, downy bromegrass (cheatgrass), chickweed, and shepherdspurse.

Biennial plants have a similar life cycle to annuals since they die after flowering and setting seed, but they require two years to complete the sequence. Growth during the first year is usually vegetative (no flowering activity) and low-growing, frequently as a rosette form. Flowering and seed production occur during the second year. Biennial weeds, though relatively few, may become established from seeds any time during the growing season. Because true biennials never produce flowers or seeds the first year, but form a rosette of leaves that usu-ally lie close to the ground, such weeds are well-adapted to lawns, pastures, hayfields, and orchards where grazing or mowing removes very few of the rosette leaves. Consequently, growth and energy storage in the roots or crowns continues until a killing frost. This large amount of stored energy supports spring growth ahead of other pasture and for-age species. If biennial weeds are not controlled before this time, there are no forage species that will crowd them out. Examples are: bullthistle, wild carrot, and common mullein.

Perennial weeds become established by seed or by vegetative parts, such as root stocks or rhizomes and once established, they live for more than two years. Perennials are usually herbaceous (top growth usually

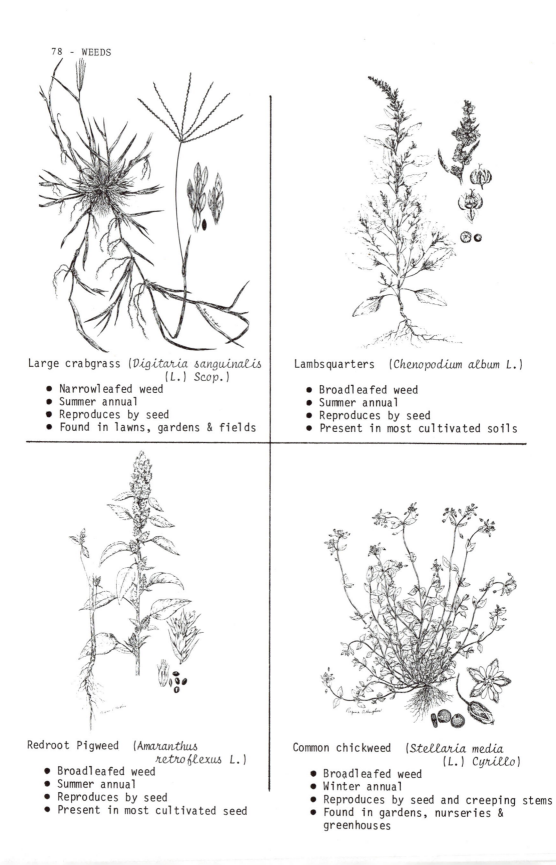

Large crabgrass (*Digitaria sanguinalis* (L.) Scop.)

- Narrowleafed weed
- Summer annual
- Reproduces by seed
- Found in lawns, gardens & fields

Lambsquarters (*Chenopodium album* L.)

- Broadleafed weed
- Summer annual
- Reproduces by seed
- Present in most cultivated soils

Redroot Pigweed (*Amaranthus retroflexus* L.)

- Broadleafed weed
- Summer annual
- Reproduces by seed
- Present in most cultivated seed

Common chickweed (*Stellaria media* (L.) Cyrillo)

- Broadleafed weed
- Winter annual
- Reproduces by seed and creeping stems
- Found in gardens, nurseries & greenhouses

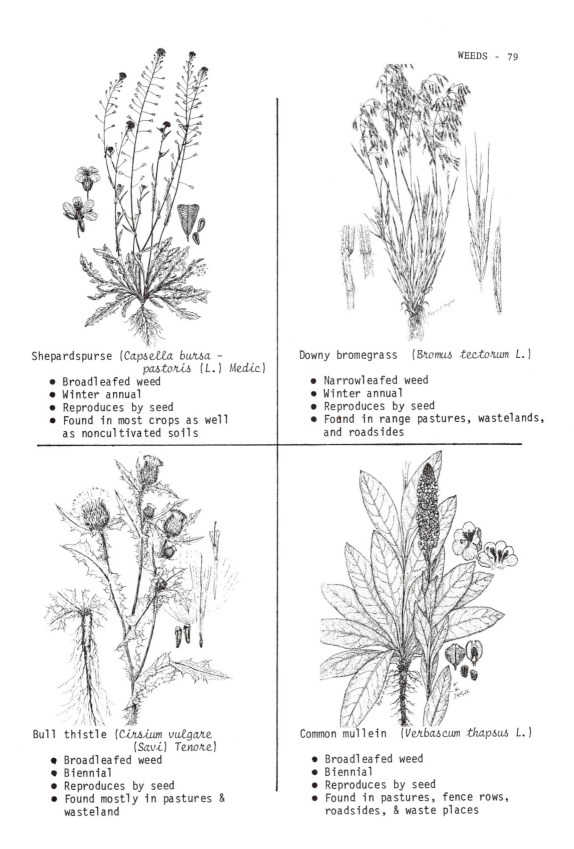

Shepardspurse (*Capsella bursa - pastoris* (L.) *Medic*)

- Broadleafed weed
- Winter annual
- Reproduces by seed
- Found in most crops as well as noncultivated soils

Downy bromegrass   (*Bromus tectorum L.*)

- Narrowleafed weed
- Winter annual
- Reproduces by seed
- Found in range pastures, wastelands, and roadsides

Bull thistle (*Cirsium vulgare* (*Savi*) *Tenore*)

- Broadleafed weed
- Biennial
- Reproduces by seed
- Found mostly in pastures & wasteland

Common mullein (*Verbascum thapsus L.*)

- Broadleafed weed
- Biennial
- Reproduces by seed
- Found in pastures, fence rows, roadsides, & waste places

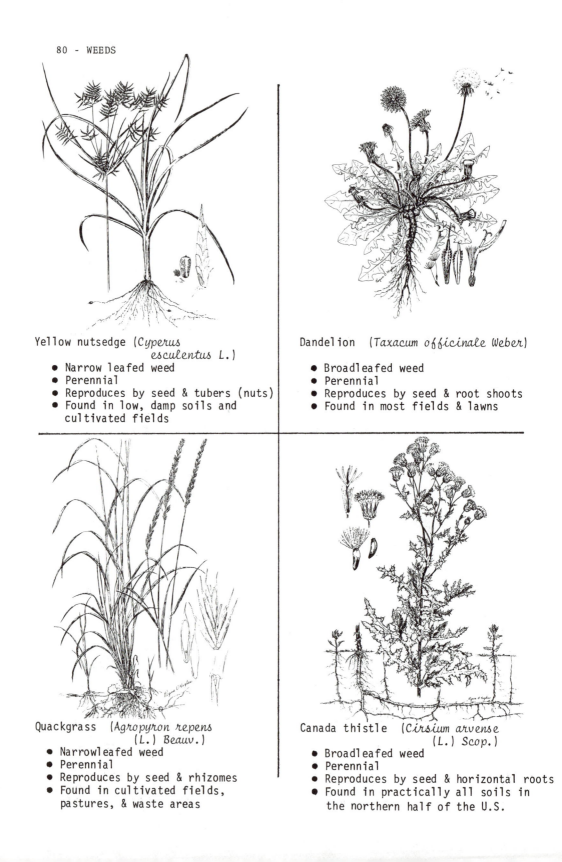

Yellow nutsedge (*Cyperus esculentus L.*)
- Narrow leafed weed
- Perennial
- Reproduces by seed & tubers (nuts)
- Found in low, damp soils and cultivated fields

Dandelion (*Taxacum officinale Weber*)
- Broadleafed weed
- Perennial
- Reproduces by seed & root shoots
- Found in most fields & lawns

Quackgrass (*Agropyron repens (L.) Beauv.*)
- Narrowleafed weed
- Perennial
- Reproduces by seed & rhizomes
- Found in cultivated fields, pastures, & waste areas

Canada thistle (*Cirsium arvense (L.) Scop.*)
- Broadleafed weed
- Perennial
- Reproduces by seed & horizontal roots
- Found in practically all soils in the northern half of the U.S.

winter kills) or woody, brush or trees. Since perennial weeds live indefinitely, their persistence and spread is not as dependent on seeds as with the other three weed groups. Seed is the primary method of introducing these weeds into new areas but perennial weeds are often spread during soil preparation and cultivation. Most perennial weeds that spread by rhizomes or root stocks will spread in circular patches if left undisturbed. In crop fields, patches of perennial weeds such as Canada thistle spread in oblong patches in the direction the field is worked.

Perennial weeds grown from seed are no more difficult to control than any other kind of weed coming from seed, but once established, perennial weeds are the most difficult to control.

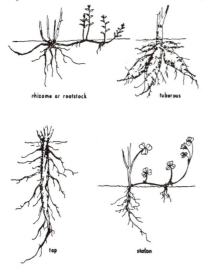

rhizome or rootstock   tuberous

top   stolon

They compete with any crop, especially row crops where competition between rows is not too great. Perennial weeds also get an early start in the spring and compete with perennial crops, such as alfalfa and other forage species. Examples are: field bindweed, Canada thistle, Russian knapweed, johnsongrass, yellow nutsedge, dandelion, quackgrass, and many tree and brush species.

Some perennials have underground structures in addition to true roots. Such structures are: Rhizomes, a modified underground stem (johnsongrass, bermuda grass); bulbs (wild onion, death camas); tubers (nutsedge); creeping roots (field bindweed, Canada thistle); and tap roots (dandelion and plantain). These underground structures serve as food storage organs and produce new shoots. A few perennials have above ground roots called stolons. Perennial plants are the most difficult to control because of their ability to reproduce in several ways.

WEED IDENTIFICATION

Correct identification of weeds in essential for an effective weed control program. It may not be necessary to identify plants down to the individual species, but it is necessary that plants be properly identified to major weed groups. There are many plant species that can create weed problems and most of them are flowering plants (the higher forms of life that produce seed).

Some herbicides are effective on some groups of weeds, while other herbicides perform best on other plant groups. Very often, poor results with herbicides can be attributed to having used an herbicide for control of a weed that a particular herbicide was not intended to control. For example,

alachlor (Lasso®) does not generally kill mustard species or lambsquarters; atrazine (AAtrex®) will not usually control puncture vine or
Russian thistle; and trifluralin (Treflan®) does not usually control
nightshade in dry beans. Often a crabgrass herbicide is applied on turf
that is infested with perennial weedy grasses. Crabgrass herbicides control annual grasses but have little effect on perennial species. Therefore,
it is important that an applicator of herbicides know the weed species he
is trying to control and use the herbicide that is most effective on that
weed.

Narrowleafed weeds -- These include grasses, sedges (bullrush, nutsedge),
rushes, cattails, and several other plants that are less often weed problems, such as iris and lillies. These plants can be identified by the
characteristic parallel veins in the leaves.

Broadleafed weeds -- This is the larger of the two groups and includes all
of the broadleafed weeds as well as most trees, and many other brush
species. In this group, the leaf veins are net-like, not all parallel.
This group includes mustard, dock, pigweed, purslain, field bindweed, and
many others.

The decision to use a pre-plant or preemergence herbicide has to be made
before weeds germinate. Therefore, it is important to know the history of
weed species in a field (what weeds were there the year before) in order
to make the right decision as to what herbicide to use.

Seedling identification is necessary to properly select post-emergence
herbicides in many cases. Wrong identification can lead to less desirable results. There are numerous publications available that will help
an applicator identify weeds. These are available from state universities
or the U.S. Department of Agriculture. Weeds can be identified by simply
looking at pictures, or a more scientific method is by using keys that
systematically describe different characteristics of plants and by deciding between two possibilities. However, the use of plant identification
keys requires knowledge of plant characteristics and experience in distinguishing between plant parts that are not always obviously different.

The following is an example of a small part of an analytical couplet key
to the major genera of one plant family:

KEY TO SELECTED GENERA

(1) Fruit a group of akenes; flowers not spurred....(2)
(1) Fruit a group of follicles; flowers spurred.....(4)
    (2) Petals none................................(3)
    (2) Petals present............................Ranunculus
(3) Sepals usually 4; involucre none...............Clematis
(3) Sepals usually 5; involucre present............Anemone
    (4) Flowers regular; spurs 5..................Aquelegia
    (4) Flowers irregular; spur 1................Delphinium

To use such a key, the specimen is examined for the characters described after
number (1). A choice must be made and this results in a referral to the next

couplet, either (2), as in this case, or sometimes a number farther down
the list.

It must be remembered that there are approximately 130 different plant
families ranging from Aceraceae to Zygophyllaceae and each family may
have hundreds of sub-families, genera and species.

WEED CONTROL METHODS

Methods that can be used for weed control include: (1) prevention;
(2) crop competition and rotation; (3) burning; (4) mechanical control;
(5) biological controls; (6) chemical control. Although chemical control is often thought of first, it is not a panacea for all weed problem
situations. Alternative control methods should be identified and corrected. Good farming or other good land use practices should always be
foremost in any weed control program.

(1) Prevention is the most practical method of controlling weeds. If
weeds are not allowed to infest an area and seed is not produced, there
is much less chance of weeds becoming established. Once weeds have become established, they are difficult and costly to control and may persist for many years if the seeds can lie dormant in the soil. Field
bindweed seed can lie dormant and remain alive in soil for 40 years or
more.

Preventative control measures should be adapted where practical and should
be the first step in any effective weed control program. These should include:

   a. Always use clean seed.
   b. Do not feed grains or hay containing weed seeds without destroy-
       ing their viability.
   c. Do not spread manure unless the viability of the weed seeds have
       been been destroyed.
   d. Livestock should not be moved directly from infested to clean
       areas.
   e. Make sure harvesting equipment is clean before moving from
       infested areas.
   f. Avoid the use of soil from infested areas.
   g. Inspect nursery stock for presence of weed seeds or other weed
       parts.
   h. Keep irrigation ditches, fencerows, roadsides, and other non-
       cropped areas free from weeds.
   i. Prevent the production and spread of weed seeds by wind when
       possible.
   j. Use weed screens for trashy irrigation water.

Obviously, the implementation of an effective weed prevention program
requires alertness and perserverance.

(2) <u>Crop Competition and Rotation</u> is probably the cheapest and easiest method of controlling weeds; it is based on the law of nature, "Survival of the Fittest". A crop will survive and flourish if it can compete more efficiently for sunlight, water, nutrients, and space than the unwanted plants. It is important to understand the growth habits of the crop in relation to the weeds. Early weed competition is usually more detrimental to a crop than later competition when the crop is well established. Crop rotations can be a means of controlling weeds; certain weeds are more common in some crops than in others and some crops are more competitive to certain weeds than others.

(3) <u>Burning</u> has been used for many years to control unwanted vegetation. Selective burning is used to kill weeds and row crops. Controlled burning is a valuable tool in the conversion of brush lands to productive grazing lands. Fire is an effective tool for removal of vegetation from ditch banks, roadsides, fencelines, and other waste areas. Intense heat can sear green vegetation which often grows abundantly in irrigation ditches. Burning can kill small weed seeds on the soil surface.

(4) <u>Mechanical Control</u> includes cultivation, mowing, hoeing, hand pulling, root plowing, chaining and tree grubbing. All of these methods involve the use of tools to physically cut off, cover, or remove undesirable plants from soil. Cultivation or tillage is the most common method of weed control. It is effective on small annual weeds, but when the plants are larger, tillage effectiveness may be reduced. Tillage can also be used to disturb perennial root systems. However, repeated tillage operations are usually required to effectively control perennial weeds.

(5) <u>Biological Controls</u> include the use of insects, diseases or parasitic plants to control weeds, but not harm the desirable crops. Biological controls will usually not eradicate weeds because the host plant is necessary as a food source for the predator of the weed species. Examples of biological controls are a moth borer from Argentina used to destroy prickly pear cactus in Australia, and the <u>Chrysolina</u> beetle used to control Klamath Weed.

(6) <u>Chemical Controls</u> or herbicides are used for killing or inhibiting plant growth. Selective herbicides date back to the turn of the century, but the first break-through with selective herbicides occurred with the introduction of 2,4-D in the early 1940's. Since that time, chemical manufacturers have developed a large number of herbicides. Since the discovery of 2,4-D, the development of organic chemicals for weed control has expanded until now there are more than 145 different herbicides on the market. With the different formulated combinations, the total number available is over 200. Herbicides are now applied to more agricultural lands than insecticides and fungicides combined.

| CLASSIFICATION OF HERBICIDES |

There are several ways to classify herbicides. One is based on chemical structures and effect on plants, and another is the separation of herbicides into families. The most practical classification for field use is based on how herbicides are used.

A. Classification of Herbicide Action

The classification discussed on the following pages is based on whether the herbicide is applied to the foliage or the soil; the pathway by which the herbicide enters the weed; and whether the herbicide is selective or non-selective.

1) *Foliage - Contact - Nonselective.* Herbicides applied to the weed's foliage, in the absence of a crop or directed underneath a crop, kills the

foliage the herbicide contacts with little or no translocation to underground or shaded parts of the weed. Since biennial and perennial weeds normally have dormant and protected buds with stored energy in the crown or root systems that produce new growth, these weeds will recover after treatment. Small annual weeds are completely and permanently controlled. For example, paraquat applied before a no-tillage corn planting, or used as a directed spray underneath fruit trees, will control seedling weeds. Since contact herbicides normally don't leave any residue beyond the initial contact, a residual herbicide is commonly mixed with the contact herbicide to control the regrowth of the biennial and perennial weeds plus weeds coming from seed.

2) *Foliage - Contact - Selective.* These herbicides kill the weeds by

contact-burning effect on the foliage. Due to differences in the waxy covering on the leaves, certain crops and weeds are not injured. These herbicides are effective only on weeds in the seedling stage and biennial and perennial weeds with dormant buds near the soil level will regrow after the foliage is killed. Generally, these herbicides are not translocated, although some may have a short residual effect in the soil. For example, Stoddard Solvent is used for annual weed control in onions; dinitro is used to control broadleaf annual weeds in new legume seedlings. Onions have a very waxy leaf that Stoddard Solvent cannot penetrate.

3) *Foliage - Translocated - Nonselective.* These herbicides are applied to the foliage of the weeds and are absorbed and translocated throughout the plant. Since they are nonselective, they cannot be applied when a crop is present but they can be used before planting or after harvesting. For example, Amitrol T which can be used on noncrop land only is nonselective and will kill anything that is growing. Amitrol T has only a short residue so dormant weeds or weeds germinating from seed after herbicide application will not be controlled. A residual herbicide, such as atrazine or simazine, is commonly mixed with this herbicide to give continued weed control for the rest of the year.

4) *Foliage - Translocated - Selective*. This group includes some of the oldest and most widely used herbicides. These herbicides are applied

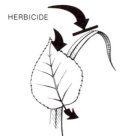

HERBICIDE

to the foliage and are absorbed primarily through the foliage and translocated throughout the plant. They are also selective, so weeds may be treated while the crop is present with little or no injury to the crop. The best example of this type of herbicide is 2,4-D. For example, 2,4-D has many uses; it takes dandelions out of lawns or broad-leaved weeds out of oats, wheat, barley, and corn. Dicamba is used for tough-to-control weeds such as thistle, common milkweed, horse nettle and hedge or field bindweed in corn, grass pastures small grains and range lands. Other common herbicides in this group are 2,4,5-T, and 2,4,5-TP. These herbicides are used for brush control along highways, railroad right-of-ways, fence lines, and in waste areas. They will selectively remove brush without killing grass so that there are no bare areas once the brush dies.

5) *Soil - Short residual - Nonselective*. Only a few herbicides belong to this group. The most common, methyl bromide,

HERBICIDE

2 DAYS

is a gas at normal air temperature and dissipates into the air if the area to be treated is not covered. A sheet of plastic or gas-tight cover must be used and the gas is released under the cover. All weeds, including weed seed, are killed by methyl bromide. The cover must be kept on the treated area for about 24 hours for best results; another 24 hours is required to aerate the soil after the cover is removed. After this 48-hour period, anything can be planted without injury, making methyl bromide one of the shortest residual herbicides available.

6) *Soil - Short residual - Selective*. Any herbicide that is applied to the soil prior to planting a crop or immediately after planting and has a residue of less than one year belongs to this group. This includes

HERBICIDE

6-8 WEEKS

almost all those herbicides used for weed control in vegetable and feed crops as well as for other specialized crops such as flowers and ornamentals. Since the crop is present at the time the herbicide is applied or it is planted soon after application, the herbicide must be selective and should not cause injury to the crop. These herbicides are often referred to as preplant incorporated or pre-emergence-type herbicides. For example, atrazine on corn, linuron on potatoes, or EPTC on ornamentals have residues that last from 6 to 8 weeks to all summer. Because these herbicides are applied to the soil, they are primarily root absorbed although some may be absorbed through the foliage if any is present at the time of application.

7) *Soil - Long residual - Nonselective.* Herbicides that are used to

HERBICIDE

MORE THAN 1 YEAR

control all vegetation for as long as possible are desirable for use near parking lots, around buildings, oil storage areas, factory supply storage areas, etc. For example, bromacil and simazine, at high rates of 10 to 25 lbs. per acre, last for 3 to 5 years. Bromacil will kill brush and trees present in a treated area. Simazine will not kill deep rooted trees since it doesn't leach as readily as bromacil.

8) *Soil - Long residual - Selective.* These herbicides can be used only on deep rooted crops, such as fruit trees, nut trees, cane fruits, and

HERBICIDE

MORE THAN 1 YEAR

grapes. They may be applied to the foliage of the weeds, although most of the herbicide is eventually absorbed through the root system. These herbicides have a low solubility in water and do not leach readily; therefore, they seldom get down to the root systems of the deeper rooted crop. They have a residue and will give weed control for more than one year. For example, simazine applied at rates of about 5 pounds per acre will control shallow rooted weeds without injuring deeper rooted crops. At higher rates, selectivity is lost and it becomes a nonselective herbicide.

B. Classification of Herbicides by Chemical Families

The following does not list all of the herbicides, but is given as an example of chemical classification:

1. Inorganic Compounds
    a. AMS
    b. Boron
    c. Copper sulfate
    d. Sodium chlorate

2. Organic Compounds
    a. Arsenicals (organic)
    b. Amino acids
        glyphosate (Roundup®)
    c. Benzoic acids
        dicamba (Banvel®)
    d. Benzonitriles
        dichlobenil (Casoron®)
        bromoxynil (Brominal®)
    e. Carbamates
        chlorpropham (Chloro-IPC®)
        barban (Carbyne®)

f. Carbanilates
    phenmedipham (Betanal®)
g. Dinitroanilines
    benefin (Balan®)
    nitralin (Planavin®)
    Trifluralin (Treflan®)
h. Dipyridyls
    diquat (Ortho Diquat®)
    Paraquat (Ortho Paraquat®)
i. Halogenated Aliphatic acids
    dalapon (Dowpon®)
j. Phenols
    dinoseb (Sinox®)
k. Phenoxy compounds
    Acetic
       2,4-D
       MCPA
       2,4,5-T
    Butyric
       2,4-DB
       MCPB
    Propionic
       2(2,4-DP)
       2(MCPP)
       Silvex (2,4,5-TP)
l. Phenyl ethers
    nitrofen (TOK®)
    bifenox (Modown®)
m. Phthalic acids
    DCPA (Dacthal®)
    endothal (Aquathol®)
    napthalam (Alanap®)
n. Pyridyls
    picloram (Tordon®)
o. Pyridazinones
    Pyrazon (Pyramin®)
p. Substituted Amides
    alachlor (Lasso®)
    bensulide (Betasan®, Prefar®)
    propachlor (Ramrod®)
q. Thio-carbamates
    butylate (Sutan®)
    diallate (Avadex®)
    EPTC (Eptam®)
    pebulate (Tillam®)
    triallate (Fargo®)
    vernolate (Vernam®)
r. s-Triazines (symmetrical)
    atrazine (AAtrex®)
    cyanazine (Bladex®)
    cyprazine (Outfox®)
    prometone (Pramitol®)
    propazine (Milogard®)
    simazine (Princep®)

s. as-Triazine (asymmetrical)
        metribuzin (Sencor®, Lexone®)
t. Triazoles
        amitrole (Weedazol®, Amino-triazole®)
        methazole (Probe®)
u. Uracils
        bromacil (Hyvar®)
        terbacil (Sinbar®)
v. Ureas
        diuron (Karmex®)
        fenuron (Dybar®)
        linuron (Lorox®)
        monuron (Telvar®)
        siduron (Tupersan®)

HARVEST AIDS AND GROWTH REGULATORS

Harvest aids and plant growth regulators are not herbicides per se, but
they are included in this section because they are used to control plant
growth and they do act on plants much the same as many herbicides.

Harvest aids -- Materials generally referred to as harvest aid chemicals
fall into two classes: (1) defoliants that induce the plant to drop its
leaves, but do not kill the plant; (2) desiccants that kill plant foliage.

These classifications often overlap, depending upon the amount of chemical applied. Before the use of chemicals to defoliate plants, some
crops such as potatoes were defoliated with machines, such as beaters with
rubber flails or chains. Contact chemical vine killers have replaced
machines and are favored by growers because they provide the means of
artificially hastening the maturity of the crop.

Desiccants applied to alfalfa grown for seed purposes is common in some
areas and has replaced mowing and wind-rowing, as was the common practice
for harvesting alfalfa seed. Desiccation for pre-harvest drying of dry
beans has been used in some years when the weather is not right for good
drying and a bean field can remain green up until the frost time.

Some of the commonly used defoliants or desiccants registered by the EPA
are: paraquat, and diquat, endothal, pentachlorophenol, sodium borate and
sodium chlorate.

Plant growth regulators -- Plant growth regulators are used to regulate
or modify the growth of plants. These chemicals are used to thin apples,
control the height of turfgrass, control the height of some floral potted
plants, promote dense growth of ornamentals, and to stimulate rooting.
These are used in minute amounts to change, speed up, stop, retard, or in
some way influence vegetative or reproductive growth of a plant.

A plant is made up of many cells with specialized functions. Plant growth regulators can change or regulate the development of cells. Maleic hydrazide is a plant growth regulator. When it is sprayed on some plants, it is translocated and restricts development of new growth by preventing further cell division. The older cells, affected by the chemical, may continue to grow. In other words, cell division, but not cell maturity is prevented.

Potted plants such as chrysanthemums, easter lillies, and poinsettias may grow too tall. Plant height can be regulated with growth regulators that are applied at the proper stage of growth. Plant growth regulators are used to control vegetative growth and promote earlier flowering and development in some species. Some materials are used to kill shoot tips which results in additional branching.

Plant growth regulators are used on apples and peaches. They are used to increase fruit color and may result in earlier and more uniform ripening of fruit. Plant growth regulators can also be used to thin the fruit, widen branch angles, produce more flower buds, prevent fruit drop, increase fruit firmness, reduce fruit cracking, reduce storage problems, and encourage more uniform fruit bearing.

Rooting hormones are used to increase the development of roots and speed up rooting of certain plants.

Gibberellic acid is the opposite of a growth regulator, it is an accelerator. This is sometimes used to initiate uniform sprouting of seed potatoes and increase the size of sweet cherries.

BULL THISTLE

6
PESTICIDE LAWS AND LIABILITY

Pesticide Laws & Liability

As the use of chemicals to control pests has increased in the United States, the scope of Federal and State laws and regulations has also increased to protect the user of pesticides, the consumer of the treated products, and the environment from pesticide pollution. In 1910, the Federal Government passed the Federal Insecticide Act, which was intended  primarily to protect farmers against adulterated or misbranded products. Prior to that time, there was no protection for the buyer. For the past thirty-five years, the continual development of new and more effective chemicals to control pests has been an exciting story for farmers and the consuming public. Synthetic organic compounds, such as chlorinated hydrocarbons, organic phosphates, carbamates, and phenoxy herbicides have been added to the early arsenal of pesticides such as arsenic, copper and sulfur compounds. During this same period, the United States Department of Agriculture laws were broadened to regulate the labeling and interstate distribution of insecticides, herbicides, fungicides, nematicides, germicides, plant growth regulators, defoliants, dessicants, and rodenticides for use by farmers, home owners, and industry.

The Federal Insecticide, Fungicide, and Rodenticide Act of 1947, added a new concept. It placed the burden of proof of acceptability of a product on the manufacturer prior to its being marketed. The Act was oriented to protect the user, the consumer, and the public from pesticides, some of which are dangerous and all of which are subject to limitations in application. In 1948, the Food and Drug Administration began establishing safe levels of residue tolerances in foods. In no case were tolerances established that exceeded a safety factor of 100-1. In addition to  this safety factor, tolerances were never approved for levels higher than necessary to accomodate registered uses. The Pesticide Chemicals Amendment (the Miller Pesticide Amendment), amended the Federal Food, Drug, and Cosmetic Act in 1954 and formalized the tolerance setting procedures of FDA, and as a matter of policy USDA registered only pesticide uses which would result in no residue or residue levels declared safe by FDA. The pesticide industry was required to submit to USDA data proving the efficacy of the chemical to control the pest and to submit to FDA proof of the safety of any measurable residue in the food produced. In 1958 a Food Additives amendment to the Federal Food, Drug and Cosmetic Act was passed which prescribed regulations for the safe use of food additives. This amendment included the Delaney Clause which prohibits any residue of carcinogenic (cancer producing) chemicals.

FEDERAL AGENCIES REGULATING PESTICIDES

Environmental Protection Agency

The Environmental Protection Agency, created in 1970, is now responsible for the pesticide regulatory functions previously delegated to the departments of Agriculture; Health, Education, and Welfare; and Interior. These responsibilities include the registration of pesticides as required under the Federal Insecticide, Fungicide and Rodenticide Act (FIFRA); the setting of tolerances as required by the Miller Amendment to the Federal Food, Drug and Cosmetic Act; and many of the research and monitoring programs relating to pesticides previously conducted by the three departments.

The Federal Environmental Pesticide Control Act of 1972, and amended in 1975 and 1978, completely revises the Federal Insecticide, Fungicide and Rodenticide Act (FIFRA) which has been the basic authority for federal pesticide regulations since 1947. The new Act regulates the use of pesticide to protect man and the environment and extends federal pesticide regulation to all pesticides including those distributed or used within a single state.

All persons using and applying pesticides should have a general understanding of the laws pertaining to pesticide use and application. The following is a section by section synopsis of the major provisions of the Federal Environmental Pesticide Control Act of 1972, including the 1975 and 1978 amendments. A copy of the complete Act may be obtained from a regional office of the Environmental Protection Agency.

Section 1. Short Title and Table of Contents
Section 2. Definitions
Section 3. Registration of Pesticides--With certain exceptions, any pesticide in U.S. trade must be registered with the Administrator.

All information except trade secrets and privileged information in support of registration must be made available to the public within 30 days of the registration.

A pesticide which meets the requirements of the Act shall be registered and the fact that it is not essential cannot be a criterion for denial of registration.

Registered pesticides will be classified for either general use or restricted use. A pesticide would be restricted because of its potential for harm to either human health or the environment.

The Administrator may require that the packaging and labeling of a pesticide for its restricted uses be clearly distinguishable from its general uses.

A restricted use pesticide which is a hazard to human health is subject to the restriction that it be applied by a certified pesticide applicator who is certified by the State according to standards set by the Administrator, and may be either a commercial or private certified applicator.

Section 4. Use of Restricted-Use Pesticides; Certified Applicators--
The Administrator sets minimum standards for the certification of
applicators, which are to be separate for private and commercial
applicators.

Certification will be accomplished by state programs whose plans are
approved by the Administrator. The Administrator may withdraw approval
of a state plan if the program is not maintained in accordance with it.
In any state for which a state plan has not been approved by the Administrator, the Administrator shall conduct a program for certification of
pesticide applicators.

The Administrator cannot require private applicators to take written
examinations.

Instructional materials concerning integrated pest management techniques
are to be made available to interested individuals.

Section 5. Experimental Use Permits -- The Administrator may issue
permits and set the terms for the experimental use of a pesticide in
order to gather data for registration, and may establish a temporary
pesticide residue tolerance level for that purpose.

The Administrator may authorize a state to issue experimental use
permits under an approved plan.

Section 6. Administrative Review, Suspension -- If the Administrator
has reason to believe a registered pesticide does not comply with the
Act or that it when "used in accordance with widespread and commonly
recognized practice, generally causes unreasonable adverse effects
on the environment," he may move to cancel the registration or change
the classification of the pesticide.

Suspensions, hearings, emergency orders, judicial reviews and condi-
tional registrations are also provided for.

Section 7. Registration of Establishments -- Pesticide producing
establishments are required to be registered with the Administrator.

Producers are required to submit upon initial registration, and annually thereafter, information on the amount of pesticides produced,
distributed and sold.

Section 8. Books and Records -- Allows for EPA officer inspection
provided a written reason is presented.

Section 9. Inspection of Establishments -- The Administrator's agent
may enter any establishment or other place where pesticides or devices
are held for distribution or sale, inspect pesticides or devices, and
take samples. The agent must state a sufficient reason for his action.

Section 10. Protection of Trade Secrets and other information--
Prohibits disclosure of information to the public by the Administrator and federal employees or to foreign producers.

Section 11. Standards Applicable to Pesticide Applicators -- No regulations under the Act may require a private applicator to maintain
records or file reports.

Private and commercial applicator certification standards set by the
Administrator must be separate.

Section 12. Unlawful Acts.

Section 13. Stop Sales, Use, Removal, and Seizure -- When it appears a
pesticide is in violation of the Act or its registration has been suspended or cancelled by a final order, the Administrator may issue a
"stop sale, use or removal" order to any person.

Pesticides in violation of the Act may be seized, as may misbranded
devices and pesticides which have an unreasonable adverse effect on
the environment even when used in accordance with requirements and as
directed on the labeling.

Section 14. Penalties -- Any registrant, commercial applicator, wholesaler, dealer, retailer, or other distributor is liable to a $5,000
civil penalty for an Act violation. Private applicators and other
persons are liable to a $1,000 civil penalty on their second and sub-
sequent offenses. Opportunity for hearing is required prior to
assessment of a civil penalty.

Any registrant, commercial applicator, wholesaler, dealer, retailer,
or other distributor is liable to a penalty of $25,000 or 1 year in
prison or both upon conviction of a misdemeanor. Private applicators
and other persons are liable to a penalty of $1,000, or 30 days in
prison or both. Persons who reveal formula information are liable
to a penalty of $10,000, or three years in prison, or both.

Section 15. Indemnities -- Persons owning a pesticide whose registration is suspended and later cancelled are eligible for payment by the
Administrator of an indemnity for the pesticide owned. Manufacturers
who had knowledge prior to the issuance of a suspension notice that
their products did not meet registration requirements and who continued
to produce the pesticide without notifying the Administrator of the
deficiency would not be eligible.

In lieu or indemnification the Administrator may authorize, under
certain limited conditions, the use or other disposal of a pesticide.

Section 16. Administrative Procedure; Judicial Review

Section 17. Imports and Exports -- The Administrator is required to notify foreign governments through the State Department whenever the registration, or the cancellation of suspension of the registration of a pesticide becomes effective or ceases to be effective.

The Administrator is authorized to examine imported pesticides or devices and have those in violation of the Act refused entry, seized, and destroyed.

Section 18. Exemption of Federal Agencies -- The Administrator may exempt any Federal or State agency from the Act if he determines that emergency conditions exist which require such exemptions.

Section 19. Disposal and Transportation -- The Administrator shall establish procedures and regulations for storage and disposal of pesticides and containers. He is required to accept at convenient locations for disposal a pesticide whose registration is suspended, then cancelled. Notification of cancellation of any pesticide shall include specific provisions for the disposal of the unused quantities of such pesticide.

Section 20. Research and Monitoring -- Authority for pesticide research and monitoring is provided, including research contracts and grants. A National Monitoring Plan is required to be formulated and carried out by the Administrator.

Section 21. Solicitation of Comments; Notice of Public Hearings -- The Administrator is required to solicit the views of the Secretary of Agriculture before publishing regulations under the Act.

The Administrator is authorized to solicit the views of all interested persons concerning any action under the Act and seek advice from qualified persons.

Timely notice of any public hearing to be held shall be published in the Federal Register.

Section 22. Delegation and Cooperation -- The Administrator shall cooperate with the USDA, other federal agencies and any appropriate state agency in carrying out the provisions of this Act.

Section 23. State Cooperation, Aid, and Training -- The Administrator may delegate to any state or indian tribe the authority to cooperate in enforcement of the Act, train state personnel in cooperative enforcement, and assist states to implement cooperative enforcement with grants.

The Administrator may assist states and indian tribes in developing and administering applicator certification programs under cooperative agreements and under contracts.

The Administrator shall, in cooperation with the Secretary of Agriculture, utilize the Cooperative State Extension Service to inform and educate pesticide users about accepted uses and other regulations under this act.

Section 24. Authority of States -- A state may regulate the sale or use of any federally registered device in the state but only if and to the extent the regulation does not permit any sale or use prohibited by this act. However, different state packaging or labeling requirements are prohibited.

A state may provide registration for additional uses of federally registered pesticides formulated for distribution and use within that state to meet Special Local Needs in accordance with the purposes of this Act.

Section 25. Authority of the Administrator -- The Administrator is authorized to prescribe regulations to carry out the provisions of this Act, after consulting with the Secretary of Agriculture and notifying the appropriate committees in the Senate and House of Representatives.

The Administrator is authorized to declare that which is a pest; determine highly toxic substances in pesticides; specify classes of devices subject to certain misbranding provisions; and under Section 7, prescribe pesticide discoloration requirements and determine suitable pesticide names.

Section 26. State Primary Enforcement Responsibility -- A state shall have primary enforcement responsibility for pesticide use vioations providing the state has adopted adequate pesticide use laws and regulations and has adequate procedures for their enforcement.

Section 27. Failure by the State to Assure Enforcement of State Pesticide Use Regulations -- Authorizes the Administrator to rescind a states' primary enforcement responsibility if he determines that the state has been inadequate in its program.

Section 28. Identification of Pests; Cooperation with Department of Agriculture Program -- The Administrator, in coordination with the Secretary of Agriculture, shall identify those pests that must be brought under control.

Section 29. Annual Report

Section 30. Severability.

Section 31. Authorization for Appropriations.

United States Department of Agriculture

The consumer and marketing services of the meat and poultry inspection program monitors the quality of meat and poultry products. Sampling is scheduled on a national basis through the use of a computer. Samples are identified as to source and questionable samples are followed up with inspections at the product's origin where appropriate action is taken to prevent the marketing of contaminated meat.

The Food And Drug Administration

The FDA still retains the responsibility to monitor food for humans and feed for animals. Any products violating pesticide residue tolerances are subject to seizure by the FDA. The Federal Food, Drug and Cosmetic Act delegates this authority.

The Federal Aviation Administration

The FAA under the Federal Aviation Regulation, part 137 -- Agricultural Aircraft Operations, January 1, 1966, regulates the dispensing of pesticides by aircraft. Under these regulations, as amended, it is a violation for an aerial applicator to apply any pesticide except according to Federally registered use.

The Federal Department of Transportation

The hazardous nature of pesticides requires strict interstate regulation for their transport. The Federal Department of Transportation (DOT) is the major agency involved in enforcing hazardous material laws. Individuals directly involved in transporting hazardous pesticides should know the following:

Hazardous materials are classified as -

Class A Poisons -- Extremely dangerous poisons, so poisonous that a small amount of gas, vapor, or liquid mixed with air would be dangerous to life.

Class B Poisons -- Less dangerous poisons, including liquids, solids, pastes, and semi-solids which are known to be toxic to man and afford a health hazard during transportation. Most pesticides fall in this group.

Class C Poisons -- Tear gas and irritating substances. Materials that upon contact with fire or exposed to air give off dangerous fumes. Does not include any material included under Class A.

Each hazardous material transported must be packaged in the manufacturer's original container. Each container must meet the DOT standards.

Each vehicle carrying a hazardous material must have the proper placard (sign). The manufacturer is responsible for placing the proper warning signs on each box or container containing Class A or B poisons.

DOT regulations prohibit carriers from hauling Class A and B poisons or other hazardous materials in the same vehicle with food products.

DOT regulations require carriers of hazardous materials to immediately notify DOT after each accident when a person is killed, receives injuries requiring hospitalization, or when property damage exceeds $50,000.

STATE AGENCIES REGULATING PESTICIDES

Various state agencies including the Department of Agriculture, State Department of Environmental Protection (or similar designation), Labor Department, and others have one or more laws pertaining to pesticide use and application. These include laws for pesticide registration and sale, laws regulating commercial applicators, and structural pest control applicators.

Individuals planning to sell or apply pesticides should obtain copies of the applicable laws from the appropriate agencies in their state. Check the Index for the addresses of the various State regulatory agencies.

RESTRICTED USE PESTICIDES

The following pesticides are those that have been classified as "Restricted Use" by the Environmental Protection Agency. Most of these pesticides have been classified by regulation, i.e., they have been reviewed by EPA and determined to fit the requirements for restricted use pesticides. Approximately one-fourth of the pesticides have been classified as restricted at the time of their registration. This list will grow as the EPA completes its review of all previous registrations, and whenever a newly registered material requires it.

The chemicals' common name is listed first, followed by a trade name (or names) if there is one, and finally the type of pesticide it is. Trade names are given only as an example and may not include all of them. No endorsement is intended of those trade names shown, nor is discrimination intended against any trade names not shown.

Common Name	Trade Name	Type of Pesticide
acrolein	Aqualin®	Aquatic herbicide
aldicarb	Temik®	Insecticide
allyl alcohol	(None)	Soil fungi & Weed Seed Killer
aluminum phosphide	Phostoxin®	Insecticidal Fumigant
amitraz	BAAM®	Insecticide/Miticide
azinphos - methyl	Guthion®	Insecticide
calcium cyanide	(None)	Insecticidal Fumigant/ Rodenticide
carbofuran	Furadan®	Insecticide
chlordimeform	Galecron®/Fundal®	Insecticide/Acaricide
chlorfenvinphos	Supona®	Insecticide/Acaricide
chlorobenzilate	Acaraben®	Miticide
chlorophacinon	Rozol Blue®	Rodenticide
chloropicrin	(Many)	Soil fumigant for all life forms
chlorpyriphos	Killmaster II®/Dursban®	Insecticide
clonitralid	Bayluscide®	Molluscicide
cycloheximide	Acti-Aid®/Acti-Dione®	Growth Regulator/ Antibiotic Fungicide
DBCP	(None)	Nematicide
demeton	Systox®	Insecticide
diclofop methyl	Hoelon®	Herbicide
dicrotophos	Bidrin®	Insecticide
diflubenzuron	Dimilin®	Insect Growth Regulator
dioxathion	Delnav®/Deltic®	Insecticide/Acaricide
disulfoton	Di-Syston®	Insecticide
dodomorph acetate	Milban®	Fungicide
endrin	(Many)	Insecticide
EPN	(Many)	Insecticide
ethoprop	Mocap®	Nematicide/Insecticide
ethyl parathion	(Many)	Insecticide

Restricted Use Pesticides (continued)

Common Name	Trade Name	Type of Pesticide
fenamiphos	Nemacur®	Nematicide
fensulfothion	Dasanit®	Insecticide/Nematicide
fenvalerate	Pydrin®	Insecticide
fluoracetamide	1081®	Rodenticide/Predicide
fonophos	Dyfonate®	Insecticide
heptachlor	Heptachlor®	Insecticide
hydrocyanic acid	Fumico hydrocanic acid®	Fumigant Rodenticide
isofenphos	Amaze®	Insecticide
magnesium phosphide	Fumi-cel®	Rodenticide
mercaptodimethur	Mesurol®/Mesrepel®	Insecticide/Bird Repellent
methamidophos	Monitor®	Insecticide
methidathion	Supracide®	Insecticide/Miticide
methomyl	Nudrin®/Lannate®	Insecticide
methyl bromide	(Many)	Fumigant for all life forms
methyl parathion	(Many)	Insecticide
mevinphos	Phosdrin®	Insecticide
monocrotophos	Azodrin®	Insecticide
nicotine (alkaloid)	Nico-fume®	Fumigant Insecticide
nitrofen	Tok®	Herbicide
paraquat	Gramoxone®	Herbicide/Desiccant
permethrin	Ambush®/Pounce®	Insecticide
phorate	Thimet®	Insecticide
phosacetim	Gophacide®	Rodenticide
phosphamidon	(None)	Insecticide
picloram	Tordon®	Herbicide
pronamide	Kerb®	Herbicide
sodium cyanide	(None)	Predicide
sodium fluoracetate	1080®	Predicide/Rodenticide
strychnine	(Many)	Rodenticide/Avicide
sulfotep	Flora-fume Dithione®	Insecticide/Acaricide
TEPP	Vapotone®	Insecticide
zinc phosphide	(Many)	Rodenticide
sulprophos	Bolstar®	Insecticide

Some states have their own "restricted use" list in addition to the pesticides shown above. Some states also have additional requirements that must be satisfied before a permit can be issued for the use of certain pesticides.

A pesticide applicator contemplating the use of Restricted Use pesticides must not only be certified to use those pesticides, but must be aware of any state laws that may be in effect regarding use and application of certain pesticides.

CANADIAN PEST CONTROL PRODUCTS REGULATIONS

All persons using and applying pesticides in Canada should have a general understanding of the laws pertaining to pesticide use and application. The following is a section by section synopsis of the major provisions of the Pest Control Products Regulations established in 1972 and amended in 1973, and again in 1977. The complete Act is published in Part II of the Canada Gazette and should be consulted for all purposes of interpreting and applying the Regulations. A copy may be obtained from Agriculture Canada or from the regulatory offices in provinces. Refer to the Appendix for the appropriate offices and addresses.

REGULATIONS MADE PURSUANT TO THE PEST CONTROL PRODUCTS ACT

A Short Title

1. These Regulations may be cited as the Pest Control Products Regulations.

Interpretation

2. Definitions of Act, active ingredient, applicant, assessed or evaluated, certificate of registration, device, Director, display panel, District Director, metric unit, Plant Products Division, registrant, residues, and seed.

Exemption of Certain Control Products

3. The following control products are exempt from the Act:

   (a) A control product that is subject to the Food and Drugs Act and is only used for:

      (i) the control of viruses, bacteria, or other micro-organisms on or in humans or domestic animals,

      (ii) the control of arthropods on or in humans, livestock or domestic animals if the control product is to be administered directly and not by topical application,

      (iii) the control of micro-organisms on articles that are intended to come directly into contact with humans or animals for the purpose of preventing or treating disease when associated with medical care,

      (iv) the control of micro-organisms in premises in which food is manufactured, prepared or kept, or

      (v) the preservation of foods for humans during cooking or processing; and

   (b) a control product that is a device other than a device of a type and kind listed in Schedule I.

Application

4. These Regulations do not apply to a control product, other than a live organism, that is imported into Canada for the importer's own use, if the total quantity of the control product being imported does not exceed 500 grams by mass and 500 millimeters by volume and does not have a monetary value exceeding $10.

Exemption from Registration

5. A control product is exempt from registration if

   (a) it is a control product, other than a live organism, that is used only in the manufacture of a registered control product and conforms to the relevant specifications of that registered control product, set forth in the register of control products;

   (b) it is for use by a person for research purposes

      (i) on premises owned or operated by that persons, or

      (ii) on any other premises not owned or operated by that person, if such use has been approved by the Director; or

   (c) it is a control product that

      (i) is a substance or a thing the primary purpose of which is not for controlling, preventing, destroying, mitigating, repelling or attracting any pest, but is represented as having such properties or contains an active ingredient possessing such properties, and

      (ii) is of a type and kind listed in Schedule II and meets the conditions relevant to that substance or thing set forth in that Schedule.

Registration of Control Products Required

6. Subject to Section 5, every control product imported into or sold in Canada shall be registered in accordance with these Regulations.

Application for Registration

7. An application for a certificate of registration shall contain the name and address of the applicant or agent; name and address of the manufacturer; brand name, if any; active ingredient; size, type, and specification of the package; guarantee statements; and other relevant information.

8. An applicant or registrant who is not resident in Canada shall appoint an agent permanently resident in Canada to whom any notice or correspondence under the Act and these Regulations may be sent.

9. (1) In addition to the information required by section 7, an applicant shall provide the Minister with such further or other information as will allow the Minister to determine the safety, merit, and value of the control product.

(2) Without limiting the generality of subsection (1), where a control product

(a) is a device that has not been previously assessed or evaluated for the purposes of the Act and these Regulations or contains an ingredient that has not been so assessed or evaluated, the applicant shall provide the Minister with the results of scientific investigations respecting effectiveness; safety to persons occupationally exposed to it when it is manufactured, stored, displayed, distributed or used; safety to the host plant, animal or article in relation to which it is used; effects of the product on non-target organisms; degree of persistence, retention and movement of the control product and its residue; suitable methods of analysis for detecting the active ingredient and measuring the specifications of the control product; suitable methods for analysis in food, feed and the environment; suitable methods for the detox- ification or neutralization of the control product in soil, water, air or on articles; suitable methods for the dis- posal of the control product and its empty packages; stability of the control product under storage conditions; and the compatability of the control product with other products; or

(b) is intended for use on living plants or animals or pro- ducts derived therefrom which plants, animals or products are for human consumption, the applicant shall provide the Minister with the results of scientific investigation respecting effects of the control product or its residues when administered to test animals for the purposes of assessing any risk to humans or animals; and the effects of storing and processing food or feed, in relation to which the control product was used, on the dissipation or degradation of the control products and any of its residues.

10. Every application for a certification of registration shall be accompanied by five copies of the label for the control product or reasonable facsimiles thereof.

11. When requested to do so by the Minister, the applicant shall provide a sample of the control product; a sample of the technical grade of its active ingredient; and a sample of the laboratory standard of its active ingredient.

Fees on Applications

12. Fees in various amounts are required for certain products or devices.

Registration

13. Requirements for registration including specifications of each control product; label for each product; registration number assigned to each product; and such other information as the Minister deems necessary.

Duration and Renewal of Registration

14. Duration of certificates of registration; expiration; and renewal requirements.

    (1) Requires re-registration of products registered prior to 1977.

    (2) Provides for renewal periods not exceeding two years.

15. Where the registrant intends to discontinue the sale of a control product, he shall so inform the Minister and the registration of that control product shall, on such terms and conditions, if any, as the Minister may specify, be continued to allow any stocks of a control product to be substantially exhausted through sales.

Temporary Registration

16. The Minister may, upon such terms and conditions, if any, as he may specify, register a control product for a period not exceeding one year where the applicant agrees to endeavor to produce additional scientific or technical information in relation to the use for which the control product is to be sold; or the control product is to be sold only for the emergency control of infestations that are seriously detrimental to public health, domestic animals, natural resources or other things.

Refusal to Register

17. The Minister may refuse to register a control product if, in his opinion the application does not comply with the Act and these Regulations; the information provided is insufficient; the merit or value of the control product has not been established in accordance with its label directions; the use of the control product would lead to an unacceptable risk of harm to things on or in relation to which the control product is intended to be used, or public health, plants, animals or the environment; or the control product is not required to be registered.

Cancellation and Suspension of Registration

18. During the period of registration of a control product, the registrant shall, when requested to do so by the Minister, satisfy the Minister that the availability of the control product will not lead to an unacceptable risk of harm to things on or in relation to which the control product is intended to be used; or public health, plants, animals or the environment.

19. The Minister may, on such terms and conditions, if any, as he may specify, cancel or suspend the registration of a control product when, based on current information available to him, the safety of the control product or its merit or value for its intended purposes is no longer acceptable to him.

20. Requires the Minister to provide information to the applicant regarding his reasons for cancellation or suspension.

21. Provides for continued dealer sales providing the dealer had the product for sale prior to the Minister's suspension order.

22. Provides for hearing for those applicants affected under section 20.

23. Provides for the appointment of a review board for hearings as provided in section 22.

24. Sets forth the conditions and requirements of the board hearing and requirements by the board and the Minister.

Records

25. Every registrant shall make a record of all quantities of a control product stored, manufactured or sold by him and the record shall be maintained for five years from the time it is made; and be made available to the Director at his request at such time and in such manner as the Director may require.

Labeling

26. No label shall be used on a control product unless it has been approved by the Minister and, unless the Minister otherwise directs, every label show the the information required by sections 27-37. This section provides for labeling of "RESTRICTED" and "DOMESTIC" pesticides as well as the requirement for "READ THE LABEL BEFORE USING"; "GUARANTEE"; "REGISTRATION NUMBER"; "PEST CONTROL PRODUCT ACT"; "FIRST AID INSTRUCTIONS"; "TOXICOLOGICAL INFORMATION"; and NOTICE TO USER".

27. The label for a control product that is a device of a type and kind listed in Schedule I shall contain the information referred to in section 26.

28. The display shall consist of one principle display panel and at least one secondary display panel.

29. Additional requirements to labeling set forth in section 26.

30. Further clarification of requirements for registration under section 26.

31. Clarification of requirements in section 30.

32. Not withstanding sections 29, 30 and 31, the Minister may, for reasons satisfactory to him, approve the inclusion of the information required by those sections elsewhere than on the display panel.

33. Additional requirements regarding the product class designation "RESTRICTED".

34. When the information required to be shown on the label is, pursuant to section 32, not included in the display panel, the display panel shall contain the words in capital letters "READ ATTACHED BROCHURE (OR LEAFLET) BEFORE USING" prominently displayed thereon.

35. Subject to the approval of the Minister, additional information relating to the control product and any graphic design or symbol may be shown on the label if it does not unreasonably detract or obscure the information required to be shown on the label.

36. Pertains to a statement regarding limitation of warranty.

37. Requirements for bulk container information.

38. The information on every label shall be printed in either the English, or the French language or both.

Units of Measurement on Labels

39. All units of measurements shown on a label shall be expressed only in metric units in accordance with the Rates and Measures Act. This section provides for additional information regarding net quantities and the metric units required in the declaration of net quantities.

40. All information shown on a label shall be printed in a manner that is conspicuous, legible and indelible.

Denaturation

41. Where the physical properties of a control product are such that the presence of the control product may not be recognized when it is used and is likely to expose a person or domestic animal to severe health risks, the control product shall be denatured by means of color, odor, or such other means as the Minister may approve to provide a signal or a warning as to its presence.

Storage and Display

42. A control product shall be stored and displayed in accordance with any condition set forth on the label, and a control product bearing the poison symbol superimposed on a danger symbol and shall be stored and displayed apart from food for humans or feed for animals in a separate room; or separated by a physical barrier so as to avert the contamination of the food or feed.

Distribution

43. A control product shall be distributed in a manner that is consistent with any special condition specified by the Minister and, when required by the Minister, the condition shall be shown on the label; and on the shipping bill respecting the control product or on a statement accompanying the shipment.

Prohibitions Respecting Use

44. (1) No person shall use a control product in a manner that is inconsistent with the directions or limitations respecting its use shown on the label.

    (2) No person shall use a control product imported for the importers own use in a manner that is inconsistent with the conditions set forth on the importer's declaration respecting the control product.

    (3) No person shall use a control product that is exempt from registration under paragraph 5 (a) for any purpose other than the manufacture of a registered control product.

Packaging

45. (1) The package for every control product shall be sufficiently durable and be designed and contructed so that it will contain the control product safely under practical conditions of storage, display and distribution.

    (2) Every package shall be designed and constructed to permit the withdrawal of any or all of the contents in a manner that is safe to the user; and the closing of the package in a manner that will contain the control product satsifactorily under practical conditions.

    (3) Every package shall be constructed so as to minimize the degradation or change of its contents resulting from interaction or from the effects of radiation or other means.

    (4) When the package is essential to the safe and effective use of the control product, it shall be designed and constructed to meet specifications acceptable to the Minister on registration of a control product.

Standards

46. Every control product shall conform to the specifications and bear the label contained in the register of control products.

47. Every control product shall have the chemical and physical composition and uniformity of mix necessary for it to be effective for the purposes for which it is intended.

General Prohibitions

48. A control product shall not contain an active ingredient unless it is present in an amount sufficient to add materially to the effectiveness, merit or value of the control product.

49. A label shall not contain any information respecting any organism or causative agent of a disease of humans mentioned in Schedule A to the Food and Drugs Act.

50. Unless otherwise authorized by the Minister a label shall not contain any information respecting any organism or causative agent of a disease of domestic animals that is required to be reported under the Animal Contagious Diseases Act.

Sampling

51. Provides for sampling of control products by an inspector in a manner approved by the Director.

Detention

52. A control product seized pursuant to section 9 of the Act may be detained by an inspector at any place by attaching a detention tag to at least one package of the control product in the lot that has been seized. This section stipulates further provisions for detention of products.

53. Where a control product has been seized and detained by an inspector, the registrant shall be entitled to a hearing if he so requests, and sections 23 and 24 apply in respect of the hearing.

Import

54. A control product may be imported into Canada if it is accompanied by a declaration, in a form specified by the Minister, which form shall be signed by the importer and shall state the name and address of the person who is shipping the control product; the name and brand, if any, of the control product; common or chemical names; total amounts being imported; name and address of the importer; and purpose of the importation of the control product using the words "for resale" or "for manufacturing purposes" or "for research purposes".

55. This section was revoked in 1977.

56. Where the collector of customs at a port of entry is not satisfied that an importer's declaration is complete and in order, he shall hold the control product at the port of entry or place the control product in bond and forthwith advise a District Director.

57. The collector of customs at a port of entry shall forward one copy of every importer's declaration to a District Director.

Transitional

58. This is a "grandfather clause" section regarding pest control products that were registered under the former Act and Regulations.

59. Additional information regarding pest control products that were not registered or required to be registered under the former Act and Regulations.

60. Sections 58 and 59 do not apply to a control product that is listed or is of type or kind listed in Schedule IV.

SCHEDULE I

1. Garment bags, cabinets or chests that are manufactured, represented or sold as a means to protect clothing or fabrics from pests;

2. Apparatuses that are manufactured, represented or sold as a means to attract or destroy flying insects, or to attract and destroy flying insects;

3. Devices that are manufactured, represented or sold as a means to repel pests by causing physical discomfort by means of sound or touch;

4. Devices for attachment to garden watering hoses that are manufactured, represented or sold as a means to dispense or apply a control product;

5. Devices that are manufactured, represented or sold as a means of providing the automatic or unattended application of a control product.

6. Devices that are sold for use with chemical products containing cyanide as a means to control animal pests.

SCHEDULE II

1. Feed for animals;

2. Fertilizer that is subject to the Fertilizers Act if the control product contained therein is registered under these Regulations;

3. Seed that has been treated with a control product registered for the purpose of treating seed if

   (a) the seed is sold and shipped in bulk and the shipping documents bear information setting forth the common name or chemical name of the active ingredient of the control product used to treat the seed; and

   (b) where the seed is packaged, the package bears a label with the words "This seed is treated with", followed by the name of the control product including the common name or chemical name of its active ingredient together with the appropriate precautionary symbols and signal words selected from Schedule III and such other statements as are required by these Regulations and are applicable to the control product used to treat the seed.

SCHEDULE III

Precautionary Symbols and Signal Words

Signal Word	Symbol	Signal Word	Symbol
Degree of Hazard		Nature of Primary Hazard	

1. Danger

2. Warning

3. Caution

1. Poison

2. Corrosive

3. Flammable

4. Explosive

SCHEDULE IV

Control products that are manufactured, represented or sold as a means to

(a) control the axillary "sucker" growth of tobacco.

(b) extend the dormancy period of vegetables or control the respiration rate of fruit or vegetables held in storage.

CANADIAN REGULATIONS IN THE PROVINCES

Each province in Canada has its own regulations regarding pesticide use and application. Individuals contemplating pesticide application in any province should contact the proper authorities regarding local laws. The Appendix contains the addresses of the Canadian provincial regulatory offices.

LIABILITY

Responsible pesticide applicators rarely have legal claims brought against them. However, most pesticides are considered hazardous material and even the most careful applicators have claims for damages brought against them. The usual claims are for nonperformance when the grower feels that a pesticide application did not do the job that was expected. More often, the claims are for injury to crops that were treated or to crops in nearby fields. It is important for all licensed pesticide applicators to be informed of the most common legal claims that can be made against them.

Drift

Drifting pesticides are a major cause of environmental contamination and damage to non-target areas. In general the courts have held that the applicator and the grower that hired him are jointly liable in drift cases. The grower is responsible when he hires or contracts for a "particularly dangerous operation" such as the application of pesticides. However, don't depend on the grower to share costs. He may file a counter suit against you claiming that you agreed not to cause drift damage. The manufacturer of the pesticide may be held liable in drift cases in certain instances. If the label doesn't clearly warn about the possibility of drift, the manufacturer may have to share the liability.

Crop Injury

Claims of injury to the crop that was treated or claims that the pesticide had not performed as expected, involve the dealer, the manufacturer, and the applicator. The courts must decide which of the three recommended or guaranteed the product for that specific use on that crop. The party in error must accept the blame and pay damages. Applicators must make sure that all the pesticides they use are recommended on the label for that purpose.

If the crop injury was not great or total, the grower must show how much damage was from the pesticide and how much was from other conditions such as weather, disease, etc. This breakdown is not required in cases where there is total injury.

Personal Injury

The application of pesticides is considered an especially dangerous or, in legal terms an "ultra-hazardous" activity. As a result the pesticide applicator is liable for any injury to a person from a pesticide. Usually the injured person can recover damages without proving negligence on the part of the applicator. The injured party must only prove he was free of any negligence and did not assume the risk of pesticide exposure. Pest control operators are in a somewhat different category. The liability in most cases involving personal injury or death depends on proving the pest control operator is negligent.

Wrong Field

If the pesticide is applied on a field, crop, or area other than the one it was intended for, serious problems can result. If the owner did not want the area treated,the applicator may be charged with trespass. In the event that the damage involves residue or over-tolerance, the applica- tor may be liable for damages involving the entire crop. Some pesticides are highly persistent and last for long periods of time, while others are not persistent and last only a few days. Naturally, the persistent pes- ticides are more likely to cause long-lasting residues. Some examples of insecticide in the environment include Sevin 7-10 days; Diazinon 2-4 weeks; Dursban 10-14 days; Parathion 3-7 days; Chlordane 2-4 weeks; and Malathion 5-10 days. The pesticide applicator must do everything he can to eliminate residue problems. Defense is very difficult. Double check on addresses, field locations, and all landmarks before you treat an area. Applying pesticides to the wrong field or area can be costly.

Honey Bees

Honey bees are very important to the farmer and to the beekeeper who makes his living raising and caring for honey bees. Bees are insects, and un- fortunately they are very susceptible to many pesticides. If the bees in hives are killed as the result of drift from nearby fields, the appli- cator is usually held legally responsible and often he must pay damages. However, if the bees contacted the pesticides while in the sprayed fields, the applicator is usually not liable. The courts have ruled that the bee is trespassing, and that the land doesn't need to be safe to uninvited animals. Pesticide applicators should know where the bee hives are lo- cated in their area. Warn the beekeeper before hand when and where you will be spraying.

Attractive Nuisance

The rulings on "attractive nuisance" usually involve cases when children are attracted to ground equipment or aircraft and injure themselves. The owner and/or applicator is held liable for leaving the "nuisance" where a child could be "attracted" to it. Therefore, be very careful. Do not leave ground equipment with exposed drive belts, drive wheels, gears, or any moving parts alone in areas where children can get to them. Air- craft should never be parked where curious children can find them. Empty containers and aerosol cans are also attractive and dangerous to children. Be sure to store or to dispose of them properly. (See Section on PESTI- CIDE TRANSPORTATION, STORAGE, DECONTAMINATION, AND DISPOSAL).

Noise

Recently claims have been brought against applicators for noise damage. Owners of specialty operations such as mink raising, poultry farms, and occasionally cattle ranches may claim injury to their animals from fright caused by noise of aircraft and ground equipment operating above or near their place of operation. They must prove direct loss of property due to noise from machinery operated carelessly or negligently. In some cases, the ranch owner may claim that the applicator made an unlawful flight over his property without permission. This is especially important in aerial applications when pull-ups over nearby property are necessary. Successful defense is possible when the applicator can show that the noise wasn't the cause of injury or that no injury occurred.

Cross Contamination

There are three main ways that cross contamination may occur.

1. The manufacturer may make a mistake in labeling or formulating a product. In these cases the pesticide container has not only the pesticide named on the label, but another pesticide also. The materials in the container may then damage crops being treated.

2. The applicator may error in mixing or filling the spray tank. Or he may not have removed from the spray tank all of the pesticide left over from a previous application.

3. Open containers of herbicides such as 2,4-D can vaporize and penetrate other pesticides which are stored nearby. When the other pesticides are applied, the 2,4-D contamination can seriously injure the crops.

The applicator must know which container of pesticide was used on the crop so that laboratory tests can be made. The lab test can show whether the contamination occurred during mixing and filling, or was the fault of the manufacturer. In cases involving herbicide contamination, it is difficult to prove whether it is a result of vaporization during storage or a manufacturer formulation error. The courts must decide who is to blame.

All applicators should keep a careful record of the location and the date that a particular pesticide was applied. They should also record the lot number of the pesticide in case of cross contamination. (See Section on RECORD KEEPING).

INSURANCE

To protect himself and his business, the applicator should have insurance
for possible pesticide mishaps. There are many different types of insurance plans ranging from bodily injury, property damage, restricted chemical
liability, and comprehensive chemical liability. The plan you choose should
fit your needs and your business. If you are in a relatively low risk position from a legal standpoint, then minimum coverage may be enough. On the
other hand, aerial applicators will probably need more insurance. Be sure
to explore the costs, benefits, and drawbacks of insurance before you buy.
You need to know exactly what you are covered for. An insurance agent, who
specializes in pesticide insurance is the best person to advise you on
your individual insurance needs.

What to do When You Are Involved

No matter how careful you are, accidents will happen. You must be prepared
for any emergency. If you are sued or become involved in any legal problem act carefully and promptly. Always be friendly and helpful. Never ad-
mit liability. Be careful who you give information to about your spray
operation. Offer to look into the matter at once.

▲ Examine your records to make sure that you were actually operating in
the area at the time of the alleged injury.

▲ Make sure that all of your records are up-to-date, particularly as to
the identity of the equipment used, temperature, and wind direction
and velocity, and all other pertinent information.

▲ Proceed to the scene immediately and make notes of all essential information when you get there.

▲ Record the presence of any adverse condition that you observed at the
time of your investigation, particularly insect infestations, disease,
plant stress, late planting, carry-over effect from other materials
or herbicides, and general condition of the crop.

▲ Photograph any adverse conditions with color film at a sufficiently
close focal length, so that the symptoms can be examined by an expert.

▲ Collect samples of crop injury including roots so that the materials
can be examined by an independent expert.

▲ If the container from which the product was removed that was used on
the job is still available, it is wise to keep it for laboratory examination. If the container is not available, check your records to
see if you at least have the lot number for that particular pesticide.

▲ Notify your insurance company immediately.

▲ If you do not have insurance for the loss involved, request permission to have an expert examine the crop or the property, in order that you may have the benefit of his opinion.

▲ If a chemical company is likely to be involved because of a particular pesticide used, notify them immediately. The company will probably want to send experts to the site also.

▲ Obtain the names and addresses of all witnesses who might testify as to the nature of the operation and conditions of the crop before and after treatment. In the event that the crop is perennial such as fruit trees or wind breaks, USDA aerial photographs may be available for your examination to ascertain the condition of the trees or other perennial plantings in years prior to the year of alleged injury.

READ THE LABEL

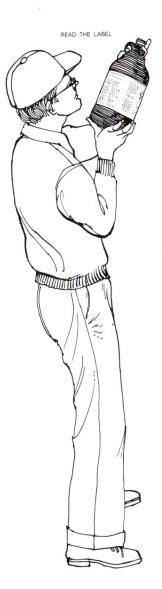

The Pesticide Label

The label is the key to the information about any pesticide you are planning to use. The label on the bottle, can or package of a pesticide tells you the most important facts you must know about that particular pesticide for safe and effective use. The information on the label gives not only the directions on how to mix and apply the pesticide, but also offers guidelines for safe handling, storage, and protection of the environment.

Information on pesticide labels has been called the most expensive literature in the world. The research and development that leads to the wording

READ THE LABEL

on a label frequently costs millions of dollars. The combined knowledge of laboratory and field scientists, including chemists, toxicologists, pharmacologists, pathologists, entomologists, weed scientists and others in industry, universities, and government is used to develop the information found on the label.

To appreciate the value of the information on the label, consider the time, effort, and money spent in gathering and documenting the label information.

Chemical companies all over the world continually make new compounds and then screen them in the laboratory and greenhouse for possible pesticide use. For each material that finally meets the standards of a potential pesticide, thousands of other compounds are screened and discarded for various reasons. When a promising pesticide is discovered, its potential use must be evaluated. If the company believes it has a worthwhile product and there is a possibility for reasonable sales volume, wide scale testing and label registration procedures are begun. From this point of development until the pesticide reaches the market, it requires from 7 to 10 years with costs reaching to 10 to 12 million dollars.

Many kinds of carefully controlled tests must be made to determine the effectiveness and safety of each pesticide under a wide range of environmental conditions. The facts below can help us better appreciate the effort that goes into the pesticide label and the value of the information on the label to the user.

   --Toxicological tests: These tests determine the possible hazards of the
     new pesticide to man, animal life and the environment. The pesticide
     is fed by mouth and applied to the skin of test rats, rabbits and
     other animals. Tests are conducted to determine if it has gases or
     vapors that would harm the skin or the eyes. Extended feeding of test
     animals is also carried out to determine if the pesticide will cause
     cancer or affect the offspring of the test animal.

--Degradation studies: These studies are made to determine how long it takes for the compound to breakdown into harmless materials under various conditions.

--Soil movement tests: Tests are performed to determine how the pesticide moves in the soil and ground water, and how long it remains in the soil where it might be absorbed by crops planted later in the same field.

--Residue tests: These tests determine how much of the pesticide or its breakdown products remains on the crops. Similar tests determine residues (if any) in the meat of cattle, swine, and poultry and also in milk and eggs. From this data, the number of days from the last pesticide application until harvest or slaughter are determined. Tolerances for the amount of pesticide (residue) may legally remain on or in food or feed are also established. A tolerance is set for the pesticide on each crop or commodity and is established by the Environmental Protection Agency as the amount that can safely remain on a food or feed commodity without hazard to the consumer. All food or feed commodities must not be above the pesticide residue tolerance limits when ready for market or for farm feeding. If a residue is over the tolerance level, the commodity can be condemned and the producer penalized.

--Wildlife tests: Tests are performed to determine immediate and long range effects of field application of pesticides on wildlife. The possibilities of residue building up in wildlife are checked on mammals, fish, and birds.

--Performance tests: The producing company must prove that the new pesticide has practical value as a crop protection tool. Performance data must be collected for each pest and from each crop or animal species on which the material is to be used. Information on crop varieties, soil types, application methods and rates, and number of applications required are also needed. This information must prove that the pests are controlled, crops or animals are not injured, that yield and/or quality has been improved, and that the pesticide definitely provides a worthwhile benefit.

--Label review: After all of the tests described above have been completed the chemical company is ready to take their data to the Pesticide Regulatory Division, Environmental Protection Agency, Washington,D.C.

The company asks for a Registered Use of the pesticide on as many pests, crops, and animals as they have test data to support their claim.

The request for label registration must now pass three reviews: (1) a review to determine the effectiveness of the pesticide for the purpose or purposes claimed; (2) a review by the tolerance section where the environmental, degradation, residue, and toxicological data are studied. This section sets the tolerance levels for the pesticide, and they are published in the Federal Register; (3) a final review by the Pesticide Regulation Division to determine if all claims on the label are supported by data.

WHAT IS ON A PESTICIDE LABEL?

The printed material on a pesticide label has all of the necessary information and instructions for the effective and safe use of the pesticide. It must, by law, include the following:

<u>Brand name, Common name and Chemical name</u> -- The brand name is the producers or formulators proprietary name for the pesticide. The common

name is the generic name accepted for the active ingredient of the pesticide product regardless of the brand name. The chemical name of the pesticide is the one that is long and difficult to pronounce unless you are a chemist. For example, Benlate ® is a brand name and the common name for this fungicide is benomyl. The chemical name of the active ingredient in Benlate ® is [Methyl 1-(butylcarbamoyl) 2-benzimidazolecarbamate].

<u>Use Classification</u> -- All pesticide labels must have a use classification statement. Pesticides are classified for general use or restricted use and the label will bear the following statement on restricted use pesticides:

> *"For retail sale and use only by Certified Applicators or persons under their direct supervision and only for those uses covered by the Certified Applicators' certification."*

<u>Ingredients</u> -- The ingredient statement tells you what is in the container. Ingredients are usually listed as active and inert (inactive). The active ingredient(s) is a chemical(s) that does the job that the product is intended to do. The chemical name must be listed for the active ingredient as well as the percent or pounds of active ingredient in the container. The inert ingredient(s) make up the balance of the contents. They may or may not be named, but the total percentage or amount of these ingredients is usually listed. Inert ingredients may be the wetting agent, the solvent, the carrier, or filler to give the product its most desirable quality and quantity for proper use.

<u>Uses of the Pesticide</u> -- The label lists only the uses for which the pesticide is registered. These are the legal uses for the product. It should not be used for any other purpose or in any other manner not listed on the label.

Some products are registered for use only in specific areas or states. The use may or may not be legal for the area in which you live.

Remember, the purpose for which you intend to use a particular pesticide must appear on the label when you buy the product.

<u>Directions for Use</u> -- These are specific directions for using the product properly and tell you when to use the pesticide, how to apply the particular formulation, where to apply it, and dosage rates or how much to use.

Directions also give you information on the number of applications that can be made over a given period of time and the length of time and days that you must allow between the last application and harvest or slaughter.

On some pesticide labels, you will find a range for the suggested dosage. For example, the label may suggest 1/2 to 2 pints/acre. In this case you will want to use the lowest rate that you can that will give the best control needed. If you have questions or doubts be sure to contact an agricultural authority for suggested rates that work well in your area.

Safety Information, Signal Words , and Precautions -- The label contains the information you need to use the product safely. Certain signal words are required on every label. These words are DANGER-POISON and the skull

and crossbones (all in red) or WARNING or CAUTION, depending on the hazard of the particular product to the user. In addition, the label must carry the statement KEEP OUT OF THE REACH OF CHILDREN. The signal word required, depends upon the toxicity and potential hazard of the active ingredient and also the formulation of the pesticide. For example, a 10% granular formulation usually will be much less hazardous to the user than a 50% emulsifiable concentrate with the same active ingredient.

The signal words DANGER-POISON and the skull and crossbones are required on labels of all highly toxic compounds. These materials all fall within the acute oral LD_{50} range of 0-50 mg/kg.

The word WARNING is required on the labels for all moderately toxic compounds. These materials all fall within the acute oral LD_{50} range of 50-500 mg/kg.

The word CAUTION is required on the labels for all slightly toxic compounds. These chemicals all fall within the acute oral LD_{50} range of 500-5000 mg/kg.

The word CAUTION is also required on the labels for all relatively non-toxic compounds. These chemicals are all in the acute oral LD_{50} range of 5000 mg/kg and above.

The following table gives the hazard indication associated with the signal words and the estimated amounts required to kill a man when the pesticide is taken by mouth.

Signal Word		Meaning	Amount Required to Kill a Man
DANGER	Poison (in red)	Highly hazardous	Taste to a teaspoon
WARNING		Moderately hazardous	Teaspoon to a tablespoon
CAUTION		Slightly hazardous	Ounce to a pint
CAUTION		Relatively non-hazardous	More than a pint

All pesticides should be considered potentially dangerous. If the material is hazardous, the label will give detailed emergency first-aid measures. There will also be information about symptoms of poisoning, first aid and a note to the physician. The pesticide label is the most important information you can take to the physician when a person is suspected of being poisoned.

Additional information on the toxicity and hazards of pesticides can be found in the SAFETY SECTION of this manual.

Net Contents -- Net contents indicates the amount of pesticide formulation in the container. It is given in gallons, quarts, pints, pounds, or other units of measure. This information helps you determine how much of the product to buy for your needs.

EPA Registration number and Establishment number -- The EPA registration number signifies that the product has been registered with the Environmental Protection Agency. This indicates the pesticide has passed the agency's review of the data and is ready for use according to the label's directions. The establishment number is a special number assigned to the plant that manufactured the pesticide for the company. Some companies manufacture pesticides in more than one manufacturing plant, and the establishment number identifies the place of manufacture.

Name and Address of Manufacturer or Registrant -- This is the name and address of the company manufacturing the product and offering it for sale.

WHEN SHOULD YOU READ THE LABEL?

There are five times when you should read a pesticide label. Many people are guilty of just reading the label when they are ready to apply the pesticide. No wonder there are so many misuses and mistakes with pesticides. Listed below are the five times that you should read a pesticide label and why it is necessary each time.

FIRST TIME -- Before you buy the pesticide.

Why? To determine:
1. If this is the chemical - the best chemical - for your job.
2. If the product can be used safely under your conditions.
3. If the concentration and amount of active ingredient is right for your job.
4. If you have the proper equipment to apply the pesticide.

SECOND TIME -- Before you mix the pesticide.

Why? To determine:
1. The protective equipment needed to handle the pesticide.
2. The specific warnings and first aid measures.
3. What it can be mixed with (compatibility).
4. How to mix it.
5. How much to use.

THIRD TIME -- Before you apply the pesticide.

READ THE LABEL

Why? To determine:
1. Safety measures for the applicator.
2. What it can be applied on.
3. When to apply (check harvest waiting periods - see important note below).
4. How to apply.
5. The rate of application.
6. Restrictions on use.
7. Special instructions.

Important Note: If you apply the chemical but you are not sure if the waiting period before harvest is up, then read the label again to make sure it is safe to harvest the crop. Remember, if you do not follow the waiting period precautions your crop may be seized due to the presence of illegal pesticide residue.

FOURTH TIME -- Before you store the pesticide.

Why? To determine:
1. Where and how to store.
2. Where it should not be stored.
3. What it should not be stored with.

FIFTH TIME -- Before you dispose of the excess pesticide in the container.

Why? To determine:
1. How and where to dispose of the pesticide.
2. How to clean (decontaminate) a pesticide container, and how and where to dispose of the container.

A TYPICAL PESTICIDE LABEL

The following pages show a representative pesticide label. Note that it consists of two parts. The center or front panel of the label gives certain basic information about the pesticide. The rear panel or side panels of the label give more detailed information about the pesticide and how to use it. All pesticide labels follow a similar style.

Let's take a look at the information on the front or center panel of the label starting at the top and reading it line by line:

1. The container contains 2 lbs. of product.
2. The product is manufactured by Mow Chemical Company.
3. The product's name is Killmite, the ® means the name is a registered trademark and the 50W means it is a 50% wettable powder (50% is active ingredient).
4. Miticide means this product is for the control of mites and will kill the mites indicated under the "Directions for Use" on the side panel.

2 LB

KILLMITE® 50W
Miticide

Wettable Powder Formulation
For Control of Plant-Feeding Mites

ACTIVE INGREDIENT
Methylethylbutyl   **50.0%**
INERT INGREDIENTS   **50.0%**
E.P.A. Registration No. 000-000-AA
EPA ESTABLISHMENT NO. 00-AA-0

CAUTION
KEEP OUT OF REACH OF CHILDREN
Read Complete Label Precautions on Side Panel

THE MOW CHEMICAL COMPANY
SNOWBALL, N.C. 27000

Front Panel of Container

KILLMITE® 50W Miticide

DIRECTIONS FOR USE

General Information: Killmite 50W miticide is recommended for control of motile forms of plant-feeding mites, including strains both susceptible and resistant to other miticides. Species controlled include European red, two-spotted spider, Schoene spider, Pacific spider, yellow spider, apple rust, and McDaniel mite. **Killmite** 50W is not highly injurious to certain predatory mites and insects and is essentially non-toxic to honeybees. **Killmite** 50W is compatible with insecticides and fungicides commonly recommended in orchard spray schedules.

Amount and How to Use: Killmite 50W miticide mixes readily with water to form a suspension that can be applied using conventional ground spray equipment. Apply as a dilute spray according to the following table.

DOSAGE RECOMMENDATIONS

CROP	AMOUNT		
	Killmite 50W per acre	Killmite 50W per 100 gallons water	Spray mixture per acre
Apples	12 to 48 oz.	4 to 6 oz.	300 to 800 gallons
Pears	12 to 30 oz.	4 to 6 oz.	300 to 500 gallons

APPLICATION NOTE: Do not apply more than 4 sprays per season. Do not apply within 14 days of harvest.

For more concentrated sprays reduce the amount of water used in accordance with instructions supplied by the application equipment manufacturer. However, do not exceed the amount of **Killmite** 50W recommended per acre, whether dilute or concentrate sprays are used.

Apply sprays containing **Killmite** 50W to obtain thorough, complete coverage of foliage and fruit. Wetting of both upper and lower leaf surfaces is essential for good mite control.

When to Apply: Make the first application when mites are active, usually at or soon after petal-fall. Repeat the treatment in 10 to 14 days and in mid-summer if mites increase in numbers.

USE PRECAUTIONS

Do not graze or feed livestock on cover crops growing in treated areas.

Avoid contamination of food, feedstuffs and domestic water supplies.

This product is toxic to fish. Keep out of lakes, streams and ponds. Apply only as specified on this label.

Rinse equipment and dispose of wastes by burying in non-croplands away from water supplies. Dispose of empty containers by burying with wastes or by burning.

CAUTION
KEEP OUT OF REACH OF CHILDREN AND ANIMALS
MAY BE HARMFUL IF SWALLOWED
MAY CAUSE IRRITATION
Avoid Contact with Eyes and Skin
Avoid Breathing Dust or Spray Mist
Wash Thoroughly after Handling and Spraying
Do Not Wear Contaminated Clothing

In case of contact flush eyes with plenty of water. If irritation persists or develops get medical attention.

NOTICE: Seller warrants that the product conforms to its chemical description and is reasonably fit for the purpose stated on the label when used in accordance with directions under normal conditions of use, but neither this warranty nor any other warranty of MERCHANTABILITY OR FITNESS FOR A PARTICULAR PURPOSE, express or implied, extends to the use of this product contrary to label instructions, or under abnormal conditions, or under conditions not reasonably foreseeable to seller, and buyer assumes the risk of any such use.

Licensed for use under U.S. Patent No. 0,000,000

LOT - 0000

2 LB

REAR PANEL OR SIDE PANELS
OF THE
PESTICIDE CONTAINER

5. Wettable Powder Formulation for Control of Plant-Feeding Mites. This further clarifies the type of formulation and tells you this miticide is for mites that feed on plants.

6. The active ingredient is methylethylbutyl. This is what kills the mites and 50% of the contents in this package is the active ingredient.

7. The inert ingredients, the ingredients that do not kill mites make up the rest, or the other 50% of the contents.

8. The EPA Registration Number 000-000-AA tells you this product is registered with the Environmental Protection Agency for the uses on this label.

9. The EPA Establishment Number 00-AA-0 is the number of the plant that manufactured this product.

10. Caution means this chemical formulation is slightly toxic. An ounce to a pint taken by mouth might kill a man.

11. Keep Out of Reach of Children means just what it says.

12. "Read complete label precautions on side panel" tells you there are more precautions to be followed on the side panel of the label.

13. The Mow Chemical Company manufactures this product and the address of its main office is Snowball, North Carolina.

14. Since there was no "Restricted Use" statement on the front panel, this product can be purchased and used by anyone.

Now let's take a look at the rear panel or side panels of this pesticide label.

1. Killmite® 50W Miticide is repeated from the front panel.

2. Directions for use. Here, you find the kinds of mites the product will control and also the amount and how to use it.

3. Dosage Recommendations: The product is for use on apples and pears and the dilution rates and amount of dilute spray to be used per acre are given.

4. Application note: Here, you see some notes on applying this chemical. These precautions should be followed.

5. When to apply: Tells you when you can safely apply this pesticide. The label should also tell you, where appropriate, when the last application before harvest can be made.

6. Use precautions: Gives you more information on the safe use of this product. Information includes precautions about grazing and feeding.

7. CAUTION: Tells you again to keep away from children, that it may be harmful if swallowed, not to breathe the dust or spray mist and what to do if it gets in the eyes.

8. Notice: Tells you the seller backs the products when used as stated on the label, but advises you not to use it for any other purposes.

9. U.S. Patent Number: It is given if the product's patent has been granted or is still in effect.

10. Lot number - 0000: It is important in checking on the pesticide if anything goes wrong.

Pesticide manufacturers publish up-dated label books every year. The specimen labels on the following pages are from 1982 label books and are representative of the toxicity category signal words CAUTION, WARNING AND DANGER-POISON. You may wish to study them for a few minutes. It is important that you be aware that labels may change annually and to be alert in knowing you are using the latest available labels. If you have doubts, contact the manufacturer.

Hopkins
7½% CAPTAN DUST

ACTIVE INGREDIENT:
Captan (N-trichloromethylthio-4-cyclohexene-1,2-dicarboximide) 7.5%
INERT INGREDIENTS: 92.5%

TOTAL............100.0%

EPA Reg. No. 2393-108 AA

EPA Est. No. 2393-WI-1

CAUTION
KEEP OUT OF REACH OF CHILDREN

AVOID CONTAMINATION OF FEED AND FOODSTUFFS. AVOID INHALATION OF DUST. AVOID CONTACT WITH SKIN. HARMFUL IF TAKEN INTERNALLY. STORE IN A COOL, DRY PLACE, PROTECT FROM EXCESSIVE HEAT.

THIS PRODUCT IS TOXIC TO FISH. KEEP OUT OF LAKES, STREAMS OR PONDS. DO NOT CONTAMINATE WATER BY CLEANING OF EQUIPMENT, OR DISPOSAL OF WASTES.

RECOMMENDATIONS

APPLE: Apple Scab, Block Rot (Frogeye Leaf Spot) Black Pox, Bitter Rot, Botryospheria, Brooks Fruit Spot, Flyspeck — 25 to 50 lbs. per acre (9.2 ozs. to 1 lb. 2.4 ozs./1,000 sq. ft.). Make first application at delayed dormant period. Repeat applications for each infection period or at weekly intervals as long as necessary.

APRICOT: Brown Rot, Jacket Rot — 50 to 60 lbs. per acre (1 lb. 2.4 ozs. to 1 lb. 6 ozs./1,000 sq. ft.). Apply at red bud stage and repeat at 75 per cent petal fall.

CHERRY: Brown Rot, Leaf Spot — 25 to 50 lbs. per acre (9.2 ozs. to 1 lb. 2.4 ozs./1,000 sq. ft.). Apply at popcorn, blossom and petal fall stages; continue at 10 to 14 day intervals.

PEACH: Brown Rot, Scab — 25 to 50 lbs. per acre (9.2 ozs. to 1 lb. 2.4 ozs./1,000 sq. ft.). Make 2 to 3 prebloom applications. Repeat at petal fall and 3 to 4 covers as necessary. Make pre-harvest application where necessary.

PEAR: Pear Scab (except Pacific Northwest) — 25 to 50 lbs. per acre (9.2 ozs. to 14.7 ozs./1,000 sq. ft.). Make first applications during early finger and petal fall stages. Repeat applications for each infection period or at weekly intervals as long as necessary. Russeting may be produced on Bosc pears. Do not use on D'Anjou pears.

RASPBERRIES: Fruit Rot — 35 lbs. per acre (12.8 ozs./1,000 sq. ft.). Make applications 3 to 5 days before harvest starts, at mid-harvest and 8 to 10 days after mid-harvest.

STRAWBERRIES: Botrytis Rot — 30 to 40 lbs. per acre (11.0 ozs. to 14.7 ozs./1,000 sq. ft.). Make application when new growth starts in spring before fruit starts to form. Repeat applications weekly. Severe Infection Conditions — Continue through harvest period, treating immediately after each picking.

BEANS, PEAS: Soil Treatment — Damp-off (Beans only), Root Rot — Apply 50 to 60 lbs. per acre (1 lb. 2.4 ozs. to 1 lb. 6 ozs./1,000 sq. ft.). Work into top 3 to 4 inches of soil before planting.

CUCURBITS (Cantaloupe, Cucumber, Melon, Squash): Angular Leaf Spot, Anthracnose, Downy Mildew — Start application when runners begin to form or at the first sign of disease. Apply at rate of 20 to 30 lbs. per acre (7.35 ozs. to 11.0 ozs./1,000 sq. ft.) on young plants, increasing to 40 lbs. per acre (14.7 ozs./1,000 sq. ft.) on mature plants.

POTATO SEED PIECE TREATMENT: To protect against seed piece breakdown and rot or decay from Brown Eye Disease, Damp-off, Seed Rot, Verticillium Wilt — Use at rate of 10 ozs. per bushel cut seed pieces (16 ozs. or 1 lb. to 100 lbs. cut seed pieces). Apply in convenient container or by duster attachment over belt. Coat each seed piece evenly with a light film of dust. Treat pieces within 6 hours after cutting. If planting is delayed, store treated seed pieces in open crates for two days before bagging. **DO NOT USE TREATED SEED FOR FOOD OR FEED PURPOSES.**

TOMATOES: Anthracnose, Early and Late Blight, Gray Leaf Spot, Septoria Leaf Spot — Apply 30 to 40 lbs. per acre (11.0 ozs. to 14.7 ozs./1,000 sq. ft.) at weekly intervals on young plants; 40 to 50 lbs. per acre (14.7 ozs. to 1 lb. 2.4 ozs./1,000 sq. ft.) on mature plants.

DO NOT USE FOR ANY PURPOSE NOT RECOMMENDED ON THIS LABEL, SINCE INJURIOUS EFFECTS MAY RESULT.

Do not reuse empty container. Destroy it by burying with waste or burning. Stay away from smoke or fumes.

NOTICE OF WARRANTY

Seller makes no warranty, express or implied, concerning the use of this product other than indicated on the label. Buyer assumes all risk of use and/or handling of this material when such use and/or handling is contrary to label instruction.

NET WEIGHT

MANUFACTURED BY

Hopkins
agricultural chemical co.

BOX 7532 • MADISON, WISCONSIN 53707

50 POUNDS

CODE 11.1.76—   3 4 5 6 7 8 9 (11.76)

Pramitol®

80 WP

SAMPLE LABEL

Herbicide

For weed control on railroad rights-of-way and industrial sites

Controls johnson-grass, bindweed, and other hard-to-kill weeds

25 Pounds
Net Weight

Active Ingredient:
Prometon: 2,4-bis (isopropylamino)-6-methoxy-*s*-triazine 80%

Inert Ingredients:	20%
Total:	100%

EPA Est. 100-AL-1
EPA Reg. No. 100-536

Pramitol 80WP is a wettable powder

Keep Out of Reach of Children.

Warning

See additional pre-cautionary statements on back of bag.

See directions for use on back of bag.

CIBA—GEIGY

CGA 130-585A

Pramitol® 80WP

DIRECTIONS FOR USE AND CONDITIONS OF SALE AND WARRANTY

IMPORTANT: Read the entire **Directions for Use** and the **Conditions of Sale and Warranty** before using this product.

Conditions of Sale and Warranty

The **Directions for Use** of this product reflect the opinion of experts based on field use and tests. The directions are believed to be reliable and should be followed carefully. However, it is impossible to eliminate all risks inherently associated with use of this product. Crop injury, ineffectiveness, or other unintended consequences may result because of such factors as weather conditions, presence of other materials, or the manner of use or application all of which are beyond the control of CIBA-GEIGY or the Seller. All such risks shall be assumed by the Buyer.

CIBA-GEIGY warrants that this product conforms to the chemical description on the label and is reasonably fit for the purposes referred to in the **Directions for Use,** subject to the inherent risks referred to above. **CIBA-GEIGY makes no other express or implied warranty of Fitness or Merchantability or any other express or implied warranty. In no case shall CIBA-GEIGY or the Seller be liable for consequential, special, or indirect damages resulting from the use or handling of this product.** CIBA-GEIGY and the Seller offer this product and the Buyer and user accept it, subject to the foregoing **Conditions of Sale and Warranty** which may be varied only by agreement in writing signed by a duly authorized representative of CIBA-GEIGY.

General Information

Pramitol 80WP is a nonselective herbicide that may be applied in water or phytotoxic oil before or after plant growth begins. When applied with water as the carrier, Pramitol 80WP has minimal activity through foliar contact. Much of its activity is through plant roots; therefore, its effectiveness is dependent on rainfall to move it into the root zone. Very dry soil conditions and lack of sufficient rainfall may result in poor weed control.

Use only in areas where complete control of all vegetation is desired, such as industrial sites, rights-of-way, lumberyards, petroleum tank farms, around farm buildings, along fence lines, etc. When applied to the soil, this product usually inhibits plant growth for a year or more. Pramitol 80WP should not be used on land to be cropped, or near adjacent desirable trees, shrubs or plants, or injury may occur.

Following many years of continuous use of this product and chemically related products, biotypes of some of the weeds listed on this label have been reported which cannot be effectively controlled by this and related herbicides. Where this is known or suspected, we recommend the use of this product in combination with other registered herbicides which are not triazines. Consult with your State Agricultural Extension Service for specific recommendations.

Do not use for weed control in greenhouses.

Pramitol 80WP is noncorrosive to equipment and metal surfaces, nonflammable, and has low electrical conductivity.

Mixing Instructions

Pramitol 80WP is a wettable powder that should be mixed with water or phytotoxic oil and applied as a spray. Make a slurry by adding Pramitol 80WP to water or phytotoxic oil. Pour the slurry into the spray tank during or after filling. Sufficient hydraulic (jet) or mechanical agitation must be provided during mixing and application to keep the material in suspension. Wash sprayer thoroughly after use.

Directions for Use

Industrial Sites and Noncrop Areas

Apply Pramitol 80WP prior to or up to 3 months after weed emergence. Application rates vary from 12½-74 lbs. of Pramitol 80WP per acre, depending on climatic conditions, soil conditions, weeds present, and the stage of growth of the weeds. In the following rate recommendations, the higher rates are intended for use on heavier soils where weed growth is heavy and where rainfall is expected to be relatively high. Use the higher rates also where longer residual control is desired in regions with a long growing season.

For best results, apply Pramitol 80WP prior to weed emergence or when weeds are young and actively growing.

For control of annual and susceptible perennial weeds (including downy bromegrass, oatgrass, goosegrass, quackgrass, puncturevine, goldenrod, and plantain): Broadcast 12½-18½ lbs. Pramitol 80WP in 50-100 gals. of water per acre. For small areas, apply 4½-6¾ oz. Pramitol 80WP per 1,000 sq. ft. in sufficient water to give thorough and uniform coverage.

For control of hard-to-kill perennial weeds and grasses (including johnsongrass, bindweed, and wild carrot): Broadcast 50-74 lbs. Pramitol 80WP in 50-100 gals. of water per acre. For small areas, use 1 lb. 2 oz.-1 lb. 11 oz. Pramitol 80WP per 1,000 sq. ft. in sufficient water to give thorough and uniform coverage.

For faster top-kill of existing vegetation, apply Pramitol 80WP in diesel oil, fuel oil, or weed oil. Use 100-200 gals. of oil per acre. Tall dense vegetation including johnsongrass will generally require a minimum of 150 gals. of oil per acre for thorough coverage.

Under Asphalt Pavement

Pramitol 80WP may be used to extend the useful life of asphalt pavement by preventing weeds from emerging through it. Pramitol 80WP may be applied to the ground before laying asphalt.

Pramitol 80WP should be used only where the area to be treated has been prepared according to good construction practices. The site should be disked, plowed, or otherwise prepared in accordance with good construction practices and recompacted to allow good penetration of the Pramitol 80WP mixture.

Pramitol 80WP does not control woody vegetation such as small trees, brush, and woody vines. Roots of such species should be grubbed and removed from the site prior to application. If rhizomes, tubers, or other vegetative parts are present in the site, they should be removed by scalping with a grader blade.

Pramitol 80WP may be applied to the ground before laying rapid, medium, or slow curing asphalt coatings such as those used in parking lots, highway shoulders and median strips, roadways, and other industrial sites. Pramitol 80WP should be applied just prior to the laying of the asphalt coating to prevent possible lateral movement of the herbicide by rainfall or other mechanical means. Sprayers should be equipped to provide continuous agitation of the spray mixture during application to ensure application of a uniform spray emulsion.

Apply Pramitol 80WP uniformly at the rate of 50-62 lbs. per acre (0.16-0.20 oz. per sq. yd.) in a minimum of 100 gals. of water. Use the lower rate to control annual and susceptible perennial weeds, including downy bromegrass, oatgrass, goosegrass, quackgrass, puncturevine, goldenrod, and plantain; use the higher rate to control hard-to-kill perennial weeds and grasses including johnsongrass, bindweed, and wild carrot.

> **Precaution:** Do not use Pramitol 80WP under asphalt coatings less than 2 inches thick.

Storage and Disposal

Do not contaminate water, food, or feed by storage or disposal. Open dumping is prohibited. Do not reuse empty container. Pesticide, spray mixture, or rinsate that cannot be used or chemically reprocessed should be disposed of according to procedures approved by federal, state, or local disposal authorities. Consult federal, state, or local disposal authorities for approved procedures.

Precautionary Statements

Hazards to Humans and Domestic Animals

WARNING

Causes eye irritation. Avoid contact with eyes. Wash thoroughly after handling. Wash clothing before reuse. Harmful if swallowed. Avoid contact with skin, inhalation of dust, and contamination of food and feed.

First Aid

In case of contact with eyes, flush with plenty of water for at least 15 minutes and get medical attention. In case of contact with skin, wash with soap and water. In case of inhalation exposure, move from contaminated area.

Note to Physician: There is no specific antidote if ingested. Induce emesis or lavage stomach. Give a saline laxative and supportive therapy.

Environmental Hazards

Do not contaminate water used for irrigation or domestic purposes.

Do not contaminate water by cleaning of equipment or disposal of wastes. Apply this product only as specified on this label.

Pramitol® trademark of CIBA-GEIGY for prometon

Agricultural Division
CIBA-GEIGY Corporation
Greensboro, North Carolina 27409
CGA 44L2C   090

RESTRICTED USE PESTICIDE
FOR RETAIL SALE TO AND USE ONLY BY CERTIFIED APPLICATORS OR PERSONS UNDER
THEIR DIRECT SUPERVISION AND ONLY FOR THOSE USES COVERED BY THE CERTIFIED
APPLICATOR'S CERTIFICATION. KEEP DRY. MOISTURE CAN INCREASE HANDLING HAZARDS.

Temik®
10% Granular

Aldicarb Pesticide

For Control of Certain Insects, Mites and Nematodes on Commercially Grown Ornamentals

Active Ingredient: Aldicarb [2-methyl-2-(methylthio)propionaldehyde $\underline{O}$ (methylcarbamoyl) oxime] 10%
Inert Ingredients: 90%
E.P.A. Reg. No. 264-322   **E.P.A. Est. 10352-GA-01**

KEEP OUT OF REACH OF CHILDREN
DANGER POISON

POISONOUS IF SWALLOWED. MAY BE FATAL OR HARMFUL BY SKIN OR EYE CONTACT OR BY BREATHING DUST. RAPIDLY ABSORBED
THROUGH SKIN OR EYES. DO NOT GET ON SKIN OR EYES. DO NOT BREATHE DUST. ANTIDOTE IS ATROPINE SULFATE. SEE STATEMENT OF
PRACTICAL TREATMENT AND OTHER DETAILED WARNINGS ON TOP PANEL. READ THE ENTIRE LABEL BEFORE USING THIS PESTICIDE.

STORAGE AND DISPOSAL

STORE UNUSED "TEMIK" ALDICARB PESTICIDE IN ORIGINAL CONTAINER ONLY IN WELL VENTILATED CLEAN DRY AREA OUT OF REACH
OF CHILDREN AND ANIMALS. DO NOT STORE IN AREAS WHERE TEMPERATURE AVERAGES 115°F OR GREATER. DO NOT STORE IN OR
AROUND THE HOME OR HOME GARDEN.

DO NOT CONTAMINATE WATER, FOOD, OR FEED BY STORAGE OR DISPOSAL.

OPEN DUMPING IS PROHIBITED.

IF CONTAINER IS BROKEN, HANDLE WITH RUBBER GLOVES. DO NOT GET DUST OR GRANULES ON SKIN OR EYES. DO NOT BREATHE
THE DUST. SWEEP UP AND BURY ANY SMALL SPILLS OR EXCESS "TEMIK" ALDICARB PESTICIDE AT LEAST 18 INCHES DEEP IN SOIL
ISOLATED FROM WATER SUPPLIES AND FOOD CROPS.

DISPOSE OF PESTICIDE CONTAINER IN AN APPROVED INCINERATOR OR LANDFILL. NEVER REUSE EMPTY CONTAINER.

CONSULT FEDERAL, STATE OR LOCAL DISPOSAL AUTHORITIES FOR APPROVED ALTERNATIVE PROCEDURES SUCH AS LIMITED OPEN
BURNING. STAY AWAY FROM AND DO NOT BREATHE OR CONTACT SMOKE.

IN CASE OF A TRANSPORTATION OR WAREHOUSE EMERGENCY INVOLVING A SPILL, FIRE OR EXPOSURE, CALL COLLECT (304) 744-3487
TWENTY-FOUR HOURS A DAY IN THE U.S.A.

UNION CARBIDE AGRICULTURAL PRODUCTS COMPANY, INC.
RESEARCH TRIANGLE PARK, NC 27709

Active Ingredient protected by U.S. Pat. No. 3217037
TEMIK* is the registered trademark of Union Carbide Corporation for aldicarb pesticide.   Printed in U.S.A.
Form No. AG82005-7.5M-11/81-TCG
UCC-2400-4110331

FOR USE ONLY BY TRAINED PERSONNEL IN COMMERCIAL PRODUCTION OF ORNAMENTAL PLANTS.

READ GENERAL DIRECTIONS AND CAUTIONS ON THIS PANEL AND ALL DETAILED WARNINGS ON TOP PANELS BEFORE USING. USE ONLY IN ACCORDANCE WITH LABEL DIRECTIONS, WARNINGS AND CAUTIONS. THIS PRODUCT CANNOT BE USED ON ANY CROP UNLISTED ON THIS LABEL AS ANY RESIDUES REMAINING MAY BE HARMFUL.

DO NOT USE IN THE HOME OR HOME GARDEN.

DO NOT STORE IN OR AROUND THE HOME.

DO NOT APPLY THIS PRODUCT IN SUFFOLK COUNTY, LONG ISLAND, NEW YORK.

PRECAUTIONARY STATEMENTS

HAZARDS TO HUMANS AND DOMESTIC ANIMALS POISONOUS IF SWALLOWED. MAY BE FATAL OR HARMFUL BY SKIN OR EYE CONTACT OR BREATHING DUST. RAPIDLY ABSORBED THROUGH SKIN AND EYES. DO NOT BREATHE DUST. ANTIDOTE IS ATROPINE SULFATE. SEE ANTIDOTE STATEMENT. INFORMATION FOR PHYSICIAN AND OTHER DETAILED WARNINGS ON THIS PACKAGE. READ THE ENTIRE LABEL BEFORE USING THIS PESTICIDE.

Wear long-sleeved clothing and protective gloves when handling. Wash hands and face before eating or smoking. Bathe at the end of work day, washing entire body and hair with soap and water. Change contaminated clothing daily and wash in strong washing soda solution and rinse thoroughly before reusing.

SYMPTOMS OF POISONING
may be one or more of the following:

Weakness	Vomiting	Pinpoint Eye Pupils
Headache	Diarrhea	Abnormal Flow of Saliva
Sweating	Tightness in Chest	Abnormal Cramps
Nausea	Blurred Vision	Unconsciousness

CONTACT A PHYSICIAN IMMEDIATELY IN ALL CASES OF SUSPECTED POISONING.

ENVIRONMENTAL HAZARDS

TOXIC TO FISH, BIRDS, AND WILDLIFE
This product is toxic to fish, birds and other wildlife. Birds feeding on treated areas may be killed. Keep out of lakes, streams and ponds. Do not contaminate water by cleaning of equipment or disposal of wastes. Apply this product only as specified on this label.

STATEMENT OF PRACTICAL TREATMENT

ANTIDOTE
ANTIDOTE is Atropine Sulfate

FIRST AID TREATMENT
If symptoms are apparent give atropine sulfate tablets as directed by the physician. Do not give antidote unless symptoms of poisoning have occurred. Transport patient to a physician or hospital, whichever is faster. Keep patient warm and quiet. If breathing is difficult, give oxygen. Start artificial respiration immediately if breathing stops. SHOW A COPY OF THIS LABEL TO THE PHYSICIAN.

IN CASE OF SKIN OR EYE CONTACT, wash skin immediately and thoroughly with soap and water and rinse well; flush eyes with plenty of water for 15 minutes. Remove contaminated clothing and wash before reuse.

IN CASE OF SWALLOWING drink 1 ot 2 glasses of water and induce vomiting by touching back of throat with finger or blunt object. Do not induce vomiting or give anything by mouth to an unconscious person. Get medical attention.

INFORMATION FOR PHYSICIAN
This product is a Carbamate Pesticide containing 2-methyl-2-(methylthio) propionaldehyde O-(methylcarbamoyl)oxime. It is a spontaneously reversible cholinesterase inhibitor causing parasympathetic nerve stimulation. Preferred treatment of poisoning in adults is atropine sulfate given intravenously. As much as 2 to 4 mg may be needed every 10 to 12 minutes until patient is fully atropinized. Dosage for children is appropriately reduced. Atropinization should be maintained for 12 hours by intramuscular administration of atropine in lower doses given at appropriate time intervals. Do not administer opiates or cholinesterase inhibiting drugs. Artificial respiration or oxygen may be necessary. Observe patient continuously for at least 24 hours. Allow no further exposure to any cholinesterase inhibitors until cholinesterase level is normal by blood test.

**IN CASE OF EMERGENCY
PHONE COLLECT (24 HOURS A DAY)
IN U.S.A. (304) 744-3487**

GENERAL INFORMATION

TEMIK 10G aldicarb pesticide is a granular product that provides systemic control of certain insects, mites and nematodes. When applied to soil, the active ingredient is rapidly absorbed by root systems and translocated throughout all parts of the plant. Do not apply to very dry soil unless treatment is followed by irrigation or rainfall. Residual activity varies with dosage and pests involved, but often lasts more than 6 weeks.

GENERAL CAUTIONS

Calibrate all mechanical devices for delivery of the required amount of TEMIK 10G (See DIRECTIONS FOR USE AND RECOMMENDED APPLICATION on this PANEL). Clean application equipment thoroughly after use. Bury any excess TEMIK 10G in soil (See instructions for Disposal of Spillage and Containers on TOP PANEL).

For outdoor nurseries, granules should be worked into the soil to a depth of at least 2 inches or covered with soil to a depth of at least 2 inches to provide minimum hazard to birds. Deep disc any spills of granules at row ends immediately to prevent birds from feeding on exposed granules.

To avoid injury to plants do not apply to open blooms. Do not mix TEMIK 10G with potting soil. Use only on standing plants.

SPECIAL PRECAUTIONS FOR HUMANS SAFETY

Apply only as directed. Do not mix TEMIK Aldicarb Pesticide with other materials before application.

Do not use applicators that will grind the granules. A respirator mask and rubber gloves are recommended while filling or cleaning equipment.

To avoid residue problems:

Do not use plant parts for food or feed purposes.

Do not market potted plants within 4 weeks after last application.

Do not plant food crops in soil previously treated with TEMIK 10G for at least six months.

Directions for use and recommended applications

It is a violation of Federal law to use this product in a manner inconsistent with its labeling.

FOR USE ON POTTED PLANTS OR PLANT BEDS IN COMMERCIAL GREENHOUSES

Dosage TEMIK 10G

Crop	Pest Controlled	Pounds Acre	Oz. 1000 Sq. Ft. of Plant Bed or Closely Packed Pots	Time of Application	Recommended Application
CHRYSANTHEMUM	Aphids, Leafminers, Thrips Spider mites, Whiteflies	50 to 75 75 to 100	20 to 30 30 to 40	3 to 5 weeks after transplanting cuttings or just after pinching. Repeat in 6 to 8 weeks if needed.	If beds are covered with leaf or compost type mulch, remove mulch or place granules beneath mulch. Work into soil and water thoroughly. Use the high rate on heavy organic or clay soils or if pest populations become severe. Apply granules evenly around base of plants. Wash off granules adhering to foliage.
ORCHIDS (Cymbidiums)	Spider mites	75 to 100	30 to 40	Apply just prior to emergence of flower spikes. Repeat if needed.	
EASTER LILIES	Aphids	50 to 75	20 to 30	After plants are established and as pests appear. Repeat in 6 to 8 weeks, if needed.	
POINSETTIA	Spider mites, Mealy bugs, Whiteflies	75 to 100	30 to 40		
GERBERA	Aphids, Leafminers Spider mites, White flies	50 to 75 75 to 100	20 to 30 30 to 40		
CARNATIONS, ROSES, SNAPDRAGONS	Aphids Spider mites	50 to 75 75 to 100	20 to 30 30 to 40		On old well-established roses, use maximum rate for mite control.

FOR USE ON COMMERCIAL FIELD GROWN AND NURSERY PLANTINGS

To provide maximum performance and to minimize hazard to birds work granules into or cover with soil to a depth of 2 inches or more. Deep disc any spills at row ends immediately to prevent birds from feeding on granules.

Dosage TEMIK 10G

Crop	Pest Controlled	Pounds Acre	Pounds/1000 Linear Row Feet (42-inch Row Spacing)	Time of Application	Recommended Application
ROSES	Spider mites	70 to 100	5 to 8	As pests begin to appear. Repeat if needed.	Apply granules as sidedress to both sides of plant or row. Work into soil at least 2 to 4 inches deep.
DAHLIA	Aphids, Leafhoppers, Leafminers, Spider mites	50 to 80	4 to 6		
LILIES, BULBS	Nematodes (root lesion)	50 to 70 (on 40 row spacing)	4 to 6 (or 50 to 70 lbs. 12,000 linear feet of row)	At planting time.	Apply in furrow with bulblet. Cover with soil.
BIRCH & HOLLY	Aphids, Leafminers	50 to 100	4.5 to 9 (or 1 to 2 ounces/ one inch diameter of trunk at soil level)	After spring growth begins and prior to occurrence of leafminer activity.	Apply granules as sidedress 3 to 4 inches deep and 10 to 12 inches to one or both sides of row. Cover with soil.

LIMITED WARRANTY AND DISCLAIMER

1. The manufacturer warrants (a) that this product conforms to the chemical description on the label; (b) that this product is reasonably fit for the purposes set forth in the directions for use when it is used in accordance with such directions, and (c) that the directions, warnings and other statements on this label are based upon responsible experts' evaluation of reasonable tests of effectiveness, of toxicity to laboratory animals and to plants, and of residues on food crops, and upon reports of field experience. Tests have not been made on all varieties or in all states or under all conditions. THE MANUFACTURER NEITHER MAKES NOR INTENDS, NOR DOES IT AUTHORIZE ANY AGENT OR REPRESENTATIVE TO MAKE, ANY OTHER WARRANTIES, EXPRESS OR IMPLIED, AND IT EXPRESSLY EXCLUDES AND DISCLAIMS ALL IMPLIED WARRANTIES OF MERCHANTABILITY OR FITNESS FOR A PARTICULAR PURPOSE.

2. This warranty does not extend to, and the Buyer shall be solely responsible for, any and all loss or damage which results from the use of this product in any manner which is inconsistent with the label directions, warnings or cautions.

3. BUYER'S EXCLUSIVE REMEDY AND MANUFACTURER'S OR SELLER'S EXCLUSIVE LIABILITY FOR ANY AND ALL CLAIMS, LOSSES, DAMAGES, OR INJURIES RESULTING FROM THE USE OR HANDLING OF THIS PRODUCT, WHETHER OR NOT BASED IN CONTRACT, NEGLIGENCE, STRICT LIABILITY IN TORT OR OTHERWISE, SHALL BE LIMITED, AT THE MANUFACTURER'S OPTION, TO REPLACEMENT OF, OR THE REPAYMENT OF THE PURCHASE PRICE FOR, THE QUANTITY OF PRODUCT WITH RESPECT TO WHICH DAMAGES ARE CLAIMED. IN NO EVENT SHALL MANUFACTURER OR SELLER BE LIABLE FOR SPECIAL, INDIRECT OR CONSEQUENTIAL DAMAGES RESULTING FROM THE USE OR HANDLING OF THIS PRODUCT.

THIS SPECIMEN LABEL IS INTENDED TO BE USED AS A GUIDE IN PROVIDING INFORMATION ON THE GENERAL DIRECTION, WARNINGS AND CAUTIONS ON THE USE OF "TEMIK" 10G ALDICARB PESTICIDE. ALWAYS READ THE LABEL ON THE PACKAGE BEFORE USING THE PRODUCT.

The importance of READING THE LABEL cannot be stressed too often. The information that appears on the label is put there for your information and your protection. If it is read, understood and all of the directions are followed, the likelihood of misusing the material or of having an accident with the pesticide is remote. This is why we so often stress that "the most important few minutes in pest control is the time spent in READING THE LABEL".

WARNING CONCERNING OUT-OF-DATE SOURCES OF PESTICIDE INFORMATION

Accepting information about pesticides from unreliable sources is the surest way to create problems for yourself and others. One mistake can cost you money, injure someone, or cause you legal problems.

Guard against the following:

1. Friends mean well but often cannot remember exactly the name of a product, or in fact, they may know very little about the problem or the product.

2. Don't be oversold by personnel in sales or others who recommend pesticides. The best source of information is from specialists who work directly with the class of chemical under consideration.

3. Interested bystanders may often offer suggestions, but if not from an authoritative source the suggestions should be disregarded.

4. Old bulletins, circulars and mimeographed information from state or federal sources should be disregarded. Only recommendations of a current year should be accepted as valid.

5. Sales catalogs may contain incomplete or out-of-date information. Doubtful recommendations should be checked against up-to-date information.

6. Memories should never be trusted. Too many pesticide chemical names sound so nearly alike.

7. Trade names may not be consistent with the actual chemical ingredient.

8. Recommendations from other states may or may not apply in your state. Never accept recommendations on any pesticide without knowledge of the source and the validity of the source.

Pesticide Safety

GENERAL CONSIDERATIONS

Pesticide technology has come a long way since the days when diesel oil, lead arsenate, bordeaux mixture, and nicotine sulfate were the mainstays of the industry. We are able to do much more with today's organic insecticides and weed killers, to increase crop and livestock yields than we were a generation ago.

Many of our modern pest control chemicals are two edged swords. We have not yet learned all of the tricks of the pharmaceutical trade, for we have not yet devised many economic poisons which are selectively toxic only to their intended victims. Some of them are highly poisonous to man and animals. This situation is true today and is likely to remain so for a long time.

Pesticides, like drugs, are beneficial to man when properly used; misused they may be extremely dangerous. Most pesticides are designed to kill something -- insects, mites, fungi, weeds, rodents or other pests. Therefore all pesticides should be handled as poisons.

With proper handling and application, pesticides rank among the safest production aids used. They undergo exhaustive manufacturer's tests and pass stringent label requirements before being approved and registered for market by the Environmental Protection Agency. Pesticide residues and their effects are carefully monitored or surveyed. Teams of scientists maintain surveillance on soils, crops, water, air, animals and mankind to insure that dangerous levels of pesticides are not accumulating in the environment.

Pesticides have an enormous safety margin when used properly. But, like automobiles, firearms, and medicines, they can be and sometimes are improperly used, causing accidents. Because of improper use, over 100 children and adults die of pesticide poisoning annually. Though these poisonings are fewer than deaths reported due to items such as aspirin, the number can and should be reduced.

It is entirely possible for a user to handle pesticides safely for many years with no obvious ill effects to himself or his environment. Illnesses resulting from overexposure to pesticides do occur among those who work with these substances, but these illnesses are preventable. They are caused by misuse of the chemicals, and misuse results from carelessness which, in turn, results from ignorance. Well informed pesticide users are more likely to take proper precautionary measures in handling toxic materials and these persons play an important part in maintaining a good safety record for pesticide usage.

There is no reason for pesticide misuse because there are many sources of information (including this manual) that are available as guidance for proper use. Manufacturers, Cooperative Extension Service, Environmental Protection Agency, United States Department of Agriculture, Public Health Service, American Medical Association, and the various state agencies all provide reliable sources of information.

PESTICIDE SELECTION

Choosing the correct pesticide to use is one of the most important segments of carrying out an effective pest control program. The pesticide you choose will not only be instrumental in the effectiveness of your control program, but will also have a direct bearing on the hazards that you subject yourself to as well as other persons and the environment.

Actually, a potential hazard may be present the moment a person purchases a pesticide. The selection of type of pesticide, the formulation to be used, and even the container type, may be factors contributing to a pesticide accident.

Before making a selection, the pest problem should be identified by a competent person. Control measures should not be undertaken unless the pest problem is of economic importance, has a potential of developing into a problem, is of health importance, or is a nuisance.

After the pest has been properly identified, select a pesticide that will control the pest with a minimum of danger to other organisms. The pesticide should be one currently registered by the EPA and the State Department of Agriculture, and is recommended by the Cooperative Extension Service.

There are considerable differences in the safety of various pesticide formulations. Select the safest pesticide formulation whenever there is a choice. Granular type formulations are safer than sprays or dusts because they drift less. Formulations with greatest drift and dispersion potential are most likely to cause damage to desirable plants under unfavorable climatic conditions. These formulations can be of greater risk to the applicator if a highly toxic pesticide is being used. (Refer to section on PESTICIDE FORMULATIONS AND ADJUVANTS).

Emulsifiable concentrate pesticides with a petroleum type carrier are generally more hazardous than water soluable ones because they penetrate the skin more rapidly and are difficult to wash off.

Finally, estimate the amount of pesticide needed and purchase only enough for the particular job or for one season so that you do not have a storage or disposal problem. Smaller containers are easier to handle and present less chance of accidental spillage and contamination.

HANDLING AND MIXING PESTICIDES

Pesticides by nature are poisonous materials and extreme caution should be used in mixing and handling them. Of the various work activities associated with pesticide use, the mixing and loading operation is considered to be the most hazardous part of the spraying or dusting job. Obviously many factors are involved that determine the degree of hazard of any operation. The mixing and loading of pesticides however, will generally result in possibilities of exposure i.e. spills, splashes, etc., when a mix is of higher concentration and in greater quantity. Although not minimizing the importance of protective measures during all work activities, the mixing and loading operation warrants special attention and care. The following general safety instructions should be observed during the mixing and loading of pesticides:

(1) Having selected the right pesticide for the job, read the label and carry out the necessary calculations for the required dilution of the pesticide. Obtain the proper equipment, including protective clothing, respirator if required, and have first aid equipment available.

(2) Never work alone when handling highly hazardous pesticides.

(3) Mix chemicals outside or in a well ventilated area. Carefully

open original concentrate containers. Never position any portion of the body directly over the seal or the pouring spout. The release of pressure may cause the liquid to be expelled from the container. Open sacks with a knife rather than tearing them because dry formulations such as dusts and powders can billow up in large concentrations. Always stand up-wind when mixing or loading pesticides.

(4) In mixing chemicals, all quantities of the active ingredient should be measured with extreme accuracy. Have clean measuring and transfer containers available. Containers to accurately measure ounces, pints, quarts, gallons, as well as scales for weighing should be kept in the area where pesticides are stored. The measuring containers should be thoroughly cleaned after each use.

(5) As concentrate containers are emptied and drained, water or other diluting material being used in the spray program should be used to rinse the container. Rinse three times allowing thirty seconds for draining after each rinse. (Use one quart for each rinse of a one gallon can or jug; a gallon for each five gallon can; and five gallons for either thirty or fifty-five gallon drums). Drain each rinse into the spray tank before filling it to the desired level. (Refer also to the section on TRANSPORTATION, STORAGE,DECONTAMINATION AND DISPOSAL).

(6) Cleanup spilled pesticides immediately. If the pesticide is accidentally spilled on skin, immediately wash it off with soap and water. Should the pesticide be spilled on clothing, change clothing as soon as possible and launder before wearing again. Do not store or wash contaminated clothing with other soiled clothing items.

(7) Protective gloves should be washed before removing them. Replace protective gloves frequently regardless of signs of wear or contamination.

(8) Persons mixing, handling, or applying pesticides should never smoke, eat, or drink until they have washed thoroughly. It can be a means of swallowing pesticides which have accumulated on the mouth or hands.

(9) Do not use the mouth to siphon a pesticide from a container.

(10) When filling the spray tank do not allow the delivery hose below the highest possible water surface to avoid back-siphoning.

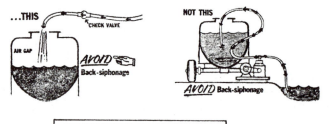

APPLICATION OF PESTICIDES

Prior to pesticide application, the label should be referred to again even though the pesticide is familar to you. Details are often forgotten and labels are frequently revised. Use protective clothing and equipment if prescribed no matter how uncomfortable it may be during hot weather. The discomfort must be endured to use the pesticide safely.

Do not apply higher dosages or rates than specified on the label or recommended by the Agricultural Extension Service. To apply the correct pesticide amount, the application equipment must be calibrated properly. Knowledge of equipment output is essential to compute the amount of pesticide needed in the mixture to treat a given area. (Refer to the section on CALIBRATION).

Be sure the application equipment is clean, in good condition, and operating properly. Poorly operating equipment can cause unnecessary hazards to the user as well as possible damage to the crop and the environment. Extra time spent during spraying to fix and adjust equipment that is in poor operating condition can cause excessive exposure to a pesticide. Do not blow out clogged hoses, nozzles, or lines with your mouth.

The application should be performed at the proper time using recommended dosages to prevent illegal residues from remaining in or on food, feed, or forage crops. Many pesticide labels state the number of days interval between the last pesticide application and the time that the crop can be

harvested. These waiting intervals should be followed closely. For livestock there is a waiting period between treatment and slaughter.

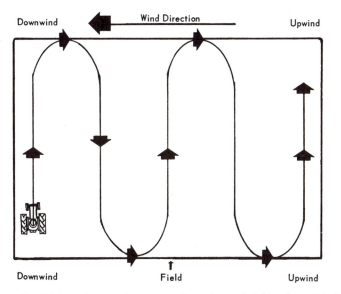

Precautions should be observed to guard against drift of pesticides on to
nearby crops, pastures or livestock. Extreme care should be observed to
prevent application from contaminating streams, ponds, lakes or wells.
If pesticides must be applied when there is a breeze, drive or spray into
or at right angles to the breeze to prevent the pesticide from blowing
into the applicators face.

If a person feels ill or notices any irregular bodily symptoms when applying
pesticides, work should cease and efforts should be made to seek medical
attention.

REENTRY INTO TREATED FIELDS

Standards for reentry into treated fields (not to be confused with number of
days from last spray until harvest) were established for certain pesticides
by the EPA in 1974. These standards apply particularly to agricultural uses
and were established to protect field workers.

Although a commercial applicator may feel that he is not directly
involved, there are some rules that must be followed. These are:

1. No unprotected person may be in a field that is being treated
   with a pesticide.
2. No pesticide application is to be permitted that will expose
   any person to pesticides, either directly, or through drift,
   excepting those involved in the application, who should be
   wearing protective clothing as required.
3. The label restrictions and directions must be followed.

The pesticides presently requiring waiting periods for reentry and workers to wear protective clothing within the 24- or 48-hour reentry period after treatment are:

24 hours		48 hours	
Guthion®	ethyl parathion		Trithion®
Zolone®	methyl parathion		Meta-Systox-R®
EPN	Systox®		Bidrin®
ethion	Azodrin®		endrin

In addition to the EPA requirements, there are a few companies that state reentry intervals on their labels. It is your responsibility to keep abreast of label information and EPA requirements.

Anytime you apply one of the above materials it is highly recommended that you notify your client in order to avoid misunderstandings. Tell him the name of the material, time of application and the required reentry period

during which protective clothing must be worn so that he may inform his employees accordingly. If workers must enter fields prior to expiration of the required interval, growers are obligated to notify workers of the date of treatment and furnish them with protective clothing. Protective clothing means at least a hat or other suitable head covering, a long-sleeved shirt and long-legged trousers or a coverall-type garment (all of closely woven fabric), shoes and socks.

Warnings to workers may be orally or by posting signs at field entrances and/or bulletin boards where workers usually assemble. If a worker does not read English, a reasonable effort must be made to ensure that he understands the warning.

Fields treated with pesticides other than those listed above, or those specified on the labels, may be reentered without protective clothing after the spray has dried or the dust has settled.

Keeping everyone involved informed of your actions will be good public relations and could keep you from possible litigation.

Some state agencies may impose and enforce standards for workers that are more restrictive than those specified above. Be sure you know the regulations for your state.

Don't confuse days-to-harvest intervals with reentry intervals. Days-to-harvest is the time that must pass between pesticide application and crop harvest to protect the consumer by allowing the pesticide residue to fall to or below a specific tolerance level. The same pesticide will have different intervals for different crops, based on dosages necessary to control pests, different surfaces treated, whether edible parts are treated, stripping or washing outer leaves, etc. Check the harvest interval on the label before an application is made.

PESTICIDE EXPOSURE - PROTECTIVE CLOTHING AND EQUIPMENT

There are three ways that toxic chemicals may enter the human body to cause poisoning. The routes of entry include oral exposure (through the mouth by swallowing); dermal exposure (absorption through the skin); and inhalation or respiratory exposure (absorption by breathing). The following chart summarizes how the accidental exposures might occur.

HOW DO ACCIDENTAL PESTICIDE POISONINGS OCCUR?

BY MOUTH	DUSTS AND SPRAYS ENTERING MOUTH DURING AGRICULTURAL APPLICATIONS
	DRINKING PESTICIDES FROM UNLABELLED OR CONTAMINATED CONTAINER
	USING THE MOUTH TO START SIPHONAGE OF LIQUID CONCENTRATES
	EATING CONTAMINATED FOOD
	TRANSFER OF CHEMICAL TO MOUTH FROM CONTAMINATED CUFFS OR HANDS
	DRINKING FROM CONTAMINATED BEVERAGE CONTAINER
THROUGH THE SKIN	ACCIDENTAL SPILLS ON CLOTHING OR SKIN
	DUSTS AND SPRAYS SETTLING ON SKIN DURING APPLICATION
	SPRAYING IN WIND
	SPLASH OR SPRAY IN EYES AND ON SKIN DURING POURING AND MIXING
	CONTACT WITH TREATED SURFACES AS IN TOO EARLY RE-ENTRY OF TREATED FIELDS, HAND HARVESTING, THINNING, CULTIVATING, IRRIGATING AND INSECT OR PEST SCOUTING
	CHILDREN PLAYING — IN DISCARDED CONTAINERS IN PESTICIDE MIXING OR SPILL AREAS
	MAINTENANCE — REPAIR WORK ON CONTAMINATED EQUIPMENT
BY BREATHING	DUST, MISTS, OR FUMES
	SMOKING DURING APPLICATION OR
	CONTAMINATED SMOKING SUPPLIES

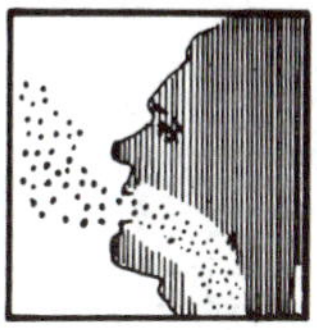

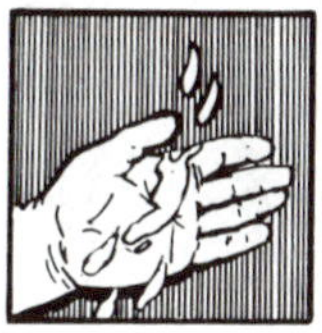

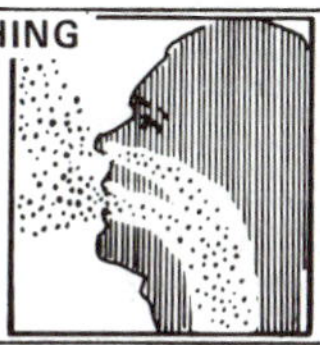

PROTECTIVE CLOTHING FOR
HAZARDOUS PESTICIDES

(Pesticides with labels marked: DANGER

☠

POISON, or WARNING)

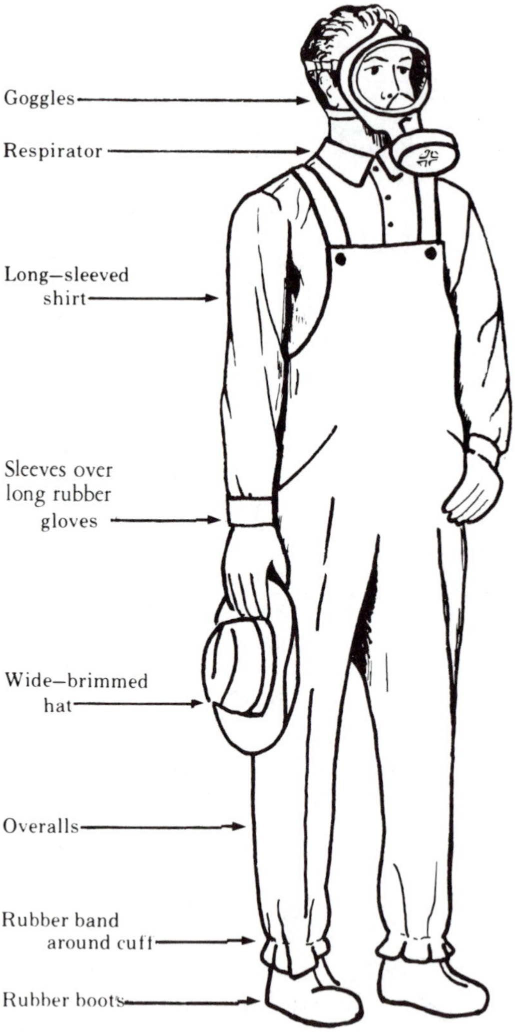

Goggles

Respirator

Long—sleeved
shirt

Sleeves over
long rubber
gloves

Wide—brimmed
hat

Overalls

Rubber band
around cuff

Rubber boots

Drawn by Dr. James R. Baker
North Carolina State University

Oral Exposure and Protective Measures

Although not considered a major source of occupational exposure to pesticides, oral exposure and subsequent absorption through the gastrointestinal tract is, with few exceptions, the most serious exposure because of the rapid internal absorption and possibility of quick death. Generally when a pesticide is taken into the mouth in amounts sufficient to cause serious injury or death, it is consumed either by accident as a result of gross negligence or by intent to do self inflicted injury. Accidental oral ingestion, in most cases, is a result of putting pesticides in unlabeled containers such as soft drink bottles or food containers for storage in an area where children or irresponsible adults may consume it.

The most likely means of ingesting pesticides is through accidental splashing of liquid into the mouth or by wiping the face with a sleeve, cuff, or hand. Other means of oral exposure of pesticides in occupational situations are food handled with contaminated hands, food exposed to pesticide sprays or dusts, contaminated drinking utensils, and attempting to clear nozzles by blowing through the nozzle openings or spray lines.

Oral exposure can be minimized by the following steps and protective measures:

1. Check the label for special instructions or warnings regarding oral exposure. Frequently pesticide labels will contain a precautionary illustration such as this one:

DO NOT SWALLOW

CAN KILL YOU IF SWALLOWED

This product can kill you if swallowed even in small amounts; spray mist or dust may be fatal if swallowed.

2. Never eat or drink while spraying or dusting.

3. Wash thoroughly with soap and water before eating or drinking.

4. Do not touch lips to contaminated objects or surfaces.

5. Do not wipe the mouth with hands, forearm, or clothing.

6. Do not expose lunch, lunch container, beverage, or drinking vessel to pesticides

7. If you are involved in operations where highly toxic pesticides are used and the label has indicated a high danger of oral ingestion, it would be advisable to wear a full face plastic shield or mask when there is a possibility of splashing of the pesticide.

Dermal Exposure and Protective Measures

Most sources of information indicate that pesticide absorption through the skin is the most common cause of poisoning in agricultural workers. The incidence of occupational poisoning by various pesticides is more closely related to their acute dermal toxicity in animals than to their acute oral toxicity, further substantiating the importance of dermal exposure as a significant route of entry into the body. Absorption through the skin may occur as a result of a splash, spillage, or drift when pesticides are being measured, mixed, loaded, or applied. It may also occur upon contact with a deposit of residue remaining on the treated crop or site.

There are several factors that affect dermal penetration of a pesticide. These factors include:

a) Physical and chemical properties of the pesticide.

b) Health and condition of the skin.

c) Temperature.

d) Humidity.

e) Presence of other chemicals (solvents, surfactants, etc.).

f) Concentration of the pesticide.

g) Type of formulation.

All of these factors can affect penetration and absorption of a pesticide by the skin. The physical and chemical properties, concentration, formulation, and presence of other chemicals are established when a specific pesticide is selected for use. Temperature and humidity are the conditions existing at the time of application, which are, at the time, selected as weather conditions most favorable for effective pest control. The health and condition of skin may be variable at the time of pesticide application. Individuals with skin problems should avoid other than minimal exposure to pesticides, unless extra precautions are observed. Skin cuts, abrasions, scratches, scuffs, or any other such damage or disruption can be sources of quick absorption of pesticides and extra care should be taken to minimize pesticide exposure if these conditions exist.

Wind, type of activity, application method, rate of application, and duration of exposure are other factors that may affect dermal exposure. All of these factors must be taken into consideration when selecting protective clothing and equipment to protect against dermal exposure of pesticides.

The type and amount of protective clothing needed depends upon the job being done and the type of pesticide being used. Several factors are important for the operator to consider when determining protective needs:

a. The toxicity, concentration, and vapor action of the pesticide being used.

b. The degree of expected exposure during application.

c. The length of expected exposure during application.

d. The extent to which the pesticide can be absorbed through the skin.

The above requirements may vary in a number of situations. The following suggested procedures should be observed depending upon an evaluation of all of the factors.

1. Observe all recommended protective measures specifically mentioned on the label of the pesticide to be used. Some pesticide labels may contain an illustration as follows:

CAN KILL YOU BY SKIN CONTACT

This product can kill you if touched by hands or spilled or splashed on skin, in eyes or on clothing (liquid goes through clothes).

2. Cover up before exposure, not after. Putting on protective clothing after you have been exposed to pesticides on the skin will only hasten the absorption rate.

3. Skin contamination can be reduced by wearing coveralls that afford the entire body protection. If the situation is such that coveralls would be wet through mist spray or spillage, a waterproof rainsuit should be used. Coveralls and rainsuits should be thoroughly washed with soap and water after each use. Contaminated protective equipment should be handled as carefully as the pesticide itself. Rain gear should be periodically checked for cracks and holes.

4. Efforts should be made to protect the hair, skin about the head, and neck from pesticides. Head gear should include one of the following: waterproof rain hat; safety hard hat; or a cap. Waterproof or repellent parkas offer good protection for the head and neck area at the same time. Old felt hats or other types of absorbent head gear should not be worn as they absorb pesticides, especially in the area of the sweat band. Once contaminated, they can provide a source of continuous and very dangerous skin contact.

5. Natural rubber gloves should generally be used when handling organophosphorous, carbamate, and other pesticides that are readily absorbed through the skin. Some fumigant type pesticides cause rubber gloves to disintegrate and polyethylene gloves must be used in this case. The type of glove to use is usually suggested by the pesticide manufacturer. One should select a light weight, unlined, natural rubber glove with long enough gauntlets to protect the wrist area, but flexible enough to allow freedom of finger movement. Several types of disposable gloves are presently being manufactured

that can be used when applying certain types of pesticides. Gloves should be checked very carefully at regular intervals for holes or other signs of wear. It is probably a good idea to replace the gloves periodically rather than chance an accidental exposure.

Leather and cotton gloves should never be used when handling or applying pesticides. They can absorb the material and become a constant source of pesticide exposure. They could then become more of a potential hazard to poisoning than if no gloves are worn at all.

6. Waterproof boots or footgear should be worn during most any type of pesticide application. Foot protection is most important when hand applications of pesticides are made. Only rubberized boots should be worn when handling or spraying pesticides on a large scale. Leather and canvas shoes can absorb the pesticide and hold it in contact with the wearer, thus making this type of footwear undesirable when applying pesticides.

7. Eye protection is most often needed when measuring or mixing pesticide concentrates as well as when spray or dust drifts might be a problem. Protective shields or goggles should be used whenever there is a chance of a pesticide coming into contact with the eyes. These pieces of equipment should be kept clean at all times and should be handled the same as protective clothing when being washed or repaired, keeping in mind that they should be handled as carefully as the pesticide itself.

Respiratory Exposure and Protective Measures

Pesticides are sometimes inhaled in sufficient amounts to cause serious damage to nose, throat, and lung tissues. The potential hazard of respiratory exposure is great, due to what is considered to be near complete absorption of pesticides through this route. Vapors and extremely fine particles represent the most serious potential for respiratory exposure.

Some means of protection for the respiratory route is needed when toxic dust and vapors or small spray droplets are prevalent. Pesticide dusts, aerosol, fog, fume, smoke, and certain mists represent a high potential for respiratory exposure.

Pesticide exposure is usually relatively low when dilute sprays are being applied with conventional application equipment. This is due primarily to the larger droplet sizes produced. When low volume equipment is being used, respiratory exposure is increased by the smaller droplets or particles being produced. Application in confined spaces also contributes to an increased potential respiratory exposure.

A respiratory device is one of the most important pieces of protective equipment both for commercial and private applicators when they are applying toxic pesticides. Respirators are sometimes uncomfortable to wear, particularly in hot and dusty conditions, but it is important that the applicator fully realize the need for protection or serious injury may occur.

Several kinds of respiratory devices are available to protect applicators from breathing dust and chemical vapors. Pesticide applicators need to know what types are available and the hazards they will protect against. Some pesticide labels may contain an illustration as follows:

CAN KILL YOU IF BREATHED

This product can kill you if vapors, spray mist or dust are breathed.

The pesticide label will contain information concerning the proper respirator to use for the pesticide you will be applying. Be sure to use the respirator designed for the pesticide and the job. Never try to use one type of respirator for all kinds of hazards.

There are primarily two types of respiratory devices for use by pesticide applicators when handling toxic pesticides. These are the chemical cartridge respirator and the canister type respirator or gas mask. Supplied air respirators and self-contained breathing equipment are also available when oxygen is deficient or highly toxic gases are so concentrated that the applicator must have his own air supply. Types of protective devices are illustrated and discussed below:

Cartridge respirator -- protects against certain pesticide dusts, mists, and fumes. Check the label to see if this type of respirator is sufficient for protection against the pesticide you are using.

CARTRIDGE RESPIRATOR

This kind usually has a half-face mask that covers the nose and mouth, but does not protect the eyes. It has one or two cartridges attached to the facepiece. An absorbent material such as activated charcoal, plus dust filters, purify the air you breathe.

DOES NOT PROTECT from lack of oxygen.

Canister type (Gas Mask) -- protects the lungs as well as the eyes against certain chemical dusts, mists, and fumes. Be sure to check the label to see if this type of protective device is required for the pesticide you are using.

CANNISTER TYPE (Gas Mask)

This covers the entire face. The face-piece may hold a cannister directly (chin-type), or connect by a flexible hose to a cannister carried on your chest or hip. The "gas mask" cannister contains more absorbing material and longer-life filters than the cartridge type respirator. *DOES NOT PROTECT from lack of oxygen.*

Supplied air respirator -- used where oxygen is deficient or highly toxic gases are so concentrated that you need your own air supply. Analyze your own spray situation to determine if you need this type of equipment.

SUPPLIED AIR RESPIRATOR

Brings in air through a hose, from a safe distant supply. The operator's hood is connected by hose to a source of out-side fresh air, usually pumped by a blower. Types are:
- Air-line respirators with constant flow or demand flow.
- Hose masks, with or without blower.

Self-contained breathing apparatus -- serves the same functions as a supplied air respirator. Determine whether you need the mobility provided with this type of equipment.

SELF-CONTAINED BREATHING APPARATUS

Lets you take your air supply with you for short term jobs. The mask is connected to an oxygen cylinder, usually carried on the operator's back.

(Certain gases, like ammonia, can cause harm through the skin. You may need protective clothing as well as oxygen.)

CARE AND MAINTENANCE OF RESPIRATORS AND GAS MASKS

No matter how well the respiratory device is designed and made, unless it is properly cared for and maintained, it may fail to provide protection. The two most common failings are said to be (1) failure to occasionally wash the face piece with soap and water and (2) neglecting to change the filter cartridges or canisters regularly.

In order to minimize respiratory exposure and to be certain that protective devices are working properly the following practices should be observed as suggested by Wilson Products Division, ESB Incorporated, Reading, Pennsylvania:

INSTRUCTIONS FOR USE

1. Remove respirator, cartridges and filters from plastic bags. Check to see that gasket is in cartridge holder before screwing in cartridges. Insert filter into retainer caps and snap onto cartridge holder or cartridges.

2. FIT RESPIRATOR ON FACE with narrow end over nose. Adjust headband around neck and crown of head, snug enough to insure a tight but comfortable seal.

3. TEST FOR TIGHTNESS: Place the palm of the hand or thumb over the valve guard and press lightly. Exhale to cause a slight pressure inside facepiece. If no air escapes, respirator is properly fitted. If air escapes, readjust respirator and test again.

4. FILTERS (a) REPLACE when breathing becomes difficult. Generally the filter discs should be changed after eight hours of dusty exposure. (b) REPLACE CARTRIDGES when any leakage is detected by taste or smell.

5. CLEAN AND SANITIZE YOUR RESPIRATOR after each day's use. First remove filters and cartridges, then wash other parts thoroughly with warm soapy water and/or sanitize with Willson's Germisol®.

6. The cartridge holders are keyed to assure their correct positioning and maintain the proper balance of the device. Make sure they are properly positioned and seated.

7. KEEP RESPIRATOR CLEAN when not in use. Store in container provided. Replace worn or faulty parts immediately, and order by part number.

8. FOR YOUR PROTECTION the DUST FILTERS and CHEMICAL CARTRIDGES must be assembled tightly, and changed frequently, according to exposure.

9. Many chemicals can be absorbed through the skin. Wear protective clothing when necessary.

TAKE CARE OF YOUR RESPIRATOR AND YOUR HEALTH

CAUTION

Respirators and canister gas masks are not to be worn in atmospheres immediately dangerous to life or health or in atmospheres containing less than 19.5% oxygen.

AIR PURIFYING RESPIRATORS (cartridge or canister (gas mask))

1. MSA Comfo® II in various sizes, MSA #460968 with GMP Pesticide Combination Cartridge.
   MSA #464025

2. MSA Belt-Mounted Respirator
   MSA #461000

3. MSA Chin Style Pesticide Mask
   MSA #448983

4. MSA Industrial Size Mask
   MSA #457100

5. MSA Fumigant Masks
   MSA #457069 for phosphine and H_2S
     #457084 for Hydrocyanic acid
     #457081 for Methyl bromide
     #457097 for sulfural fluoride (Vikane)

6. American Optical (AO Quantifit)
   in small #R4000, medium #R5000, or large #R6000 with cartridge #R58

7. H.S. Cover (H_SC)
   Model 1482 with G100 and F104 pesticide cartridge

8. Willson Chin Style Gas Mask #2100/2200 with #61F pesticide canister

9. Willson #1600 and 1700 with #R15 Filter, #R683 retainer and #R21 organic vapor cartridges

10. Willson #1200 with #R15 filter, #R683 retainer and #R21 cartridges

11. Scott Model 64 and 65 with cartridge #65-OVP (full facepiece) or #652-L
    Cartridge #642-OV
    Filter #642-F
    Retainer #642-FR

SUPPLIED-AIR RESPIRATORS

1. Scott #900034 Supplied Airline Respirator with Self-Contained Air Supply

2. Scott #801548 Type C Supplied-Air Pressure-Demand Respirator

3. Scott #802280 Half-Mask Supplied-Air Respirator

4. Willson #1810 and 1820 half-mask respirator

5. Willson #1850 and 1860 full-facepiece respirators

6. Bullard Free-Air supplied-air respiratory system

7. Bullard System 999 supplied-air system

8. Robertshaw Ram-15

9. 3M #W-2804 pesticide helmet and other Whitecap® systems

SELF-CONTAINED BREATHING APPARATUS

1. Robertshaw RAM-30

2. Scott Air-Pak 11a #900000 and Presur-Pak 11a #900014

3. Scott Air-Pak 4.5 #900450 and Presur-Pak 4.5 #900455

SUPPLIERS OF RESPIRATORY PROTECTION EQUIPMENT

1. Bullard Safety
   2105 West Amherst Avenue
   Englewood, CO 80110

2. 3M Occupational Health and
   Safety Products Division
   220-7W 3M Center
   Saint Paul, MN 55101

3. Willson Division
   INCO Safety Products Company
   P. O. Box 622
   Reading, PA 19603

4. Scott Aviation
   A Division of A-T-O Inc.
   Lancaster, NY 14086

5. HSC Corporation
   P. O. Box 192
   Buchanan, MI 49107

6. American Optical Corporation
   Safety Products Division
   South Bridge, MA 01550

7. Mine Safety Appliance Company
   600 Penn Center Boulevard
   Pittsburgh, PA 15235

8. Robertshaw Controls Company
   333 N. Euclid Way
   Anaheim, CA 92803

The preceeding listings are not all inclusive. They are intended as
a general guide to equipment that is believed to be generally available
nationwide. No endorsement is intended of equipment listed, nor is
discrimination intended toward those products or companies that may
not be listed.

SYMPTOMS OF PESTICIDE POISONING

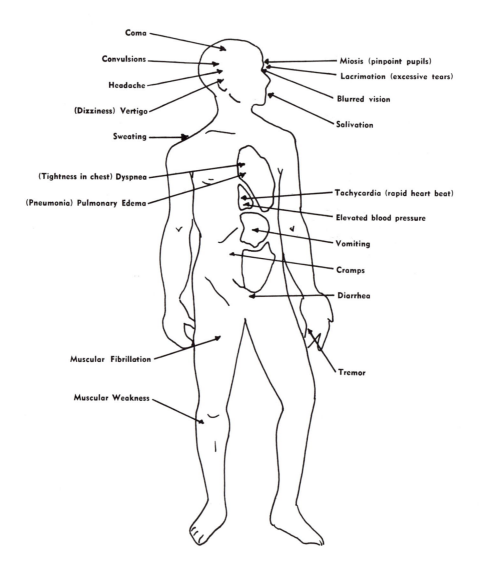

Coma

Convulsions

Headache

(Dizziness) Vertigo

Sweating

(Tightness in chest) Dyspnea

(Pneumonia) Pulmonary Edema

Muscular Fibrillation

Muscular Weakness

Miosis (pinpoint pupils)

Lacrimation (excessive tears)

Blurred vision

Salivation

Tachycardia (rapid heart beat)

Elevated blood pressure

Vomiting

Cramps

Diarrhea

Tremor

PESTICIDE POISONING EFFECTS AND SYMPTOMS

IMPORTANT! The information in this section may help to save a life -- yours. or your friend's.

Even if this manual is close by when an accident occurs, there may not be time to find and read this section before you are obliged to take action.

This information should be in your head -- not just in the manual.

The ways in which pesticides affect humans and other mammals are commonly referred to as modes of action. The modes of action of a number of pesticides in use today are either not known, or in some instances are only partially understood. Even though it may not be known exactly how a pesticide poisons the body in all cases, some of the signs and symptoms resulting from such poisoning are quite well known. These warnings(signs and symptoms) of the body in response to poisoning, should be recognized by those who use pesticides. It must be remembered, however, that everyone is subject to various sicknesses or diseases. Therefore, it cannot be assumed that because an individual is in the vicinity where pesticides are or have been used that certain signs or symptoms are the result of pesticide poisoning. But if any signs appear after contact with pesticides, call your physician and advise him of the nature of the pesticide involved.

Pesticide applicators should be alert to the early stages of signs and symptoms of poisoning and immediately and completely remove the source of exposure, such as contaminated clothing. Quick action may prevent additional exposure and will help to minimize injury. In some instances, early recognition of the signs and symptoms of poisoning and immediate complete removal of the source of exposure may save the applicators life.

Learn how to recognize poisoning symptoms. Listed below are some of the more important hazardous pesticide groups and some of the ways they affect humans or animals.

1) Organophosphorous Pesticides -- The organophosphorous chemicals are one of the largest groups of pesticides presently being used. The group includes insecticides such as parathion, malathion, phorate, mevinphos, diazinon, TEPP, and others. Toxicity values of these pesticides range from high toxicity for parathion, to low toxicity in the case of malathion. The organophosphorus pesticides can be absorbed dermally, orally, or through inhalation of vapors.

The pesticides in this group attack a chemical in the blood, cholinesterase, that is necessary for proper nerve functioning. Since this group of pesticides affect the enzyme cholinesterase, they are sometimes referred to as "Cholinesterase Inhibitors", or "Anticholinesterase Compounds". Organophosphates can inhibit, or poison, the cholinesterases by forming chemical combinations with them which prevent them from doing their work in the nervous system. (Refer to pages 20 and 21 for a graphic explanation of the role of cholinesterase and how it is affected by organophosphates).

PHYSIOLOGICAL ACTION OF ACETYLCHOLINE

ACETYLCHOLINE,

LIBERATED BY NERVE

IMPULSE, ACTS DIRECTLY UPON

EFFECTOR CELLS TO PRODUCE THEIR

CHARACTERISTIC RESPONSES

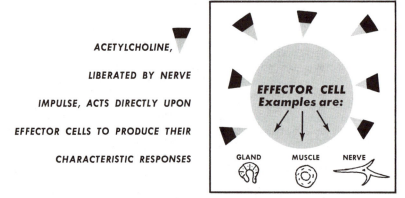

PHYSIOLOGICAL ROLE OF CHOLINESTERASE

CHOLINESTERASE

TERMINATES THE RESPONSE

BY HYDROLYZING ACETYLCHOLINE

TO CHOLINE

AND THE ACETATE ION

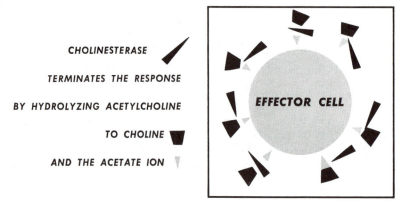

MECHANISM OF TOXIC ACTION OF PHOSPHATE ESTER INSECTICIDES

PHOSPHATE ESTER

COMPOUNDS ATTACH

A PHOSPHORYL GROUP

TO CHOLINESTERASE AND THEREBY

RENDER THE ENZYME UNABLE

TO PERFORM ITS FUNCTION

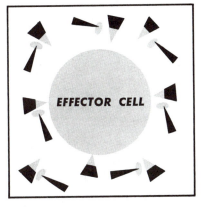

MECHANISM OF PROTECTIVE ACTION OF ATROPINE

ATROPINE

BLOCKS THE ACTION OF

ACETYLCHOLINE BY INTER-

FERING WITH THE ABILITY

OF THE CELL TO RESPOND

TO THIS CONTINUING STIMULUS

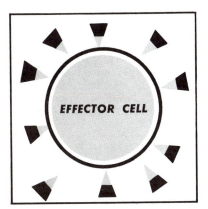

When the enzymes are poisoned, nerve impulse transmission races out of control because of a build up of acetylcholine at the ends of nerve fibers. Muscle twitchings referred to as tremors or fibulations then become noticable. Convulsions or violent muscle actions result if the tremors become intense.

Additional symptoms and signs of organophosphorus poisoning include: headache; giddiness; nervousness; blurred vision; dizziness; weakness; nausea; cramps; diarrhea; and chest discomfort. Others might be sweating, pin-point eye pupils, watering eyes, excess salavation, rapid heart beat, excessive respiratory secretions, and vomiting. Advanced stages of poisoning usually result in convulsions, loss of bowel control, loss of reflexes, and unconsciousness. Quick action and proper medical treatment can still save persons in the advanced stages of poisoning, even though they may be near death.

2) Carbamate Pesticides -- This group of pesticides is relatively new and includes such insecticides as aldicarb (Temik ®), carbaryl (Sevin®), and carbofuran (Furadan®); herbicides such as cycloate (Ro-Neet®), diallate (Avadex®), and such fungicides as benomyl (Benlate®) and ferbam (Fermate®).

The mode of action of these compounds is very similar to that of the organophosphorus compounds in that they inhibit the enzyme cholinesterase. However, they differ in action from the organophosphorus compounds in that the effect on cholinesterase is brief, because the carbamates are broken down in the body rather rapidly. Carbamates are therefore, referred to as "rapidly reversing inhibitors" of cholinesterase. The reversal is so rapid that, unless special precautions are taken, measurements of blood cholinesterase of human beings or other animals exposed to carbamates are likely to be inaccurate and always in the direction of appearing to be normal. Carbamate pesticides can be absorbed through the skin as well as by breathing or swallowing.

Symptoms and signs of carbamate poisoning are essentially the same as those caused by organophosphorus pesticides.

3) Chlorinated hydrocarbon pesticides -- This group of pesticides is one of the older categories of organically synthesized compounds and includes insecticides such as DDT, aldrin(Aldrite®), dieldrin (Dieldrite®), endrin, and chlordane. These insecticides were used quite extensively until recently when many of them have been restricted or banned from use. These compounds act on the central nervous system, but the exact mode of action is not known. It is known that these compounds or their degradation products can be stored in the fatty tissues as a result of a single large dose or repeated small dose exposures. Fat storage of these pesticides, however, appears to be virtually inactive and of no immediate consequence to the individual. The chlorinated hydrocarbon pesticides can be absorbed by breathing fumes, through the mouth, or from exposure on the skin.

Symptoms and signs of chlorinated hydrocarbon pesticide poisoning include nervousness, nausea, and diarrhea. Convulsions may result from an exposure to a large dose. Liver and kidney damage has been demonstrated in laboratory animals when administered in repeated large doses, but these signs have not been demonstrated in man.

4) Nitrophenol pesticides -- This group of pesticides is represented as different formulations in the fungicide, insecticide, and herbicide categories. DNOC (Elgetol 30 and Sinox®) is representative of the nitrophenol pesticides. All of the nitrophenol compounds can be absorbed in toxic amounts either orally or by inhalation. Some formulations can be absorbed dermally.

Symptoms and signs include fever, sweating, rapid breathing, rapid heart beat, or unconsciousness due to the speeding up of certain body processes and functions. In cases of a single, large exposure to some of the nitrophenols, signs and symptoms may occur rapidly. If death occurs, it probably will occur within twenty-four to forty-eight hours. However, if the patient receives adequate medical attention and exposure to the chemical has not been severe, he probably will recover completely. Nervousness, sweating, unusual thirst, and loss of weight have been observed in chronic or extended cases of poisoning.

5) Arsenic pesticides -- This group of pesticides includes compounds of high toxicity values, such as sodium arsenite (Atlas A®), and those that have low toxicity values such as DSMA. Other examples in this group include paris green and MSMA (Ansar and Daconate®). These compounds are used as insecticides, herbicides, defoliants, and rodenticides. Arsenic is absorbed primarily through oral exposure, but some exposure can be gained through the respiratory route.

Arsenic poisons cells in certain body tissues and also affects certain enzymes, thereby slowing down normal body functions. Signs of arsenic poisoning include stomach pain, vomiting, diarrhea, and a severe drop in blood pressure, which may cause dizziness.

6) Mercury Pesticides -- Mercury compounds were at one time frequently used in preparations for seed treatment compounds. Because of their toxic and persistent properties, many of the uses of these compounds have been banned and are no longer recommended for use in Agriculture. Mercury can accumulate in the body and may cause permanent damage to the nervous system. Severe skin problems may also occur.

Symptoms or signs of mercury poisoning may be delayed and later show up first as tingling of the fingers, tongue, lips, headache, and shakiness. Inability to think clearly, write, speak, or walk may occur. Changes in personality may follow.

7) Bipyridyliums -- This group of herbicides which includes diquat and paraquat may be fatal if swallowed and harmful if inhaled or absorbed through the skin. Lung fibrosis may develop if these materials are taken by mouth or inhaled. Prolonged skin contact will cause severe skin irritation.

Signs and symptoms of injury may be delayed, but there are no adequate treatments and effects are generally irreversible.

8) <u>Anticoagulants</u> -- This group of pesticides is commonly used as rodenticides. <u>Warfarin</u> (Prolin®), coumafuryl (Fumarin®), and diphacinone (Diphacin®) are representative examples of this group. These materials are of danger only when taken orally and large doses would be required to cause human death.

Anticoagulants reduce the body's ability to produce blood clots and sometimes damage capillary blood movement in the body. Bleeding can occur with the slightest body damage such as a severe nose bleed or massive bruises. Severe back and abdominal pains have been noted in persons attempting suicide.

9) <u>Botanical pesticides</u> -- Botanical pesticides are manufactured from plant derivatives. They vary greatly in chemical structure and toxicity to man, ranging from pyrethrum, one of the least toxic to mammals of all the insecticides in use currently, to strychnine, which is extremely toxic.

a. Pyrethrum and allethrin (Pynamin®) - These botanical materials are used against a wide variety of insect pests. Most symptoms in man as a result of exposure to pyrethrum have been reported to be skin allergies, sneezing, runny nose, and in some cases, stuffiness of the nose. Most have been reported to be minor, however, some individuals are known to be more sensitive than others.

b. Strychnine - The mode of action of this pesticide is not completely understood. However, it is known to act on the nervous system. Symptoms and signs are nervousness, stiff muscles in the face and legs, cold sweats, and fits. From one to ten violent attacks may occur before the patient either recovers or death follows due to his inability to breathe. Without medical attention, death commonly occurs within three to five hours after exposure.

c. Nicotine (Black Leaf 40®) - This is one of the most toxic of all poisons and its action is very rapid. Poisoning can be caused by oral, dermal, and respiratory exposure. Signs and symptoms of nicotine poisoning include local skin burns and irritations. If nicotine is absorbed through the skin or taken in through the mouth, the patient becomes highly stimulated and excitable. This is generally followed by extreme depression. In fatal cases of nicotine poisoning, death is usually rapid, nearly always within one hour and occasionally within five minutes due to paralysis of respiratory muscles.

d. Rotenone - Relatively few cases of poisoning to man from rotenone have been reported. Direct contact with rotenone will result in mild irritation of the skin and eyes in some individuals. Most of the symptoms reported have come from animal studies. In these studies, the animals exhibited numbness of the mouth, vomiting, and pains in the stomach and intestines. The inability of a test animal to coordinate its activities may result in muscle tremors with chronic convulsions following. Breathing is often very rapid initially,followed by extremely slow breathing. In cases of animal death, it has generally been due to the inability to breathe. If rotenone dust is inhaled, severe irritation of the inside of the nose, throat, and the lungs may occur.

10) Fumigation materials -- Most fumigation materials are highly toxic and of course are extremely dangerous when inhaled. Because of the nature of their use, they are extremely hazardous in enclosed areas.

a. Methyl bromide (Bromo Gas®) - The mode of action of this compound is to affect the protein molecules in certain cells of the body. The signs and symptoms produced by this compound include

severe chemical burns of the skin and other exposed tissue, chemical pneumonia which produces water in the lungs, and severe kidney damage as well as extreme nervousness. Any of these effects can be fatal. If a person inhales smaller amounts of methyl bromide, he may exhibit mental confusion, double vision, tremors, lack of coordination, and slurred speech. This sometimes produces effects that give the appearance of alcoholic intoxication and victims have been jailed or sent to mental hospitals by mistake.

b. Chloropicrin (Picfume®) - This is also a highly hazardous chemical, but unlike methyl bromide, it gives off a gas with an odor and it is very irritating to the eyes. It is sometimes mixed with methyl bromide as a warning agent.

c. Carbon tetrachloride - This chemical effects the nerves and also severely damages the cells in the kidneys and liver.

FIRST AID FOR PESTICIDE POISONING

It is essential that pesticide poisoning incidents be recognized immediately because prompt treatment may mean the difference between life and death. Do not substitute first aid for professional treatment. First aid is only to relieve the patient before medical help can be reached.

If you are alone with the victim . . .

First -- See that the victim is breathing; if not, give artificial respiration.
Second -- Decontaminate him immediately i.e., wash him off thoroughly.
Third -- Call your physician.

If another person is with you and the victim . . .

Speed is essential; one person should begin first aid treatment while the other calls a physician.

The physician will give you instructions. He will very likely tell you to get the victim to the emergency room of a hospital. The equipment needed for proper treatment is there. Only if this is impossible should the physician be called to the site of the accident.

- GENERAL -

1. Give mouth-to-mouth artificial respiration if breathing has stopped or is labored. (Refer to page 29.)

2. Stop exposure to the poison and if poison is on skin, cleanse the person, including hair and fingernails. If swallowed, induce vomiting as directed on page 28.

3. Save the pesticide container and material in it if any remains; get readable label or name of chemical(s) for the physician. If the poison is not known, save a sample of the vomitus.

- SPECIFIC -

Poison on Skin
   --Drench skin and clothing with water (shower, hose, faucet)
   --Remove clothing.
   --Cleanse skin and hair thoroughly with soap and water; speed in washing
      is most important in reducing extent of injury.
   --Dry victim and wrap in blanket.

Poison in Eye

--Hold eyelids open, wash eyes with gentle stream of clean running water
immediately. Use large amounts. Delay of a few seconds greatly increases extent of injury.
--Continue washing for 15 minutes or more.
--Do not use chemicals or drugs in wash water. They may increase the
extent of injury.

Inhaled Poisons (Dusts, vapors, gases)

--If victim is in enclosed space, do not go in after him without air-
supplied respirator.
--Carry patient (do not let him walk) to fresh air immediately.
--Open all doors and windows, if any.
--Loosen all tight clothing.
--Apply artificial respiration if breathing has stopped or is irregular.
--Call a physician.
--Prevent chilling (wrap patient in blankets but don't overheat him).
--Keep patient as quiet as possible.
--If patient is convulsing, watch his breathing and protect him from
falling and striking his head on the floor or wall. Keep his chin
up so his air passage will remain free for breathing.
--Do not give alcohol in any form.

Swallowed Poisons

--Call a physician immediately.
--Do not induce vomiting if:
  1. Patient is in a coma or unconscious.
  2. Patient is in convulsions.
  3. Patient has swallowed petroleum products (that is, kerosene, gasoline,
     lighter fluid).
  4. Patient has swallowed a corrosive poison (strong acid or alkaline
     products); symptoms; severe pain, burning sensation in mouth and
     throat. A corrosive substance is any material which in contact with
     living tissue will cause destruction of tissue by chemical action
     such as lye, acids, lysol, etc.
--If the patient can swallow after ingesting a corrosive poison, give the
following substances by mouth.

For acids: milk, water, or milk of magnesia (1 tablespoon to 1 cup
of water).
For alkali: milk or water; for patients 1-5 years olds, 1 to 2
cups for patients 5 years and older, up to one quart.

Chemical Burns of Skin

--Wash with large quantities of running water.
--Remove contaminated clothing.
--Immediately cover with loosely applied clean cloth, any kind will do,
depending on the size of the area burned.

Chemical Burns of Skin (continued)
  --Avoid use of ointments, greases, powders, and other drugs in first aid
    treatment of burns.
  --Treat shock by keeping patient flat, keeping him warm and reassuring
    him until arrival of physician.

HOW TO INDUCE VOMITING WHEN A NON-CORROSIVE SUBSTANCE HAS BEEN SWALLOWED

  --Induce vomiting by placing the blunt end of a spoon(not the handle), or
    your finger at the back of the patients throat, or by use of this emetic
    --2 tablespoons of salt in a glass of warm water.
  --When retching and vomiting begin, place patient face down with head
    lowered, thus preventing vomitus from entering lungs and causing fur-
    ther damage. Do not let him lie on his back.
  --Do not waste excessive time in inducing vomiting if the hospital is
    a long distance away. It is better to spend the time getting the pa-
    tient to the hospital where drugs can be administered to induce vom-
    iting and/or stomach pumps are available.
  --Clean vomitus from person. Collect some in case physician needs it for
    chemical tests.

ARTIFICIAL RESPIRATION*

In many conditions where breathing has ceased, or apparently ceased, the
heart action continues for a limited time. If fresh air is brought into
the lungs, so that the blood can obtain the needed oxygen from the air,
life can be sustained. This can be accomplished in many instances by artificial respiration.

Certain general principles must always be kept in mind in applying any method of artificial respiration.

Time is of prime importance; seconds count. Do not take time to remove the
victim to a more satisfactory place unless the place is unsafe for victim
or rescuer; begin at once. Do not delay resuscitation to loosen the victim's clothes, or warm him. These are secondary to the main purpose of
getting air into the victim's lungs.

If possible, place victim so the head is slightly lower than the feet to
permit better drainage of fluid from respiratory passage.

Remove from the victim's mouth all foreign bodies, such as false teeth,
if loose,tobacco, gum, etc., and see that the head is tipped back to maintain an open airway,and loosen any tight clothing about the victim's neck,
chest, or waist.

Keep the victim warm, by covering him with blankets, clothing, or other
material; if possible, his underside should also be covered.

---

*From "Industrial First Aid Manual", 1971. State of Washington, Dept of
Labor & Industries, Safety Education Section, as modified.

Continue artificial respiration rhythmically and uninterruptedly until spontaneous breathing starts or a doctor pronounces the patient dead.

If the victim begins to breathe of his own accord, adjust your timing to his. Do not fight his attempt to breathe.

A brief return of natural respiration is not a signal for stopping the resuscitation treatment. Not infrequently a patient, after a temporary recovery of respiration, stops breathing again. He must be watched; if natural breathing stops, resume artificial respiration at once.

Always treat the victim for shock during resuscitation, and continue such treatment after breathing has started. Do not give any liquids whatever by mouth until a patient is fully conscious.

If it is necessary (due to extreme weather or other conditions) to move a patient before he is breathing normally, continue artificial respiration while he is being moved.

Mouth-to-Mouth or Mouth-to-Nose Method of Resuscitation (Adult)

If there is foreign matter visible in the mouth, or if the victim vomits, roll him on his side, prop him up with your knee behind his shoulders, turn the head slightly downward and wipe the mouth out quickly with your fingers or a cloth wrapped around them.

Prop the body on side with knee, with head resting on extended arm--hold head with one hand and clean foreign matter from mouth if in evidence.

    1. Tilt the head back so chin is pointing upward(A). If this does not open airway, pull the jaw into a jutting-out position(B and C). These maneuvers should relieve obstruction of the airway by moving the base of the tongue away from the throat.

Mouth-to-Mouth Air Exchange

2. Open your mouth wide and place it tightly over the victim's mouth. At the same time pinch the victim's nostrils shut (D) or close the nostrils with your cheek(E). Or close the victim's mouth and place your mouth over the nose. Blow into the victim's mouth or nose. (Air may be blown through the victim's teeth, even though they may be clenched). The first blowing efforts should determine whether or not obstruction exists.

3. Remove your mouth, turn your head to the side, and listen for the return of air. Watch rise and fall of chest. Repeat the blowing effort. For an adult, blow vigorously at the rate of about 12 breaths per minute. (For infants--20 breaths per minute).

4. If you are not getting air exchange, recheck the head and jaw position (A or B and C). If you still do not get air exchange, quickly turn the victim on his side and administer one or two sharp blows between the shoulder blades in the hope of dislodging foreign matter (F).

Again sweep your fingers through the victim's mouth to remove foreign matter.

Those who do not wish to come in contact with a person may place a cloth over the victim's mouth or nose and breathe through it. The cloth does not greatly affect the exchange of air. If the aider has false teeth, they should be removed for obvious reasons.

Manual Means of Artificial Respiration

If nature of the injury is such that mouth-to-mouth resuscitation cannot be used, the rescuer should apply the manual method.

It has already been pointed out that the base of the tongue tends to press against and block the air passage when a person is unconscious and not breathing. This action of the tongue can occur whether the victim is in a face-down or face-up position.

Following is an accepted manual method of artificial respiration.

Operator kneels on one or both knees at victim's head. Place one hand on each side of spine(midback) approximately two inches from spine (according

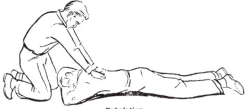

Exhalation

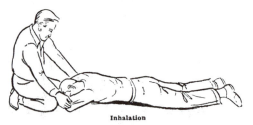

Inhalation

to size of victim), just below the shoulder blades. Do not have thumbs on spine. Rock forward and exert slow,steady, moderate pressure until firm resistance is met. Gradually release pressure, rock backwards, place hands beneath arms of subject close to elbows slowly drawing elbows toward you and upward as you rock backwards, until firm resistance is met. Gently lower arms to ground. Repeat process at the rate of approximately 12 complete cycles per minute. Back pressure decreases the chest cavity, producing active exhalation. The armlift and stretch increases the chest cavity and induces active inhalation.

➤ DO'S AND DON'TS -- ARTIFICIAL RESPIRATION ➤

DO

1. Start artificial respiration as quickly as possible,
2. Use method best suited in each particular case.

DO's -- Artificial Respiration (continued)
3. Maintain airway in all cases. (Clean out air passages and tip head back.)
4. Give respiration--apply 12 to 15 times per minute for adults, 16 to 20 times for children, and 20-24 times for infants under one year of age.
5. Treat for shock.
6. Loosen tight clothing.
7. Administer oxygen if available.
8. After victim starts breathing, stand by in case needed.

DO NOT
1. Move victim unless necessary to remove from danger.
2. Wait or look for help.
3. Do not give up.

Administration of Oxygen

It is advisable to supplement artificial respiration by administering oxygen, especially when the quantity of fresh air supplied to the patient by artificial respiration is inadequate, as is likely if the patient has been breathing poisonous gases. However, no time should be lost waiting for oxygen before artificial respiration is begun. When oxygen is administered, the manual treatment should not be stopped but should continue as long as there is hope of reviving the patient or until he begins to breathe normally. Oxygen frequently is available in compressed form in an oxygen bottle or cylinder. The valve should be opened slightly, with the flow directed away from the patient. After a moderate flow has been established, the oxygen should be directed into a cap, hat or piece of cloth used as improvised mask to confine the oxygen to the face of the victim; an oxygen inhalator aids in its administration. To eliminate explosive hazards, regulator control valves should be retained on oxygen bottles.

First Aid for Gas Poisoning

The steps in first-aid treatment of poisoning by toxic or noxious gases are:

Always remove the person to fresh air as quickly as possible.

Obtain medical aid.

If breathing has stopped, is weak or intermittent, or is present in but occasional gasps, start artificial respiration at once, and continue until normal breathing is resumed or a doctor pronounces the victim dead. When giving artificial respiration, always administer oxygen if available.

Keep the patient at rest, lying down, to avoid any strain on the heart. Later give him plenty of time to rest and recuperate.

Inhalations of oxygen, when administered immediately, will greatly reduce the severity of carbon monoxide poisoning, as well as decrease the possibility of serious after effects. Give oxygen whether the patient is conscious or unconscious.

All industries in which gas poisoning commonly is present should provide apparatus for efficient administration of oxygen. Such apparatus should be placed at convenient points and employees should be trained in its use.

ANTIDOTES -- THEIR USE BY NONMEDICAL PEOPLE

An antidote is a remedy used to counteract the effects of a poison or prevent or relieve poisoning.

Therefore, good judgment and safety practices, including the use of protective clothing, safety devices, and knowledge of first aid, are in a sense antidotes since they can and frequently do prevent or relieve poisoning. However, the general public is inclined to think of antidotes almost exclusively in terms of special chemicals that must be purchased from a druggist or prescribed by a physician.

The brief discussion of some of the more common antidotes that follows will group these materials as to internal or external use and general mode of action.

Antidotes For External Use

Clean water dilutes and washes away poisons. Recommended for poison in eyes or on skin and other tissues. Always have readily available at least several gallons of clean water for emergency use when you are handling pesticides.

Soaps, detergents, or commercial cleansers and water dilute and washes away poisons. Recommended for poison on skin, hair, under fingernails and other external tissues not irritated by soap.

Always have soap and water readily available for emergency use when handling pesticides.

Decontamination of the immediate work area can be accomplished by placing three (3) cups Arm and Hammer soda wash (available from a chemical or hardware store) into a plastic bucket. Add one-half (1/2) cup Clorox to the bucket. Fill the rest of the bucket with water. Wear rubber gloves and a respirator and wash down contaminated area. Several washings may be required. Check with the State Health Department if in doubt about decontamination.

NOTE: In an emergency use any source of reasonably clean water such as irrigation canals, lakes, ponds, water troughs, etc. Don't let victim die while you worry about how dirty the water is.

Antidotes For Internal Use

CHECK FIRST AID INSTRUCTIONS BEFORE GIVING ANYTHING TO A PERSON BY MOUTH!

--Clean water dilutes poison.
--Milk dilutes poison and helps counteract acid or alkali poisons.
--Syrup of ipecac is used to promote vomiting. Use only as directed by
  a physician or poison control center. Poisonous if improperly used.

MEDICAL ANTIDOTES FOR PESTICIDE POISONING

Antidotes such as those described below should be prescribed or given only by a qualified physician. They can be very dangerous if misused.

Medical Doctors Should be Warned Ahead of Time

Most medical doctors are not well informed as to the symptoms and treatment of pesticide poisoning. This is due to the few cases which they treat.

Pesticide poisoning symptoms are similar to those of other illnesses and poisonings. You, the pesticide applicator, should tell your doctor which chemicals you use. Then he will know the symptoms and treatment and have the antidotes on hand.

WARNING: Neither atropine nor 2-PAM should be used to prevent poisoning. Workers should not carry either antidote for first-aid purposes. They should be given only under a doctor's directions.

Group I: ORGANOPHOSPHATES

For Poisons Such as: monocrotophos(Azodrin®), dicrotophos (Bidrin®), (Bomyl®), carbophenothion(Trithion®), coumaphos(Co-Ral®), fensulfothion (Dasanit®), DDVP (Vapona®), demeton(Systox®), diazinon (Spectracide®), dimethoate (Cygon®), dioxathion(Delnav®), disulfoton(Di-Syston®), chlorpyrifos (Dursban®), fonofos (Dyfonate®), ENP, ethion, famphur (Warbex®), fenthion (Baytex®), azinphos-methyl (Guthion®), oxydemeton-methyl (Meta-Systox-R®), methyl parathion, parathion, phorate (Thimet®), mevinphos(Phosdrin®), phosphamidon(Dimecron®), schradan (OMPA®), methidathion(Supracide®) and TEPP(Vapotone®).

Antidotes:

1. Atropine Sulfate is used to counteract the effects of cholinesterase inhibitors. Injections should be repeated as symptoms recur.

2. Protopam Chloride (2-PAM) should also be injected to counteract organophosphate poisonings. It is given intravenously.

3. Do Not Use morphine, theophyllin, aminophyllin or barbiturates.

Group II: CARBAMATES

For Poisons Such as: formethanate hydrochloride (Carzol®), mexacarbate (Zectran®), aldicarb (Temik®), carbofuran (Furadan®), methomyl (Lannate®)and carbaryl (Sevin®).

Antidotes:

1. Atropine Sulfate is used to counteract the effects of cholinesterase inhibitors. Injections should be repeated as symptoms recur.

2. Do Not Use Protopam Chloride (2-PAM).

Group III: CHLORINATED HYDROCARBONS

For Poisons Such as: endrin, aldrin(Aldrite®), dieldrin (Dieldrite®), endo-sulfan (Thiodan®) and lindane (Lintox®).

Antidotes:

1. Barbiturates for convulsions or restlessness.

2. Calcium Gluconate give intravenously.

3. Do Not Use epinephrine (adrenalin).

Group IV: INORGANIC ARSENICALS

For Poisons Such as: sodium arsenite (Atlas®), paris green.

Antidotes:

1. BAL (dimercaprol) is specific for arsenic poison. Inject intra-muscularly.

Group V: CYANIDES

For Poisons Such as: hydrogen cyanide and calcium cyanide (Cyanogas®).

Antidotes:

1. Amyl Nitrite through inhalation.

2. Sodium Nitrite given intravenously.

3. Sodium Thiosulfate given intravenously.

Group VI: ANTICOAGULANTS

For Poisons Such as: warfarin(Prolin®), coumafuryl (Fumarin®), pindone (Pival®), PMP (Valone®), diphacinone (Diphacin®).

Antidotes:

1. Vitamin K by mouth, intramuscularly, or intravenously.

2. Vitamin C useful adjunct.

Group VII: FLUOROACETATE

For Poisons Such as: sodium fluoroacetate(Compound 1080®).

Antidotes:
1. Monacetin (glycerol monoacetate) intramuscularly.

Group VIII: DINITROPHENOLS

For Poisons Such as: DNOC (Elgetol 30®), dinoseb (Dow General®).

Antidotes:
1. Do Not Use atropine sulfate.
2. Maintain life supports.
3. Sodium Methyl Thiouracil may be used to reduce basal metabolic rate.

Group IX: BROMIDES AND CARBOXIDES

For Poisons Such as: methyl bromide (Dowfume®, Bromo Gas®), ethylene di-bromide(EDB) (Bromofume®) and carboxide.

Antidotes:

1. BAL(dimercaprol) may be given before symptoms appear.
2. Barbiturates for convulsions.

Group X: CHLOROPHENOXY HERBICIDES, UREAS, MISCELLANEOUS

For Poisons Such as: 2,4-D, 2,4,5-T, silvex (2,4,5-TP and Ded-Weed®), monuron (Telvar®), diuron (Karmex ®), diquat (Diquat®), paraquat (Gramoxone®),endo-thall(Endothal®), and bromacil (Hyvar®).

Antidotes:
1. None.
2. Maintain life supports.

Anticoagulants -- Vitamins have been used successfully for treatments.

1. Vitamin K administered by mouth, intravenously or intramuscularly.

2. Vitamin C is a useful adjunct.

ADDITIONAL INFORMATION -- POISON CONTROL CENTER

All states have some form of consultant service for diagnosis and treatment of human illness resulting from toxic substances. Some states have
regional offices or at least several locations that can be contacted for
information. Make sure you and your physician know the telephone number
of the Poison Control Center nearest you.

MEDICAL SUPERVISION FOR PESTICIDE APPLICATORS

Persons handling or applying cholinesterase inhibiting pesticides such as
organophosphates or carbamates throughout the application season may wish
to obtain regular cholinesterase blood tests under medical supervision.
It is valuable to have pre-season cholinesterase tests performed at a time
when contact with these insecticides has been minimal. This way a "base
line" value may be obtained for each person at a time when he should be
relatively free from evidence of exposure. Cholinesterase tests should
be obtained at weekly intervals during spray season to see if the cholinesterase levels are normal. If cholinesterase values for any person drop
below 50% of the base line value that individual should be removed from
contact with organophosphate and carbamate insecticides until his work
habits can be reviewed carefully and his cholinesterase levels have returned to normal. Many commercial companies follow the above precautions
with their chemists, biologists, and other employees that are constantly
exposed to pesticides.

AERIAL APPLICATION SAFETY

Aerial application of pesticides has developed into an extensive skilled profession. Various hazards are associated with this type of work, but these can be minimized by observing safety precautions. Although the previous sections provide general pesticide safety information common to all types of pesticide use, the trend to more pesticides being aerially applied each year creates a need for additional safety information for pesticide applicators engaged in agricultural aviation.

Equipment

The aircraft is the most important item to consider. The type of aircraft selected should be dependent upon the work it will be used for. Large areas with long runs are generally more suited to fixed wing aircraft application. Helicopters or rotary aircraft are generally safer and more maneuverable for treating small areas, particularly those which have obstacles.

Aircraft should be designed and built specifically for aerial application

to protect the pilot and increase operating efficiency. Aerial application aircraft should be equipped with numerous safety features not generally found in conventional aircraft. These features should include: a) ease of control during slow flight and partial aileron control during stalls; b) special ventilation that reduces pilot exposure to pesticides; c) provisions for pilot comfort to keep fatigue at a minimum; d) good visibility in all directions.

Fixed wing aircraft that have boom and nozzle equipment should not have booms that are longer than 3/4 of the wing span. Longer booms which extend near the wing tips result in spray being picked up by the wing tip vortices causing uneven spray patterns and in some cases increasing drift potential.

Aircraft should be equipped with a positive cut-off valve to prevent leaking when the spray is shut off at the end of swath runs. Some valves are equipped with a vacuum feature reducing the risk of dribbling nozzles.

Ground Crew

Members of ground crews including mixers, loaders, and flag men should have access to protective clothing and equipment. In addition, they should have immediate access to an adequate supply of soap and water.

Ground crew personnel should bathe and change clothes each day or immediately if a concentrate or toxic pesticide is accidentally spilled on their clothing or skin.

It is important that ground crews be able to recognize symptoms of pesticide poisoning. Many times, poison symptoms are more obvious to others than the victim himself. Any abnormal reaction may indicate the onset of pesticide poisoning.

When water is drawn from streams or ponds for pesticide application purposes, ground crews should be sure that no pesticide flows back into the water source.

Accidentally spilled pesticides should be immediately cleaned up by ground crews to protect persons and animals from coming into contact with them. See section on PESTICIDE TRANSPORTATION, STORAGE, DECONTAMINATION, AND DISPOSAL.)

Pilot Safety

The conduct of a safe aerial application program is primarily the responsibility of the pilot. The pilot must be aware of any particular condition that may affect him personally as well as drift and other hazards that could be detrimental to humans, livestock, crops, wildlife, and any other environmental situation.

With proper precautions, the Ag. pilot should experience less acutal exposure to pesticides than the ground crew. The following points should be considered for pilot safety:

1) The pilot should not assist the ground crew when toxic pesticides are being mixed or loaded. He should remain on the windward side of the loading operation to avoid inhalation or exposure to toxic materials. He can remain in the aircraft if closed system loading is being used.

2) The pilot should anticipate greater than normal hazards when applying a very toxic or highly concentrated pesticide.

3) The pilot should anticipate increased potential for hazards when fatigue results from flying for long periods.

4) The pilot should never turn or fly through the path of the previous swath.

5) The pilot should be constantly alert to detect pesticide leaks or accidental spillage inside of the aircraft. The airplane cockpit should be tight to prevent pesticide spray or dust particles entering during application. Cockpits should be checked and cleaned frequently.

6) A safety helmet and shoulder harness should always be worn during flight.

7) When applying toxic pesticides, the proper type of protective equipment should be used.

8) It is of utmost importance that the pilot be able to recognize symptoms of pesticide poisoning such as dizziness, blurred vision, watering of the eyes, and nausea. Flying should cease as soon as possible if any of these symptoms are evident.

9) Clothing worn during application should be laundered each day.

10) Two-way radio equipment to enable the pilot to keep in touch with the ground crew is a valuable asset.

Drift Control

Complete elimination of drift during aerial application is nearly impossible, but there are several factors that help to keep it at a minimum.

- Wind velocity is of utmost importance. When there are excessive winds pesticides cannot be applied without drift. When wind velocity creates a drift hazard the application operation should cease immediately.

- When possible, use sprays rather than dust applications if there is a choice. Granular materials are even less likely to drift than sprays or dusts.

- Fine spray droplets will drift easier and farther than coarse spray droplets. Pesticides applied in oil will tend to drift farther than those applied in water.

- The extent of drift can be reduced by keeping the altitude of flight as low as possible. Precautions should be observed to maintain effective pesticide dispersal and to consider the safety of the operation.

- Dispersal equipment should be calibrated precisely to produce the desired rate of application. Considerations should be given to height of flight and swath coverage. Sometimes, under some conditions, temperature variations between air and ground will cause inversions which will prevent the pesticide application from settling to the ground. If this occurs, pesticide applications should be postponed until conditions are more favorable.

- When aerial pesticide applications are made over rough terrain, the downhill movement of surface air can carry a spray a considerable distance outside of the target area. The pilot should make allowances for this air movement. Aerial applications should not be attempted on very small fields where hazards of drift to adjoining fields is unavoidable.

- Treatment of fields near canals, streams, lakes or ponds should be made with extreme caution. The swaths should be parallel to the water so that turns will not be made across it.

- Pilots should not prime the application equipment or test the flow rate while ferrying between the air strip and the area to be treated.

• It is the pilot's responsibility to see that the aircraft and its equipment is cleaned daily or after each use to prevent pesticide spray from building up.

Safety Management and Supervision

The overall responsibility of a safe pesticide application operation belongs to the managers or owners of the business. They should have all safety devices needed in the aircraft to protect the pilot. Aircraft dispersal equipment should be in good working order, delivering the desired rate and swath pattern. Loading equipment should be up-to-date to provide efficiency and safety for the ground crew operation. Pilots and workers exposed to organophosphate pesticides throughout the spray season should be placed under a medical program and required to take blood cholinesterase tests frequently.

Immediate supervisors should be sure all workers understand the nature of the pesticide being used and what to do if poisoning occurs. The names of physicians and the hospital to contact should be immediately obtainable if an emergency arises. All workers, including the pilot, should be required to wear protective equipment when needed. At least two men should work together when toxic pesticides are being mixed and loaded. It is the supervisor's responsibility to clear the area being treated, and if the material is toxic to bees, to notify bee-keepers.

CHECKLIST FOR THE SAFE USE OF PESTICIDES

There are three periods during a chemical application season which present the greatest likelihood for over-exposure among chemical handlers and users:

1) Early spring when new and inexperienced crews begin handling chemicals for the first time. Overexposures are more likely to occur among new crew members until they gain some training and experience.

2) The first prolonged hot spell of the summer. Many men may become tired of using uncomfortable protective clothing and respirators during hot weather, and thus may run greater risk of overexposure. In addition, workers are more likely to become dehydrated during hot weather, and then may be more susceptible to the effects of some chemicals.

3) Late in the season, just before or during harvest time, when the work load has been prolonged and heavy for both men and equipment. At this time, individuals who have experienced repeated small exposures to pesticides sufficient to cause cholinesterase depression, but not severe enough to cause symptoms, may acquire small additional exposures whose effects are accumulative with the previous ones and result in overt symptoms. Equipment and protective devices which were functioning well earlier in the season, may become faulty by the season's end, and further contribute to this danger period.

GREENHOUSE OPERATIONS

Applying pesticides in greenhouses presents special problems. In normal greenhouse operation, employees must work inside. Space is generally limited and personal contact with plants and other treated surfaces is almost a certainty. In addition, unauthorized persons may attempt to enter the premises. Ventilation in greenhouses is frequently kept to a minimum to maintain desired temperatures and as a result fumes, vapors, mists, and dusts may remain in the air for considerable periods of time, creating hazards.

Post warning signs in conspicuous places when using pesticides inside greenhouses or other enclosed areas.

Pesticides applied to plants and other surfaces in greenhouses do not generally break down as rapidly as they do outside. This is due to a reduced amount of moving air and lack of rain to wash the chemical off, dilute them, or combine with them chemically to break them down further. In addition, the glass in the greenhouse filters out the ultraviolet light which would normally contribute to the degradation of certain chemicals.

Checklist for Greenhouse Applications

- Select chemicals that are most effective for pest control and present the least hazard to humans and animals.

- When using toxic pesticides, especially fumigants, applicators should use gas masks and waterproof protective clothing.

- Post warning signs on the outside of all entrances to the house when fumigants or other highly toxic pesticides are being applied. Follow label instructions.

- Do not enter the building without a gas mask or permit others to do so until it has been aired for the length of time recommended on the pesticide label.

- All possible skin contact with treated plants and other surfaces should be avoided by workers and others to minimize skin irritation, sensitization, and absorption of chemicals through the skin. Where this is impossible workers should wear clean dry protective clothing and wash frequently.

- It is suggested that handles and special wrenches for steam lines be kept in custody of authorized persons so that unauthorized persons will not inadvertently use the wrong line. These lines may be labeled by name with water insoluble ink or color coded blue for water, green for fertilizer, red for pesticide and black for steam.

TOXICITY OF PESTICIDES

One must keep in mind, when considering the toxicity of pesticides, that any chemical substance is toxic or poisonous if absorbed in excessive amounts; therefore, the poisonous effect of a chemical is dependent on the amount consumed or absorbed. If enough common salt is consumed at one time it is quite poisonous.

The assumption can be made, therefore, that all pesticides in sufficient amounts can be toxic or poisonous. They are designed to be poisons to kill the pest for which they are used against. There are, however, great ranges in the level of toxicity among different pesticides. It is important that persons working with pesticides have a broad general knowledge of the relative toxicity of at least the most common pesticides used.

Pesticide users should be concerned with the hazard associated with exposure as well as the toxicity of the chemical itself. Toxicity is the inherent capability of the substance to produce injury or death. Hazard is a function of toxicity and exposure. The hazard can be expressed as the probability that injury will result from the use of the pesticide in a given formulation, quanitity, or manner. The hazard might be specific in nature, as posing a toxicity problem to humans, or to animals, or in another instance cause injury to some plants.

A pesticide can be extremely toxic as a concentrate and possibly pose very little hazard to the user if: a) used in a very dilute formulation; b) used in formulation (granule) that is not readily absorbed through the skin or inhaled; c)used only occasionally under conditions of no human exposure; or d) used by experienced applicators that are properly equipped to handle the pesticide safely.

In another example, however, a pesticide may have a relatively low mammalian toxicity, but present a hazard because it is used in the concentrated form, which may be readily absorbed or inhaled. It could be hazardous to the non-professionals who are not aware of the possible hazards to which they are being exposed.

The toxicity value of a pesticide is at best just a relative measure to estimate its toxic effect on humans or other animals. Human toxicity ratings would be the best guide to a pesticide toxicity, however, no actual scientific tests can be conducted in which humans are subjected to lethal doses of pesticides. There are some fragmentary data in regard to human toxicity ratings based on accidental exposures and suicides, caused by some pesticides, but essentially all pesticide toxicity ratings are based on animal tests.

Toxicities of pesticides are generally expressed as LD_{50} or LC_{50} values, which means lethal dose or lethal concentration to 50% of a test population. To determine LD_{50} or LC_{50} values, the dosage of a particular pesticide necessary to kill 50% of a large population of test animals under certain conditions is computed. An example might be a pesticide that has a LD_{50} of 10 which would indicate that 10 milligrams of this pesticide given to animals that weigh 1 kilogram each, would kill 50% of the population.

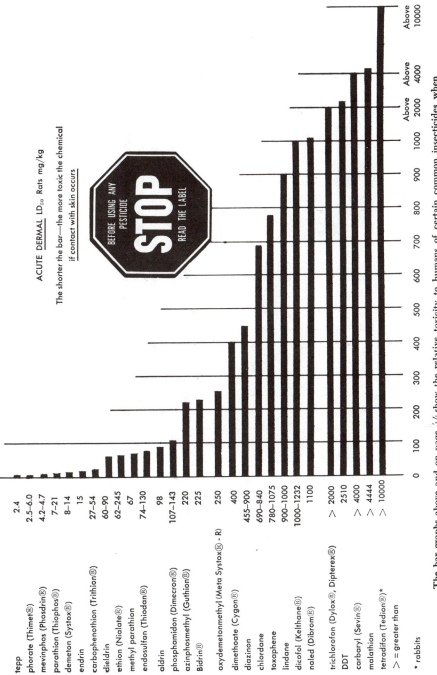

ACUTE DERMAL LD$_{50}$ Rats mg/kg

The shorter the bar—the more toxic the chemical
if contact with skin occurs

tepp	2.4
phorate (Thimet®)	2.5–6.0
mevinphos (Phosdrin®)	4.2–4.7
parathion (Thiophos®)	7–21
demeton (Systox®)	8–14
endrin	15
carbophenothion (Trithion®)	27–54
dieldrin	60–90
ethion (Nialate®)	62–245
methyl parathion	67
endosulfan (Thiodan®)	74–130
aldrin	98
phosphamidon (Dimecron®)	107–143
azinphosmethyl (Guthion®)	220
Bidrin®	225
oxydemetonmethyl (Meta Systox® - R)	250
dimethoate (Cygon®)	400
diazinon	455–900
chlordane	690–840
toxaphene	780–1075
lindane	900–1000
dicofol (Kelthane®)	1000–1232
naled (Dibrom®)	1100
trichlorofon (Dylox®, Dipterex®)	> 2000
DDT	2510
carbaryl (Sevin®)	> 4000
malathion	> 4444
tetradifon (Tedion®)*	> 10000
> = greater than	
* rabbits	

The bar graphs above and on page 44 show the relative toxicity to humans of certain common insecticides when
exposure is by the skin, or when swallowed.

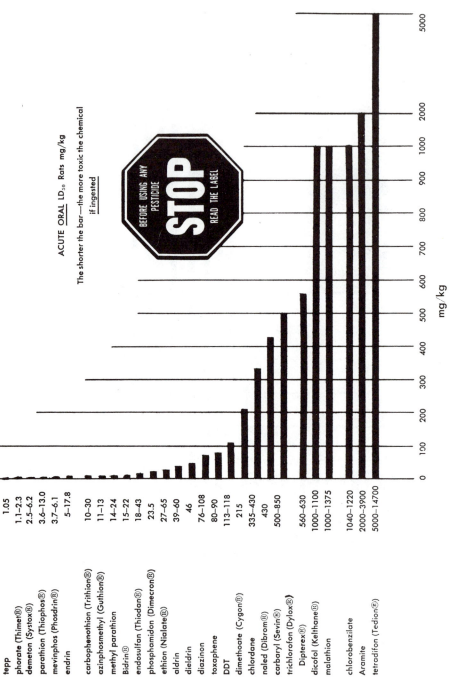

ACUTE ORAL LD₅₀ Rats mg/kg

The shorter the bar—the more toxic the chemical if ingested

Toxicity data based on LD$_{50}$ values with animals involve several problems and should not be interpreted as exact values for human toxicity, but these values can be used as guides if one uses caution in considering the following points.

1) The accidental hazard represented by any pesticide depends more upon how it is used, than how toxic it is.

2) Pesticide toxicity levels may vary according to the species of test animal, test method, sex of the species, whether the animals have been fasted or not, state of animal health, purity of the chemical tested, carrier in which the chemical is administered, route of administration and length of time and frequency of exposure.

3) LD$_{50}$ values are only a statistic. They provide no information on the dosage that will be fatal to a specific individual in a large population of animals. Statistically however, the LD$_{50}$ value is the most accurate means available to indicate the level of pesticide toxicity.

4) It must be remembered that LD$_{50}$ values are usually expressed in terms of single dosages or exposures only such as:

a) Acute Oral - referring to a single dose taken or ingested by mouth.

b) Acute Dermal - which refers to a single dose applied directly to and absorbed by the skin.

c) Inhalation - referring to exposure through breathing or inhaling.

Thus LD$_{50}$ values give little or no information about the possible cumulative effects of the pesticide.

TOXICITY TABLES

As an aid to those unfamiliar with toxicity values, the following chart groups and translates LD$_{50}$ values into practical terms.

ORAL, DERMAL AND INHALATION RATINGS OF PESTICIDE GROUPS

EPA Category-Meaning	Pesticide Label Signal Word*	Oral LD$_{50}$ (mg/kg)	Dermal LD$_{50}$ (mg/kg)	Inhalation LC$_{50}$** (μg/1)	Lethal Oral Dose For A 150 lb Man
I. Highly Hazardous	Danger Poison	0-50	0-200	0-2,000	few drops to teaspoonful
II. Moderately Hazardous	Warning	50-500	200-2,000	2,000-20,000	teaspoonful to one ounce
III. Slightly Hazardous	Caution	500-5,000	2,000-20,000	over 20,000	one ounce to one pint or pound
IV. Relatively Non-Hazardous	Caution	over 5,000	over 20,000	———	over one pint or pound

* All pesticide labels carry a "Keep Out of Reach of Children" warning.
** Vapor or gas inhalation value may also be expressed in parts per million (ppm).

Toxicity data for insecticides, herbicides, fungicides, rodenticides, nematicides, and fumigants is presented in the appendix pages, giving both the common and trade name as well as the producer of the particular chemical. Acute oral LD$_{50}$'s are shown along with a column for skin irritation indications. Dermal toxicity data for chemicals is not always available and therefore, some pesticides may not show a skin irritation rating.

9
PESTICIDES
AND
ENVIRONMENTAL
CONSIDERATIONS

Pesticides & Environmental Considerations

The quality of the environment has become a major issue, and with it an urgency for protecting the nation's air, water, land, and wildlife. Many who use pesticides regard them as a means of preserving the environment, but those who are most often quoted by the popular press place such materials near the top of the list of pollutants. The point at which a pesticide is considered a beneficial tool or a pollutant is at times hard to distinguish. Perhaps, as a weed is a "plant out of place", a pesticide pollutant is simply a "resource out of place".

Not only will problems of pollution grow with our expanding population, but sensitivity to and awareness of these problems will also increase. It appears inevitable that as our agriculture becomes more intensive, pollution will become a greater hazard to water and air and thus affect the sensitivity of more people. It is equally apparent that agriculture will be bound by rules and regulations to safeguard the environment -- regulations based not so much on the need for a pesticide as on that for preventing pollution.

PROTECT THE ENVIRONMENT!

Pesticides become problems when they are dispersed beyond the target area. The characteristics of nonbiodegradability and accumulation of some pesticides have emphasized the problem. Conditions under which pesticides can become pollutants are: leaching and contamination of ground water from treated soil; when they are carried in surface runoff from agricultural land and urban areas; misuse or accident; pesticide drift; and careless disposal of surplus materials and empty containers.

Pesticides are a help to the environment when they are used carefully and wisely. For years they have been used to control pests which are harmful to man. Rats carrying plague or mosquitos carrying malaria are good examples. These control programs are still necessary today, especially in crowded cities and countries with large numbers of people. With the help of pesticides, more food per acre can be produced. Diseases, insects, and other plant pests can be greatly reduced. There can be higher yields and better crop quality. Less land can be tilled to feed people, thus allowing more land for wildlife and recreational purposes.

Pesticides can be a big help in our environment when used to protect outdoor activities in our parks and camping areas. Fly and mosquito control programs give relief from these annoying insects. Pesticides also protect livestock from harmful and annoying pests. The quantity and quality of livestock products, milk, meat, etc. is improved when the pests are controlled. Pesticides also aid in controlling insects or diseases that get into an area for the first time. Often these pests have few natural enemies in a new environment. Without a good pesticide program, they can rapidly overrun an area. Gypsy moth and Japanese beetles are good examples of this type of pest.

There are six major areas in our environment, besides man, that require protection. These areas are water, soil, air, beneficial insects, plants, and wildlife. Each of these areas will be explored in more detail in the following paragraphs.

Soil -- Soil has become more important as the need for food increases. In order to feed the numbers of people, fertile and healthy soil is an absolute requirement. Poor soil practices and misuse can cause poor yield and second class crops, especially if root vegetables or forage crops are being planted.

Pesticides often become attached to fine soil particles and when the surface soil erodes, the chemical is carried along. With some pesticides this attachment is so strong that the pesticide does not move down through the soil layers for any important distance. Overdoses of pesticides which remain for long periods in the soil may limit planting to only a few crops which will not be harmed by the chemicals. Most pesticides that become attached to soil probably stay near the point of application of the material. Here bacteria and other microscopic organisms in the soil are important in the chemical process of breaking down the pesticide into other chemicals and eventually into the simplest chemical building blocks.

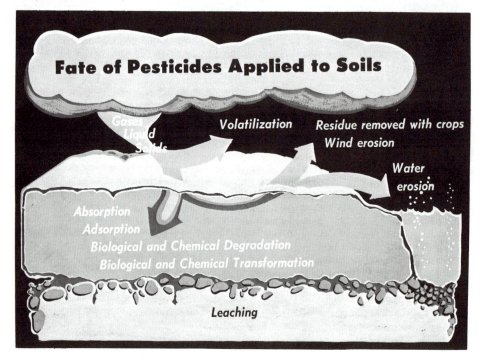

Water -- Water is one of our greatest resources. Its unusual properties and abundance make it necessary for all life. Polluted water can fill many of our needs but not the most basic ones. Man and wildlife need clean water for drinking and bathing. Most fish and other marine life can survive only slight changes in their water environment. Farmers must use uncontaminated water for their livestock and irrigation practices to prevent plant or animal poisoning or illegal residues. Clean water is essential.

Pesticides get into water in several ways. They may be applied directly for control of aquatic animals or plants; they may reach the water by accident when the nearby land is being sprayed; or they may enter as spray drift from nearby applications. Pesticides attached to soil may wash into streams and a certain amount of some chemicals may enter the water by being washed out of the air by rain. Pesticides that are dissolved in the water can be picked up directly by fish and other aquatic animals as the water passes over their gills. Therefore, it is essential that we be very careful in the application of pesticides where they may eventually find their way into our water sources or systems.

Air -- Air must be available for any plant or animal to live. It is the source of oxygen for breathing and receives the carbon dioxide waste. Air has the ability to move particles a long way before letting them go. Most of the time this ability aids the farmer. Unfortunately for the pesticide applicator, however, this same ability is the cause of drift. Pesticides carried by the air may be harmful to man and wildlife's health and safety. Pesticides in the air are not controllable and may settle into waterways, wooded areas, or heavily populated areas. Pesticide drift cannot be tolerated and must be avoided.

Plants -- Pesticides are often used on plants to protect them from pests. Insecticides and fungicides are often used on ornamental and forest trees to control serious insect pests and diseases. These chemicals aid in keeping forests, parks and lawns green and enjoyable. However, pesticides can injure plants. Injury from volatility and drift of herbicides to sensitive crops has led to specific legislation regulating the use of those materials in some states. The injury can range from slight burning or browning of leaves to death of the whole plant. This injury is called phytotoxicity. Phytotoxicity accidents can result from carelessness or from the use of a pesticide which is highly hazardous to plants and trees.

All kinds of pesticides (insecticides, fungicides, and herbicides) may injure or kill plants. Herbicides are especially hazardous because they are designed to kill or control plants. Non-target plants can be severely damaged from drift or misuse of herbicides. Always be very careful when herbicides are applied near desirable plants. In addition, some pesticides and some formulations tend to move off target readily. These chemicals can be a great threat to desirable plants and trees. Some are carried off target by rain and runoff water and may injure plants in the water's path. Other pesticides may move through the soil to surrounding areas and cause phytoxicity there. If plant injury could be a possible problem in your spray operation, try to use a pesticide and formulation which tends to remain on the target area. Be especially careful not to overdose when plant injury could be a problem. Remember that persistent pesticides can be very useful for long-term insect, disease, or weed control programs. But follow label directions carefully if future crops or other plants and trees will be planted in the area.

Beneficial Insects -- Honey bees and other pollinators are necessary for good farming and food production. In many cases when there is no pollination there is no crop. Honey bees produce honey and beeswax valued between 130 and 140 million dollars annually. Much of this honey comes from cultivated crops. The annual value of crops benefited by insect pollination, most of which is performed by honey bees, exceeds 10 billion dollars. The farmer and the beekeeper are, therefore, dependent upon each other.

Honey bees may be killed when crops are treated with pesticides. When this occurs, both the farmer and the beekeeper suffer a loss. For this reason, they need to cooperate fully in protecting the bees from pesticide damage.

Research to resolve the problem of bee losses due to pesticides has been under way since 1881 when damage to bees by lead arsenate was first

reported. A century later, there still is no solution to this problem, although intensive study is continuing. Modern agriculture is dependent upon bees for crop pollination and chemicals for pest control. Unfortunately, in the United States, pesticides annually destroy or damage about one million bee colonies or 20-25% of our honey bee colonies each year. This destruction has a significant economic impact on beekeepers and farmers alike. Yield from crops that require bees for pollination are frequently reduced due to the loss of bees. The absence of bees or inadequate use of pesticides reduces agricultural productivity which openly results in higher food costs for consumers.

Whenever it is determined that a pesticide application is necessary the following precautions should be taken:

o Use the proper dosage of the safest material (on bees) that will give good pest control.

o Tell the beekeeper what will be used and when it will be applied.

o Read the label and follow approved local, state, and federal recommendations.

o Remember that the time the pesticide is applied, depending on the blooming period and attractiveness of the crop, makes a big difference in the damage to the bees; so, treat the fields when the plants are least attractive to bees.

o Do not spray or dust chemicals over colonies, especially in hot weather when the bees cluster outside the hive.

o Apply chemicals at night or during early morning hours before bees forage.

o Do not spray or dust bee-visited plants in bloom and do not let insecticides drift to plants in bloom.

o Remember that treating a non-blooming crop, when weeds and wild-flowers are in bloom in the field or close by, can cause bee losses.

o Make as few treatments as possible because repeated applications greatly increase the damage to colonies.

o Do not treat an entire field or area if local spot treatments will control harmful pests.

o Sprays do not drift as far as dust and, consequently, are less likely to harm bees.

o Granules are usually the safest and least likely to harm bees.

We must strive for a balance between the use of pesticides and the preservation of bees. Bee losses are frequently due to the inappropriate or careless application of pesticides, improper timing of application, and dumping of unused materials. Such losses can be minimized by keeping pesticide applicators and crop producers informed of the need to protect bees and the methods of doing so. While it is unlikely that bee losses can be totally avoided when pesticides are used (many of the most effective insecticides are also very toxic to bees), they can be reduced.

In general, dusts are more toxic to bees than liquid sprays and wettable powders are more toxic than emulsifiable concentrate sprays.

Microencapsulated pesticides pose the greatest hazard to bees. The tiny capsules, adhering to foraging bees, are packed along with pollen into the pollen baskets and carried back to the hive where they are eaten by bees rearing the brood.

Granular formulations and soil applications of pesticides are usually not harmful to bees. The hazard of some systemic pesticides is reduced because they are rapidly absorbed by the plant. At least one formulation, systox, has tentatively been reported to be a bee repellant.

Pesticides Grouped According to their Relative Hazards to Honey Bees

Group 1 - Highly Hazardous:

Severe losses may be expected if these pesticides are used when bees are present at treatment time or within a day thereafter.

Group 1 - <u>Highly Hazardous (cont.)</u>:

Acephate (Orthene®)
Aldicarb (Temik®)
Aminocarb (Matacil®)
Arsenicals
Azinphosethyl (Ethyl Guthion®)
Azinphosmethyl (Guthion®)
Benzene hexachloride (BHC)
Bufencarb (Bux®)
Carbaryl (Sevin®)
Carbofuran (Furadan®)
Chlorpyriphos (Lorsban®, Dursban®)
Crotonamide (Azodrin®)
Diazinon (Spectracide®)
Dichlorvos (DDVP)
Dicrotophos (Bidrin®)
Dimethoate (Cygon®, DE-FEND®)
EPN
Famphur (Famaphos®)
Fenitrothion (Sumithion®)
Fensulfothion (Dasanit®)

Fenthion (Baytex®)
Heptachlor
Lindane
Malathion (Cythion®)
Malathion ULV
Methyl parathion
Methamidophos (Monitor®, Tamaron®)
Methidathion (Supracide®)
Methiocarb (Mesurol®)
Methomyl (Lannate®, Nudrin®)
Mevinphos (Phosdrin®)
Monocrotophos (Azodrin®)
Naled (Dibrom®)
Parathion
Phosmet (Imidan®)
Phosphamidon (Dimecron®)
Propoxur (Baygon®)
Resmethrin (Pyrethroid)
Tepp
Tetrachlorvinphos (Gardona®)

Group 2 - <u>Moderately Hazardous</u>:

These can be used around bees if dosage, timing, and method of application are correct, but should not be applied directly on bees in the field or at the colonies.

Carbophenothion (Trithion®)
Carbanolate (Banol®)
Chlordane
Coumaphos (Co-Ral®)
Counter
Demeton (Systox®)
Disulfoton (Di-Syston®)
Endosulfan (Thiodan®)
Endrin
Ethoprop (Mocap®)
Formetanate (Carzol®)
Hexaflurate

Leptophos (Phosvel®)
MAA
Oxamyl (Vydate®)
Oxydemeton Methyl (Metasystox-R®)
Perthane®
Phorate (Thimet®)
Phosalone (Zolone®)
Pyrazophos (Afugan®)
Ronnel
Temephos (Abate®, Biothion®)
Terbufos (Counter®)
Trichloronate (Agritox®)

Group 3 - <u>Relatively Nonhazardous</u>:

These can be used around bees with a minimum of injury.

-- Insecticides and Acaracides --

Allethrin
Bacillus thuringiensis
Binapacryl (Morocide®)
Chlofenvinphos (Birlane®)
Chlorbenside (Chlorparacide®)
Chlordecone (Kepone®)

Chlordimeform (Fundal®, Galecron®)
Chlorobenzilate (Acaraben®)
Cryolite
Dibromochloropropane (Nemagon®)
Dichlone (Phygon®)
Dicofol (Kelthane®)

Group 3 - <u>Relatively Nonhazardous</u> - Insecticides and Acaracides (cont.):

Dimilin®
Dinobuton (Dessin®)
Dinocap (Karathane®)
Dioxathion (Delnav®)
Ethion
Heliothis polyhedrosis virus
Menazon (Sayfos®)
Methoxychlor
Oxythioquinox (Morestan®)
Pentac®
Plictran®

Propargite (Omite®, Comite®)
Pyrethrins (natural)
Rotenone
Ryania
Sabadilla
Tetradifon (Tedion®)
Tetram®
Thioquinox (Eradex®)
Toxaphene
Tichlorfon (Dylox®, Dipterex®)

-- Herbicides, Defoliants and Desiccants --

Alachlor (Lasso®)
Amitrole
AMS (Ammate®)
Atrazine (AAtrex®)
Bifenox (Modown®)
Bromacil (Hyvar®)
Cacodylic acid (Phytar 138®)
CDAA (Randox®)
CDEC (Vegadex®)
Chloramben (Amiben®)
Chlorbromuron (Maloran®)
Copper sulfate (monohydrated)
Cyanazine (Bladex®)
2,4-D
2,4,-DB
Dalapon
Dazomet (Mylone®)
DEF®
Dicamba (Banvel®)
Dichlobenil (Casoron®)
Dichlorprop (2,4-DP)
Diquat
Diuron (Karmex®)
DMTT (Mylone®)
DSMA (Methar®)
Endothall (Endothal®)
EPTC (Eptam®)
EXD (Herbisan®)

Fluometuron (Cotoran®)
Fluorodifen (Preforan®)
Folex®
Linuron (Lorox®)
MCPA (Weedar®)
Methazole (Probe®)
Metribuzin (Sencor®)
Monuron (Telvar®)
MSMA (Daconate®)
Naptalam (Alanap®)
Nitrofen (TOK®)
Paraquat
Phenmedipham (Betanal®)
Picloram (Tordon®)
Prometon (Pramitol®)
Prometryn (Caparol®)
Pronamide (Kerb®)
Propachlor (Ramrod®)
Propanil (Rogue®)
Propazine (Milogard®)
Propham (IPC®)
Silvex
Simazine (Princep®)
Terbacil (Sinbar®)
2,4,5-T
2,3,6-TBA (Trysben®)
Terbutryn (Igran®)

-- Fungicides --

Anilazine (Dyrene®)
Benomyl (Benlate®)
Bordeaux mixture
Captafol (Difolatan®)
Captan
Carboxin (Vitavax®)
Copper oxychloride sulfate

Dithianon (Delan®)
Dodine (Cyprex®)
Fenaminosulf (Dexon®)
Fentin hydroxide (Du-Ter®)
Ferbam
Folpet (Phaltan®)
Glyodin (Glyoxide®)

Group 3 - Relatively Nonhazardous - Fungicides (cont.):

Mancozeb (Dithane M-45®, Fore®)   Oxycarboxin (Plantvax®)
Maneb (Dithane® M-22)   Sulfur
Metiram (Polyram®)   Thiram (Arasan®)
Nabam (Dithane® D-14, Parzate®)   Zineb
Nabam (Dithane® A40)   Ziram

Symptoms of Bee Poisoning by Pesticides

The following are common symptoms of pesticide poisoning in bees. Usually not all are evident at any one time.

-- Common Insecticide Poisoning Symptoms --

o Excessive numbers of dead bees in front of hive.
o Dead bees on top bars in hive.
o Lack of house cleaning by hive bees.
o Dying larvae crawling out of cells.
o Dead and dying bees, including many young (fuzzy) bees, from feeding on contaminated pollen.
o Break in brood cycle.
o Honey and pollen stored in brood area.
o Queen supersedure.
o Bees dying after being installed in hives stored over winter. (Caused by pesticides, Sevin® and Penncap-M®, retained in stored pollen for several months.)
o Queenlessness withing 30 days.

-- Organophosphate Insecticide Poisoning --
(i.e. parathion, Cygon®, imidane)

o Bees regurgitate; dead bees stick together.
o Tongues of dead bees extended.
o Bee abdomen distended.
o Erratic attempts at self-grooming.
o Gentle bees become aggressive.
o Dying bees appear disoriented or paralyzed; bees are often spinning or crawling awkwardly.
o Wings held away from the body but remain hooked together.
o Dead bees are primarily field bees except when microencapsulated materials are responsible. Then the dead bees are primarily young bees and a break in the brood rearing cycle results.
o Loss of ability to maintain hive temperature.

-- Carbamate Insecticide Poisoning --
(i.e. Furadan®, Lannate®, Sevin®)

o Affected bees become aggressive, erratic and often are unable to fly.
o Bees often die slowly; they may live up to 3 days.
o Bees appear stupefied as if chilled, paralysis is usually evident.
o Dead brood in front of hive.

Remember -- Cooperation between pesticide applicators and beekeepers is essential. Cooperation will help to prevent many unnecessary bee losses as well as law suits and hard feelings.

The information presented in this section concerning honey bees has been adapted from USDA leaflet No. 563 "Pesticide and Honey Bees" and from University of Wisconsin leaflet No. A3086 "Protecting Honey Bees from Pesticides in Wisconsin."

Wildlife -- Fish, birds, and mammals are assets to man. Land which is used

only as farmland does not have to be a wildlife refuge. However, care should be taken to protect surrounding wooded areas and waterways when applying any pesticide. Fishing and hunting are very popular sports. If pesticides are carelessly used, these sports can disappear. Pesticide-kills of mammals, birds, and fish in large numbers, have been reported on occasion and have hampered fishing and hunting activities in some areas. Pesticides can be helpful to wildlife by controlling dangerous or annoying pests which could harm the animals. The toxicity of every chemical to every animal is not known. A pesticide that is only slightly toxic to one living thing may be very toxic to another. Check the label for specific instructions concerning toxicity of a pesticide to wildlife.

Birds and mammals can be killed outright by insecticide applications. Although they may absorb chemicals through their skin or inhale them in sufficient quantities to be affected, the usual means of poisoning is by eating contaminated food. Plants and seeds or areas treated with insecticides can be hazardous to wildlife through direct exposure to the chemicals. Animals that eat other animals for food may absorb pesticides from their prey and thus develop large loads of pesticides in their own systems. Wildlife may be killed or affected in other ways. Changes in behavior in birds and mammals, hatching-failure in birds, and reduced reproduction in mammals may occur. The growth and survival rates of the young produced may be reduced. For mammals, these changes in reproduction may affect both males and females. For birds in the field, reproductive effects usually occur in females.

Most animals store certain kinds of pesticides in their body fat. Some animals such as ducks and bats use their fat quickly when they go without food for any period of time. They may then be poisoned by pesticides that were stored in their fat systems several months earlier. Over a long period of time, the most affected bird populations have been the bird eaters, such as falcons and some hawks, and the fish eaters, such as eagles, ospreys, and pellicans. Local and regional populations of these birds have on occasion declined because of the effect of certain insecticides. Similar widespread and long-term effects are not know to have occurred in mammals, although they are often killed by heavy applications of the more toxic insecticides.

Because the importance of wildlife losses cannot be expressed in economic terms, personal values must be used in evaluating losses. The loss of even one animal may be considered great by some. No estimate has been made of the annual loss of wildlife due to pesticide use. According to Cornell University "Miscellaneous Bulletin 109" by James W. Caslick:

"Since no estimate has been made of the annual national loss of wildlife due to insecticides, a direct comparison with losses from other human-made causes cannot be made. However, biologists agree that the loss of habitat is the most serious threat to wildlife in the United States today. Changes in land use, such as for housing developments and highways, have greatly altered local wildlife populations. In the early 1970's, an estimated 57,000,000 birds were killed each year by motor vehicles on U.S. roads, and 3,500,000 birds died each year as a result of flying into picture windows. In the last 200 years, the activities of people have caused the extinction of about 60 species of birds, mammals, fishes, reptiles and amphibians in the United States.

Affects of such environmental pollutants as heavy metals, radiation, and air pollution are being discovered. PCBs are known to be a serious threat to wildlife. Contamination with oil killed about 1,500,000 birds per year in the United States in the early 1970's."

Accidental losses have occurred in the past, although probably fewer than 100 such cases have been reported in the United States. Of greater concern are the possible long-term effects to wildlife populations, since some chemicals are known to reduce reproduction or affect defensive behavior.

Individual citizens making everyday choices may affect wildlife in unintended ways. When deciding about pesticide use, we are also deciding about adding to the pesticide burden of the environment. Each person's decisions about pesticides are, therefore, important.

In areas where wildlife welfare is of primary concern, the following insecticides should not be used: aldrin, Bidrin®, diazinon (granules), dieldrin, endrin, fenthion (Baytex®), heptachlor, and sodium arsenite.

The following insecticides appear to be among the safer insecticides for wild birds and mammals: allethrin, Aspon®, carbaryl (Sevin®), diflubenzuron (Dimilin®), dormant oil, ethylene dichloride, Gardona®, malathion, trichlorfon (Dylox®, Dipterex®, Neguvon®), methoxychlor, Morestan®, naphthalene, Perthane®, pyrethrum, resmethrin, ronnel, and Ryania®.

Hazard is related to conditions existing at the time and place of application, and to the fate of the chemical in the environment. The degree of exposure of wild animals is usually difficult to control, yet may be as important a factor in wildlife safety as using one of the insecticides having a low toxicity rating. Some chemicals may move in the environment and be accumulated by animals in amounts sufficient to destroy their reproductive capacity, make them more vulnerable to predators, kill them, or kill their predators.

There are reasons for optimism about the future concerning wildlife and pesticides. Some chemicals harmful to wildlife are beginning to lessen in the environment. Some scientists believe that environmental contamination by pesticide users is decreasing. New pesticides are being tested more carefully for effects on other animals.

Where wildlife values take priority over other values, avoid using pesticides. Determine whether the treatment is really necessary. If you feel you must use a pesticide, choose it carefully to pose the minimum hazard under your conditions of use, and follow label directions carefully. Select a pesticide that will last only a short time in the environment. Avoid using insecticides during bird migration and nesting periods. Whenever possible, insecticide treatment should be avoided during the nesting season of pheasants, quail, prairie chickens, grouse and doves that nest on the ground at field edges or in fields (fallow, stubble, alfalfa, etc.). Always avoid treating over or near streams, lakes, and ponds; fish are often times more sensitive to pesticides than are warmblooded animals. Do not puddle sprays during application or when cleaning spray equipment because birds may drink or bathe there within minutes. Be very careful about safe storage and safe disposal of unused chemicals and their containers. Never throw "empty" containers into water or discard them where animals have access to them.

The encouragement and maintenance of wildlife on farms where various cropping practices are conducted is dependent upon the attitude of the pesticide applicator and his appreciation of various wildlife and his willingness to provide a habitat for their coexistence with his farming practice. It is possible to maintain a fairly large variety of wildlife within a farm setting providing the pesticide applicator is aware of various options that will help to protect wildlife as well as his willingness to time some operations so that they will be the least disruptive to certain species, especially pheasants.

Some pesticides are beneficial to wildlife in that they are used to eliminate undesirable species or to enhance a setting making it more condusive to the propagation and growth of a desirable species. A farmer who is conscientious about his cropping practices is usually always careful to ensure that his practices are complimentary to and not competitive with the wildlife species that coexist on his farm.

PESTICIDE PERSISTENCE

All pesticides degrade or break down chemically into other related chemicals and, eventually, into the simple building blocks of which the whole world is made. This process occurs at very different rates for different pesticides. In some, the changes occur rapidly (in hours or in a few days), and these materials are referred to as short lived. In others, the changes are slower and the pesticides may be present for relatively long periods of time, and these are known as persistent pesticides. The rates of breakdown or degradation of any chemical may change with differing conditions of temperature, sunlight, air and location.

The environmental difficulty with persistent pesticides is that once released into the ecosystem they remain in the original chemical form long enough so that if they have the other properties of moving readily and of being stored in animal tissues, they can spread to a distance and be concentrated at some other place than where they were applied. While it is true that the persistence of a pesticide may give it practical advantages for the control of the pest against which it is used, that same persistence means that once such a material is released, its environmental damage cannot be controlled.

PESTICIDE ACCUMULATION

Some pesticides can build up in the body of animals (including man). These are called accumulative pesticides. The chemicals can build up in an animal's body until they are harmful to it.

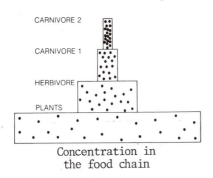

Concentration in the food chain

These pesticides also accumulate in the food chain. Meat-eaters feeding on other animals with built up pesticides may receive high doses of pesticides. If they feed on too many of these animals, the meat-eater can be poisoned without ever directly contacting the pesticide. The build up through the food chain can injure animals which aid man. In fact, man as one of the meat-eaters at the top of this food chain could get very high doses of pesticides in this way.

Some pesticides do not build up in the body of animals or in the food chain. These are called nonaccumulative pesticides. These chemicals usually break down rapidly into other, relatively harmless materials. Organophosphate pesticides, for example, have high toxicity at first and could be potentially dangerous to wildlife, but they do not accumulate so they are not as dangerous to the environment in general. Usually pesticides which break down quickly in the environment are the least harmful to it.

WHAT PESTICIDE APPLICATORS CAN DO TO PROTECT THE ENVIRONMENT

Commercial pesticide applicators are usually hired to apply a specific chemical to a particular area at a given rate within a given time period. Therefore, it may not be practical for you to assure yourself that there is a real need for pesticide use or to select a chemical that is the least dangerous to reduce hazards to a minimum. However, there are procedures that applicators can follow to help prevent damage to the total environment:

‡ Calibrate equipment carefully. A small increase in dosage rates may mean the difference between severe effects and no effects on fish and wildlife.

‡ Mix pesticides at the correct rate. Too much pesticide in the spray tank can be more harmful than a poorly calibrated spray machine.

‡ Be sure that you hit the designated target. Use care in developing ground application and flight patterns. Avoid any overlap in spray swaths, especially near water areas. Avoid applying spray materials directly to water areas if at all possible. When streams and other waterways are included in an area to be sprayed, hazards to aquatic organisms will be greatly reduced if applications are made at right angles or across these waterways rather than parallel to them.

‡ Spray after irrigating if possible. Spraying while irrigating or just before may result in the spray materials entering waters containing fish.

‡ Spray under favorable weather conditions to prevent drift into water courses or wildlife habitats.

‡ Learn what you can about the possible affects of the chemicals which you are using upon fish and wildlife.

‡ Do your part to aid the environment -- your surroundings are worth protecting.

10
PESTICIDE
FORMULATIONS
AND
ADJUVANTS

Pesticide Formulations & Adjuvants

Types of Formulations

A pesticide chemical can only rarely be used as originally manufactured. The pesticide must usually be diluted with water, oil, air, or chemically inactive solids so that it can be handled by application machinery and

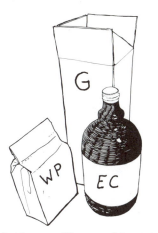

spread evenly over the area to be treated. Usually the basic chemical cannot be added directly to water or mixed in the field with solids so the manufacturer must modify his product by combining it with other materials such as solvents, wetting agents, stickers, powders, or granules. The final product is called a pesticide formulation and is ready for use, either as packaged or diluted with water or other carriers.

A single pesticide is often sold in several different formulations. For example, diazinon, an insecticide, can be purchased as 25% (2 lbs/gal) or 48% (4 lbs/gallon) emulsifiable concentrate; 25% or 50% wettable powder; a 4% dust; and as 5, 10 or 14% granules. Most pesticides however, are not available in such a wide range of form-

ulations. When applicators have the opportunity to select from several formulations, they should choose the formulation that will best meet the requirements for a particular job. Considerations in making a choice include effectiveness against the pest; habits of the pests; the plant, animal, or surface to be protected; application machinery; danger of drift or runoff; and possible injury to the protected surface.

Aerosols:

Aerosols or "bug bombs" are pressurized cans which contain a small amount of pesticide or a combination of pesticides that is driven through a fine opening by a chemically inactive gas under pressure when the nozzle is pressed. These containers are usually small and easy to handle.

Principal Uses: Pressurized cans are most often used in households, backyards and other small areas, and by licensed pest control operators. Aerosols may be used either as space sprays for flying insects but some are designed for plant diseases or weed killers. There are commercial models available for use in greenhouses, barns, and for pest control operations in warehouses and other large buildings. These are larger models holding five to ten pounds of materials, and they are usually refillable.

Advantages: Pressurized cans are very convenient in that they are always ready to use. They are also a convenient way to buy small quantities of a pesticide. They are easily stored and the pesticides do not lose their strength while in the can during their normal period of use.

Disadvantages: Pressurized aerosol cans are only practical for use in small areas. There is not much active ingredient in any one can. Because of this it is an expensive way to buy pesticides. They are also attractive play things for small children and are a hazard if left within reach. Pressurized aerosols can be dangerous if punctured or overheated and may explode and injure someone.

Dusts (D):

Dusts are finely ground dry mixtures combining a low concentration of a pesticide with an inert carrier such as talc, clay, or volcanic ash. There is a wide range in size of the dust particles in any one formulation.

Principal Uses: Because they drift badly, dusts are rarely used nowadays for large scale field use. They are now principally used for spot treatments and smaller areas such as home gardens. They work best when applied to damp surfaces such as dewy foliage in the early morning. Inside they are used in cracks and crevices for roaches and other domestic insects. Dusts can also be used to control lice, fleas, and other parasites on farm animals.

Advantages: Dusts are ready to be used as purchased so require no mixing. They can be applied with simple, light-weight equipment that is inexpensive and easy to use.

Disadvantages: Because dust particles are finely ground they may drift long distances from the treated area and may contaminate crops, pastures, and wild areas. While drifting they are highly visible and may cause public criticism. When used outside, they are easily dislodged from the treated surface by wind and rain and soon become inactive. Dusts should never be applied on windy days.

Emulsifiable Concentrates (EC):

These are liquid formulations in which the active ingredient is dissolved in one or more water insoluable solvents. An emulsifier is added to ensure mixing with oil or water. Emulsifiable concentrates are manufactured in two principle categories: low concentrate liquids containing from one to ten percent of inactive ingredient and high concentrate liquids containing from ten to eighty percent or more active ingredient.

Principal Uses for Low Concentrate Liquid: These may be used in the household for flying or crawling insects and for moth-proofing clothes. They are used in livestock areas for control of flies as a space spray in barns. In the field they are used as prepared sprays for mosquito control and shade tree insect control.

Advantages: Low concentrate liquids are designed to be sprayed as purchased. Because of this no mixing is necessary and lessens the chances for making mistakes. Household formulations have no unpleasant odors, and usually the liquid carrier evaporates quickly and does not stain fabrics or furniture.

Disadvantages: Low concentrate formulations are usually fairly expensive for the amount of actual pesticide in the container and the uses for such materials are specialized and few in number.

Principal Uses for High Concentrate Liquid: High concentrate liquids can be diluted and used in many ways. They can be used on fruit, vegetables, shade trees, for residual sprays, on farm animals, and for structural pests. They are adaptable to many types of application equipment including hydraulic sprayers, low volume ground sprayers, mist blowers, and aircraft sprayers.

Advantages: These formulations contain a high concentration of pesticides so the price per pound of active ingredient is comparatively low. Only moderate agitation is required in the tank so they are especially suitable for low pressure, low volume sprayers and mist blowers. They are not abrasive and do not settle-out when the sprayer is not running. There is little visible residue which generally allows use in populated areas. Because of the high pesticide content, the applicator is not required to transport and handle a large amount of chemical for a particular job.

Disadvantages: Because of the high concentration of pesticide it is easy to under-dose or over-dose if directions are not carefully followed. Mixtures of emulsifiable concentrates may be phytotoxic and they are also easily absorbed through the skin, thereby presenting a possible hazard to the applicator. Because of their solvents, most liquid concentrates cause rubber hoses, gaskets, and pump parts to deteriorate rather rapidly unless they are made of neoprene rubber. Some formulations can cause pitting and damage to paint on automobiles.

Flowables (F):

Some pesticides can be manufactured only as solid materials, not as liquids. Often these pesticides are formulated as flowables. They are made from finely ground wettable powder formulations which are sold as a thick suspension in a liquid to make its addition to water in the spray tank easier.

Principal Uses: Flowables are similar to high concentrate liquids and are used in the same way.

Advantages: Flowables do not usually clog nozzles and require only moderate agitation. As in the case of high concentrate liquids, the applicator is not required to transport and handle a large amount of chemical for a particular job.

Disadvantages: Flowables have the same disadvantages as high concentrate liquids in that they require very careful following of directions and careful mixing in order to get the correct amount of pesticide on a required area. Hazards and problems in handling are quite similar to those listed under high concentrate liquids.

Fumigants:

Fumigants are pesticides in the form of poisonous gases that kill when absorbed or inhaled.

Principal Uses: There are two important needs for fumigants in agriculture -- to rid stored grain of insect pests and for soil treatment to control nematodes and certain plant diseases. Fumigants are also used by pest control operators inside dwellings or other buildings to control pests that cannot easily be reached by other pesticide formulations. Soil is often fumigated in greenhouses to sterilize pests before planting.

Advantages: A single fumigant may be toxic to many different forms and types of pests; therefore a single treatment with one fumigant may kill insects, weed seeds, nematodes, and fungi all at the same time. Fumigants penetrate into cracks, crevices, burrows, partitions, soil, and other areas that are not gas tight and expose hidden pests to the killing action of the pesticide.

Disadvantages: The area to be fumigated almost always must be enclosed. Even in outdoor treatments the area must be covered with a tarpaulin or the fumigant incorporated into the soil so it doesn't escape. Frequently fumigants are highly toxic, so proper techniques and all recommended protective clothing and equipment must be used when applying them. Most fumigants can severely burn the skin.

Granules (G):

Like dusts, pesticide granules are dry, ready to use, low concentrate mixtures of pesticides and inert carriers. However, unlike dust almost all the particles in a granular formulation are about the same size and are larger than those making up a dust. Granular pesticides pour like ordinary table salt or the larger granules are similar to prilled nitrogen fertilizers.

Principal Uses: Granular pesticides are often used for soil treatments to control pests living at ground level or underground. They may be used as soil systemics, applied to soil where they are absorbed into the plant through the roots and carried throughout the plant. Granular herbicides or insecticides, or both, are frequently applied in combination with fertilizers on turf, thereby saving labor. Granular formulations are sometimes used for aerial applications when drift is a problem.

Advantages: Granules are ready to use as purchased, with no further mixing necessary. Because the particles are large, relatively heavy, and more or less the same size, granulars drift less than most other formulations. There is little toxic dust to drift up to the operator's face and present a hazard to him. They can be applied with simple, often multipurpose, equipment such as seeders or fertilizer spreaders. They will also work their way through dense foliage to a target underneath.

Disadvantages: With a few exceptions granulars are not suitable for treating foliage because they will not stick to it, therefore their use is generally confined to soil applications.

Poisonous Baits:

A poisonous bait is a food or other substance mixed with a pesticide that will attract pests and be eaten by them thereby causing their death.

Principal Uses: Baits are used inside buldings for pests such as ants, roaches, flies, rats and mice. They may be used outside in gardens for control of slugs, in dumps and similar areas for rat control, and in fields to control certain insects and pest birds.

Advantages: Baits are useful for controlling pests that range over a large area. Often the whole area need not be covered; just those spots where the pests gather. Baits may be carefully placed in kitchens, gardens, graneries, and other agricultural buildings so that they do not contaminate food or feed, and can be removed after use. Usually only small amounts of pesticides are used in comparison to the total area treated, so environmental pollution is minimized.

Disadvantages: Inside the home, baits are often attractive and dangerous to children or pets and must be used with care. Outside they may kill domestic animals and wildlife as well as the pests. Often the pests will prefer the protected crop or food rather than the baits, so the baits may be ineffective. When larger pests are killed by baits the bodies must be disposed of, or they may cause odor problems in houses. Other animals feeding on the poisoned animals may also be poisoned.

Wettable Powders (WP) and Soluable Powders (SP):

Wettable powders and soluable powders are dry preparations containing a relatively high amount of pesticide. Wettable powders are mixed with water to form suspensions. Soluable powders dissolve in water to form solutions. The amount of pesticide in these powders varies from 15% to 95%.

Principal Uses: Liquid concentrates and wettable powders are the formulations most widely used by commercial applicators. Like liquid concentrates, wettable powders can be used for most pest problems and in most spray machinery. Where toxicity to the plant or absorption through the skin of an animal is a problem, wettable and soluable powders may be used to avoid this problem.

Advantages: The pesticides in wettable and soluable powders are relatively low in cost and easy to store, transport, and handle. They are safer to use on tender foliage and usually do not absorb through the skin as rapidly as liquid concentrates. They are easily measured and mixed when preparing spray suspensions.

Disadvantages: Wettable and soluable powders may be hazardous to the applicator if he inhales their concentrated dust while mixing. They require a constant agitation in the sprayer tank and will settle quickly if the sprayer is turned off. They cause some pumps to wear out quickly. The residues are more subject to weathering than liquid concentrates, and being more visible, may require washing of automobiles, windows, and other finished surfaces where the pesticide is not intended.

Miscellaneous Formulations:

Some other formulations of pesticides which are available but less commonly used are water soluable concentrates, oil soluable concentrates, pastes, oils, and invert emulsions.

Water soluable concentrates are materials that form solutions, not suspensions, when water is added as a diluent. They are sometimes used in place of emulsifiable concentrates. Oil soluable concentrates are similar to emulsifiable concentrates except they do not mix with water but can be diluted with fuel oil, or kerosene.

Pastes are similar to wettable powders but are in a slurry form and difficult to handle. Oils have the advantage of being low in cost, spreading easily over a surface and being easy to mix and handle.

Invert emulsions are water in oil mixtures. Each spray droplet is surrounded by oil instead of water. This material is difficult to apply due to its high viscosity, but is less likely to drift and is being used in some areas for low volume and air application.

Encapsulated materials are fairly new and are still being experimented with. The active ingredient is encased in an inert material for a slow, sustained pesticidal release, resulting in decreased hazard.

Dissolvable bags have been on the market for several years. This technique involves the use of a dissolvable material in which the correct dose for a specified unit is encased. The applicator merely drops the bag into a specified amount of water and the bag dissolves releasing the material into the water. This technique is designed to protect the handler as much as possible!

AGRICULTURAL SPRAY ADJUVANTS*

The search for practical and economic means of improving our efficiency in the use of agricultural chemicals has been a continuing effort for the past several decades.

We knew, many years ago, that calcium caseinate added to the spray tank would improve deposits of lead and calcium arsenates, that nicotine applied with soap in the spray solution was more effective than without, and that blood albumin improved the performance of both oil and cryolite sprays.

Even with the advent of the miracle chemicals, such as DDT and 2,4-D, the utility of spray tank additives, generally small quantities of wetting or depositing agents, was not wholly discounted.

But in spite of a long history of limited use of tank additives there was no broad acceptance of the general concept.

The discovery, a few short years ago, that spectacular improvement in the performance of many foliage applied herbicides was possible when certain surfactants were included in the spray solution, firmly established at least one role of the agricultural spray adjuvant in improving the efficiency of pesticide chemicals.

Since then, we have been besieged by a whole gamut of surfactants and other additives, of varying effectiveness, from which the investigator, applicator, or grower must choose the proper product for his particular application.

It is from this mass of confusion over what surfactants are, what adjuvants are, and which one do you use when and where, that we must try to provide some order and understanding.

It is particularly timely now with avid public interest in, and Federal scrutiny of, chemical usage and its relationship to the environment, providing additional pressure to improve our efficiency in the use of agricultural chemicals.

The remainder of this discussion will be presented to include definitions, functions of spray adjuvants, and a product list of adjuvants suitable for providing these functions.

---

* This section is adapted from an article written in 1971 by Mr. S.M. Woogerd, Vice President for Research, Colloidal Products Corporation and reprinted from the Washington Pest Control Handbook, Washington State University, Pullman, Washington.

-DEFINITIONS-

Adjuvant: The dictionary defines "adjuvant" as a substance added to a prescription to aid the operation of the main ingredient. A spray adjuvant performs this function in the application of an agricultural chemical.
An effective spray adjuvant may be formulated to contain one or more
surfactants, solvents or co-solvents, solubilizers, buffering agents,
film formers, and other components to provide the properties listed under
"Functions".

Surfactant: A surfactant is a "surface active agent". Its primary function
is that of a wetting agent or as a component of an emulsifier or a
spray adjuvant. Some surfactants have been used successfully to enhance
herbicidal activity.

Ions: Many water-soluble materials, when dissolved in water, split apart
into electrically charged atoms or groups of atoms called ions. The
ions with the negative charge are called anions and those with the positive charge are cations.

Anionic: Those surfactants, whose negatively charged ion provides the surface active properties, are called anion-active or anionic. e.g. Sodium+
(lauryl sulfate).

Cationic: Those surfactants whose positively charged ion provides the surface active properties, are called cation-active or cationic. e.g.(Coco
amine)$^+$ acetate$^-$.

Nonionic: As the name implies, nonionic surfactants do not ionize in water
and are, therefore, non-ion active or nonionic. e.g. ethylene glycol,
alcohol, alkylarylpoly (ethylene oxy) ethanol.

Functions of Spray Adjuvants:

A spray solution may have one or more of the following functions to perform
in order to provide a safe and effective application:

1. Wetting of foliage and/or pest.

2. Modifying rate of evaporation of spray.

3. Improving weatherability of spray deposit.

4. Enhancing penetration and translocation.

5. Adjusting pH of spray solution and deposit.

    a. prolong life of alkaline sensitive pesticides.
    b. reduce re-entry time following application of hazardous chemicals.

6. Improving uniformity of deposit.

7. Improving compatibility of mixtures.

8. Providing safety to the treated crop.

9. Reducing the drift hazard.

10. Complying with FDA requirements.

The following brief discussion of each of the above ten points may be helpful in clarifying the many functions performed by the proper spray adjuvant.

1. <u>Wetting of foliage and/or pest.</u> Adequate wetting is required to provide good retention and coverage of spray solution. A suitable surfactant, at the proper concentration, will normally suffice, although certain plants and pests may have special requirements.

2. <u>Modifying rate of evaporation of spray.</u> The need for reducing the rate of evaporation of a spray solution applied at two to three gallons per acre in a hot dry area is obvious. The need, however, may be equally great in the application of a concentrate spray in an orchard. Once the spray has been applied, it may be desirable to have the spray dry as rapidly as possible. Both functions can be performed by a proper adjuvant.

3. <u>Improving weatherability of spray deposit.</u> Resistance to heavy dews, rainfall, and sprinkler irrigation can mean the difference between successful control and failure of a fungicide application, for example.

4. <u>Enhancing penetration and translocation.</u> Many chemicals perform most effectively when they have been absorbed by the plant and transported to areas other than the point of entry. "Systemic" pesticides have this ability. Their absorption can be enhanced and certain non-systemic chemicals can be made to penetrate plant cuticles through the use of a suitable adjuvant.

Translocation is included as part of the systemic performance, although I'm aware of no documented evidence to show that translocation is enhanced through the use of an adjuvant.

5. <u>Adjusting pH of spray solution and deposit.</u>

   a. Many currently used pesticides (primarily organic phosphates and some carbamates) degrade rapidly under even mildly alkaline conditions, found in some natural waters and on certain leaf surfaces. Buffering adjuvants can prolong the effective life of alkaline sensitive chemicals under these conditions.

   b. Experimental adjuvants are currently under test for reducing re-entry time following application of highly toxic pesticides.

6. <u>Improving uniformity of deposit.</u> It is almost axiomatic that, with non-systemic pesticides, the quality of performance of a pesticide can be no better than the quality of the spray deposit. This is particularly true of most fungicides which require complete and uniform coverage.

7. <u>Compatibility of mixtures.</u> With the savings in labor costs to be obtained from doing more than one job with a single application, the effort is made frequently to mix various combinations of pesticides, and pesticides with liquid fertilizers in the same spray tank for simultaneous application. The attendant compatibility problems can frequently be corrected with the proper adjuvant.

8. Safety to crop. Certainly we do not wish to harm the crop which we are trying to protect. This often happens, however, with chemicals that are potentially phytotoxic. The hazard can be increased through the use of the wrong adjuvant or substantially reduced through the choice of a proper one.

9. Drift Reduction. No method, currently in use for reducing drift of pesticide sprays, is entirely satisfactory. The most promising of the new approaches to drift reduction is the use of special foaming adjuvants, applied through foam generating pumps or nozzles, often from conventional aerial or ground equipment. Special application problems may still favor spray thickeners or invert emulsions.

10. FDA Approval. The Code of Federal Regulations, Title 21, Part 121.102, exempts from the requirement of a tolerance, those adjuvants, identified and used in accordance with 120.1001 (c) and (d) which are added to pesticide use dilutions by a grower or applicator prior to application to raw agricultural commodities. All spray adjuvants must comply with these requirements.

The functions and properties of spray adjuvants listed above can contribute substantially to safe and effective pest control. Any one of the functions may be important in a given application. It is not likely that all would be of concern in a single application.

Although a single adjuvant may provide more than one of the above properties, no single product can provide them all. As a result, there are a variety of spray adjuvants available which have been formulated to encompass those functions which are important to a particular type of application.

In an effort to provide the reader with a few practical suggestions for choosing the proper adjuvant for a given application, the following table is presented as an abbreviated guide to currently accepted usages. It must be kept in mind that all of the available products, whether or not they have utility, cannot be listed here. The table is meant only to present examples of products that have satisfactory use history, recognized recommendations, FDA approval, and are available nationally. Obviously, the inclusion of all effective regional products would be meaningless outside their marketing area.

Function	Product
1. Wetting of Foliage and Pest	Tween 20, Triton X-100, Multi-Film X-77
2. Rate of Evaporation	Bio-Film
3. Weatherability	Nu-Film 17, Plyac, Bio-Film Triton B-1956
4. Penetration and Translocation	X-77, Tronic, Surfactant WK, Atplus 300 Regulaid, Oils
5. pH of Spray and Deposit	
a. Prolong life	Buffer-X, Sorba Spray, Nutrex
b. Reduce re-entry time	Experimental compounds only

Function	Product
6. Deposition of Chemical	Bio-Film, Triton B-1956, DuPont Spreader Sticker
7. Compatability of mixtures	
a. Pesticides	Buffer-X
b. Pesticide Fertilizer	Compex, Sponto 168 D
8. Safety to Crops	Experimental Products
9. Drift Reduction	Fomex, Stull's Bi-Vert, Dacagin, Norbak
10. FDA Approved	All above

The agricultural spray adjuvant is one more useful tool for improving the effectiveness of pest control and the safety of chemical application and as with any chemical, only the proper spray adjuvant will do the proper job.

The inclusion of a trade named product does not constitute an endorsement by the author. Since the effect of any adjuvant on a chemical's performance is not predictable under all conditions of weather, plant vigor, application and water quality, the user should consult the manufacturer, his authorized or local agricultural authority to determine the optimum rate or conditions for use in his area.

PESTICIDE COMPATIBILITY

Two or more chemicals frequently are combined:

1. To increase the effectiveness of one of the chemicals. The manufacturer usually wants to increase the activity of his pesticide against specific pests. He does this by adding a chemical referred to as a "synergist", which is not an active pesticide but which increases the effectiveness of the pesticide. Synergists usually increase the toxicity of the pesticide so that a smaller amount is needed to bring about the desired effect. This may reduce the cost of the application, and also reduce the hazard, as less of the active material is used.

2. To provide better control than that obtained from one pesticide. Applicators sometimes combine two or more active pesticides to kill a pest that has not been effectively controlled by either chemical alone. Many combinations are quite effective, but in most cases it is not known if the improved control is a result of a synergistic action or an additive effect of the several chemicals on different segments of the pest population.

3. To control different types of pests with a single application. Frequently, several types of pests need to be controlled at the same time; e.g. insects, diseases, and mites. Generally, it is more economical to combine the pesticides needed and make a single application -- in this case an insecticide, a fungicide, and an acaricide. However, the compatability of the various chemicals must be known before the materials are combined.

Compatibility Problems in Certain Combinations of Chemicals

When two or more pesticides can be safely mixed together, or used at the same time, they are said to be compatable. When they cannot be mixed or used together, or used at the same time, they are said to be incompatable. Some pesticides are incompatable because chemically they will not mix. Some pesticides will mix together well but the results are not the same as when they are used alone. Some combinations of chemicals result in mixtures that produce the opposite effect of synergism, known as "antagonism". Antagonism, or "incompatability", may result in chemical reactions which cause the formation of new compounds or a separation of the pesticide from the water or other carrying agent. If one of these reactions occurs, one of the following may result:

⊙ Effectiveness of one or both compounds may be reduced.
⊙ Precipitation may occur and clog screen and nozzles of application equipment.
⊙ Various types of plant injury (phytotoxicity) may occur; e.g. russeting of fruits or vegetables, stunting of plants, and reduction of seed germination and production.
⊙ Excessive residues.
⊙ Excessive runoff.

Potentiation

Another less familiar but extremely important undesirable effect of combining certain pesticides is potentiation. Some of the organophosphorous pesticides potentiate or activate each other as far as animal toxicity is concerned. In some cases, the combination increases the toxicity of a compound that is normally of very low toxicity to one that is highly toxic to man and other animals.

Phytotoxicity, excessive residue, or poisoning of livestock can also occur when one chemical is applied several days after the application of a different chemical. Check the pesticide label carefully for such warnings.

When a combination of chemicals is to be made refer to the compatability charts that are available through your pesticide salesman or from various other sources. There are two pesticide compatability charts (spray compatability chart and herbicide compatability chart) available from:

Meister Publishing Company
37841 Euclid Avenue
Willoughby, Ohio 44094.

These charts are available annually and the publishing company can be contacted for current prices and availablity. Before you mix pesticides, or a pesticide and another material such as a fertilizer, you should be certain that once they are mixed you will get the best benefit from both. One reason for combining pesticides is to save time. However, if the two materials are incompatable you may lose valuable time. Besides time, you may also lose the chemicals (which are costly) or you may get poor results or injure the crop.

The pesticide label will sometimes indicate incompatability problems. Some pesticide formulations are prepared for mixing with other materials and are registered for pre-mixes or for tank mixes. If this is true, it will be so indicated on the label. Make sure materials are compatable and cleared for the combination before mixing or using them together.

Because of the risks involved, combinations of pesticides and/or other aggricultural chemicals should not be used unless a specific combination has been proven to be effective, tested for side effects, and accepted for registration by the EPA and the State Department of Agriculture.

11
PESTICIDE
APPLICATION
EQUIPMENT

Pesticide Application Equipment

Proper application is the key to success of any pesticide treatment. Simply stated, the application process is getting the pesticide to the target. This process usually involves a carrier which may be liquid, dry, or air to transport the pesticide to the intended surface or target. Application may range from the simple act of spraying a repellent on our skin or painting a surface with a brush to the use of very elaborate and expensive application equipment.

Many factors affect our ability to place a pesticide on the target in the manner and amount for most effective results and with the least undesirable side effects and at the lowest possible cost. Certainly, the selection and use of equipment is of utmost importance and deserves major emphasis when considering pesticide application. However, without proper consideration to formulation, adjuvants, compatibility, and use records, successful application is not very probable. Successful application must also involve principles of proper timing and drift control.

Drift

The control of drift has become an important item with the custom applicator as well as the private applicator. In order to be effective, the pesticide must be applied precisely on target at the correct rate, volume, and pressure.

Why Be Concerned with drift?

There are many reasons why the pesticide applicator should be concerned with drift and ways to minimize damage that it may cause.

- Chemicals are expensive. Do not waste them by letting them get off target.

- Chemicals are undesirable in the general environment for many reasons -- smell, appearance, and danger to wildlife and non-target plants.

- Chemicals can damage sensitive crops in surrounding areas. Recent Federal laws place rigid penalties on both private and custom applicators for misapplication. Thus, an applicator who permits chemicals to drift off the target area to injure another crop, may be faced with a law suit that requires many hours of litigation, or may even put him out of business.

What is Chemical Drift?

Drift is the movement of spray particles or droplets away from the spray site before they reach the target plant or ground surface. This is usually physical movement, that can also be volatile action -- the evaporation or vaporization at ordinary temperature or on exposure to air. Physical drift is normally due to wind movement. However, temperature inversion or convection currents, which usually occur in temperatures above 85° F can prevent the settling of very small droplets and allow their movement away from the spray site during relatively calm conditions.

In agriculture, drift is only of concern when chemicals are deposited other than where we intended them to be.

Drift is influenced by many factors. Spraying pressure, particle size, specific gravity, nozzle design, evaporation rate, height of release, horizontal and vertical air movements, temperature, and humidity are among the important considerations. Since so many variables play an active role in drift, we will discuss some of them in detail to provide insight into drift control.

○ Pressure - Some equipment operators attempt to correct drift by varying pressure. Droplet size is influenced by pressure, but rate and pattern configuration are affected even more. The higher the pressure, the smaller the droplet size. With smaller droplets, better coverage is gained, resulting in higher chemical performance at the expense of drift control. Reducing pressure will enhance drift control because of larger droplets, but it will also reduce coverage. Evaluations have demonstrated that a constant pressure setting produces the best results.

○ Nozzles - In terms of drift control, this is probably more important than pressure. Nozzle tips that give the largest droplet size and application rate acceptable with the proper pressure range are generally recommended. The larger droplet size will help reduce the drift and deliver chemicals on target. Position and orientation of nozzles on the spray boom are also important. Droplet size distribution can be greatly affected by nozzle orientation.

○ Specific Gravity - Small, light weight particles fall much slower than large droplets. For example, a five micron (1 micron = 0.00004 inch) drop of water would require 5 1/2 hours to fall 50 feet in still air as compared to only 55 seconds for a 100 micron drop under the same conditions. It is the small droplets of lower specific gravity that will drift from the target area. Furthermore, water droplets and oil droplets react differently in the presence of wind movement. Because they are much lighter, oil spray droplets remain airborne longer.

Although a coarse spray will drift less than a fine spray, the coarse spray results in narrower swaths with fewer drops per unit of area. In applying certain herbicides, this may be beneficial as thorough coverage may not be critical as long as the material reaches the target area. However, for more complete coverage, a finer spray may be necessary. When the average drop size is doubled, the number of drops in a gallon will be reduced to 1/8 the original number. Generally spray droplet size should be no finer than is necessary to do an effective job.

⊙ Height of nozzles - Nozzles positioned too high will disperse spray over a wider area, but increase the likelihood of drift. Spray particles must fall a greater distance. The applicator must determine the desired swath width by striking a balance among nozzle size, pressure and height above the target. For example, if an operator increases the application pressure, and thereby increases the rate, but maintains the swath width, then he should lower the nozzle to compensate for the increased pressure. If the pressure is increased correspondingly, there will be an increase in the application rate, and also an increased probability of drift.

⊙ Horizontal and Vertical Air Movement - Unless it is calm, most fields are subjected to constant air movement. Unpredictable changes in this air movement may occur at any time to cause drift of spray being applied by ground or aerial sprayers. Thus, weather conditions directly affect the direction, amount, and distance of drift. Furthermore, as the ground warms up during the middle of the day, ground-air temperatures increase to about that of the air at 10-20 feet. This warmer air rises and sets up convection and thermal air currents to lift small spray particles suspended above the target area. Horizontal air movement then carries these droplets to some distant point before they can settle out. The air temperature differential between the ground and ten to twenty feet up is considerably less during early morning and late evening hours than during the middle of the day. On the other hand, during temperature inversion, the air near the ground is cooler than the air above, and minute spray droplets remain suspended in the layer of cold, undisturbed air. Eventually, these mists move out of the area before coming to rest. This kind of drift is difficult to control.

⊙ Temperature and Humidity - The rate of droplet evaporation is determined to a great degree by temperature and humidity. As the diameter of the droplet decreases, the ratio of surface area to volume increases and evaporation occurs at a faster rate. The time of exposure to evaporation conditions also increases with decreasing droplet size because of slower fall.

Vapor drift, unlike spray or dust drift is related directly to the chemical properties of the pesticide. Most problems are the result of accidental or unintentional misuse. If the applicator understands the chemical properties, he can control and avoid damage from vapor drift.

Vapor drift can be due to vapor leakage. Vapor leakage can be stopped by properly sealing fumigant or other volatile materials after they are applied and by applying these materials with vapor tight equipment. Many volatile pesticides cause safety problems from vapor drift. For example, ester formulations of phenoxy ester herbicides may volatilize and drift under high temperature conditions. Amine or acid formulations should be used where vapor drift might cause problems.

The careful use of pesticides is of prime concern to everyone today. Many factors interact to influence the distance material will drift from the target area. Even when common sense and good application technology are adhered to, drift continues to present a problem for the applicator. Label instructions must be followed and strict attention must be given to the control of pesticide drift.

TYPES OF APPLICATION EQUIPMENT

Most application equipment can be used for several different kinds of pest problems. By choosing the type of equipment best suited for his type of operation, the applicator can be assured that he will do a good job. The agricultural applicator's equipment differs greatly from that of a structural pest control operator. Even when he specializes in a specific type of pest control, the pesticide applicator will need to make a choice of equipment. The choice will depend on his **working conditions**, pesticide formulations, type of area treated, and possible **problems that he will** encounter. While large power equipment may be desirable for some jobs, other problems may be best handled by **using small** portable or hand equipment.

Aerial Equipment *

1. Fixed Wing Aircraft - Approximately 7,500 aircraft are used each year

in the United States for aerial application. Most of them are single engine aircraft and are either high wing mono-planes, low wing mono-planes, or bi-planes. These type of aircraft are used on smaller jobs. Multi-engine aircraft are being used on large areas such as forests and range lands and also for forest fire control.

Advantages - Aerial application offers a fast convenient method for pest control, especially when quick action must be taken. Aerial application also allows for treatment of pests when a field is too wet or muddy to allow ground application equipment to operate.

Disadvantages - Fixed wing aircraft generally cannot treat small fields and are difficult to operate in areas where many hazards are present such as power lines and tall trees. Application costs are generally higher with aircraft than with ground equipment, but the speed of application and timeliness may offset this cost differential.

2. Helicopters - The helicopter has shown substantial increase in use in

recent years but the total flight time in comparison to other aircraft is still quite small.

Advantages - Helicopters in agriculture offer certain advantages over fixed wing aircraft, such as slower speeds, safety, accuracy of swath, coverage and placement of chemical, and ability to operate without an airport.

Disadvantages - The complicated construction means higher initial cost and maintenance and therefore a higher per acre application charge. Here again however, this may not be a disadvantage if quick control of the pest problem is essential.

---

* See additional information at end of this Section.

Ground Equipment

1. <u>Low Pressure Boom Sprayers</u> - These sprayers are usually mounted on

tractors, trucks, or trailers. They are designed to be driven over fields or large areas of turf, applying the pesticides in swaths to the crop. Low pressure sprayers generally use a relatively low volume of dilute spray ranging from ten to forty gallons/acre applied at 30-60 lbs. pressure. These sprayers are designed to handle most of the spraying needs on general farms and there are perhaps more low pressure boom sprayers in use than any other type of equipment. They usually have roller type pumps that limit their pressure to about 80 lbs/sq. in. Handguns can be attached for remote spraying for spot treatment and patches of weed infestations.

<u>Advantages</u> - Low pressure sprayers are relatively inexpensive, light weight, adapted to many uses, and can cover large areas rapidly. They are usually low volume so that one tankful will cover a large area.

<u>Disadvantages</u> - Low pressure sprayers cannot adequately penetrate and cover dense foliage because of their low pressure and gallonage rate. Because most rely on bypass systems and return flow agitation, wettable powder formulations often settle out. However, if mechanical agitators are used, this is not a problem.

2. <u>High Pressure Sprayers</u> - High pressure sprayers are often called "hy-

draulic sprayers." They operate with dilute sprays and pressures can be regulated up to several hundred pounds. They are used for spraying shade trees and ornamentals, livestock, orchards, farm buildings, and unwanted vegetation where dense foliage requires good penetration.

<u>Advantages</u> - High pressure sprayers are useful for many different pest control jobs. They have enough pressure to drive a spray through heavy brush, thick cow hair, or to the tops of tall shade trees. Because they are strongly built they are long lasting and dependable. Piston pumps are standard and resist wear by gritty or abrasive materials. Mechanical agitators are also standard and keep wettable powders well mixed in the tank. With a long hose and handgun, trees, shrubs or other targets in hard to get at places can be treated. This kind of attachment is commonly used by commercial sprayers for applications to "ornamentals."

<u>Disadvantages</u> - High pressure hydraulic sprayers have to be strongly built and so are heavy and costly. They usually use large amounts of water and thus require frequent filling.

The National Aborist Association, Inc. has formulated suggestions for the proper and most efficient use of high pressure sprayers. Their suggestions are offered here:

High pressure, high volume hydraulic sprayers are used to apply pesticides to tall shade trees as well as landscape plants. When properly adjusted, and when used with the appropriate hose and nozzle, such equipment is an effective and efficient delivery system.

The tops of tall trees are reached by maximizing the volume of material that comes from the nozzle, not by increasing the pressure. If the pressure is increased beyond the capacity of the hose and/or nozzle, the spray will come out as a fine mist which will not travel very high and which will be subject to any breeze.

If the inside diameter of the hose, its fittings, or the orifice of the nozzle are too small for the volume of material that the pump is delivering, the same things will occur. If a swivel is used between the nozzle and the hose, its inside diameter must be the same as the spray hose.

Use of the proper combination of pump, pressure, hose, fittings and nozzle will provide maximum volume of delivery at the nozzle permitting a higher, straighter trajectory of the material. This will result in a large droplet size pattern, avoiding premature atomization (small droplet size) of the material, thereby minimizing drift.

The following recommendations should be considered:

a. Hydraulic Sprayers

   1) Hydraulic sprayers to be used for spraying trees over 60 feet in height should have a pump capacity of not less than 35 gallons per minute.

   2) Hydraulic sprayers to be used for spraying trees less than 60 feet in height must have a pump capacity sufficient to reach the tops of the trees to be sprayed.

   3) Tank capacities should not be less than 400 gallons to provide sufficient material at the job site.

   4) Spray tanks must have either mechanical or jet agitation.

   5) To provide maximum height and a minimum of spray drift, pump shall operate at such pressure (may be up to 800 P.S.I.) so as to produce a pressure of about 400 P.S.I. when the gun is open and equipped with the largest size spray tip (disc) that the pump will support at this pressure.

b. Spray Hose

   1) Spray hose shall have sufficient inside diameter to deliver a minimum of two thirds of the gallons per minute capacity of the pump.

2) Spray hose shall have a minimum burst pressure of not less than two times the maximum operating pressure of the pump.

c. Spray Nozzles

1) Spray nozzles (guns) should have a sufficient capacity to deliver up to the gallons per minute rating of the pump.

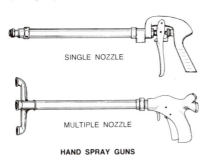

SINGLE NOZZLE

MULTIPLE NOZZLE

HAND SPRAY GUNS

2) Spray nozzles (guns) should be adjustable from a straight stream, to a fan, to shut off.

3) Spray nozzles should have interchangeable tips to allow for increasing or decreasing volume of material.

4) Tips should be checked for wear and replaced as required.

Pesticide Application Techniques for Hydraulic Sprayers

a. The applicator should: position himself to take advantage of air movement with each tree or shrub; treat only designated trees or shrubs; treat so that the material carries into the tree being treated; when spraying a shade tree, start at the top working downward from side to side and gradually reduce the volume of spray as the lower extremities are reached; apply a sufficient amount of material to provide thorough coverage; avoid drenching; be sure that the pesticide reaches all of the leaves or bark, top or bottom, as the case requires; when treating plants near buildings, position yourself so as to spray away from the building.

b. When spraying a tall shade tree there are two very important techniques which must be considered.

1) In order to compensate for the force of gravity, the proper distance can be determined by first aiming at the top of the tree and gradually raising the nozzle until the moving leaves indicate the top has been reached.

2) Air movement can be used to drive the column of material up into the top of a tree rather than through it. By pointing the nozzle back over your shoulder and slightly away from the target, the air will push the material up and into the crown of the tree. (This principle is similar to that which applies when you are trying to water a garden that is 20 feet away and you only have 15 feet of hose. Raising the nozzle a few degrees enables you to reach the target.)

3. Air Blast Sprayers - Practically all spraying in commercial orchards and much of the spraying on shade trees is with air-blast sprayers.

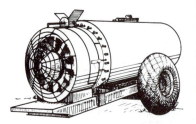

TYPICAL AIR BLAST ORCHARD SPRAYER

Air-blast sprayers are primarily designed to carry pesticide-water mixtures under pressure from a pump through a series of nozzles into a blast of air that blows into the tree by means of a fan. High volume fans supply the air which is directed to one or both sides of the sprayers as it moves between rows of trees. Nozzles operating at low, moderate or high pressure deliver the spray droplets into the high velocity air stream. The high speed air aids in breaking up larger droplets and transporting these smaller droplets for thorough coverage. Agitation of the spray material in the tank is usually accomplished with a mechanical agitator.

Advantages - A small amount of water covers a large area and very little operating time is lost in refilling. They are usually less tiring to operate than hydraulic sprayers and are particularly adapted to applying sprays over a large area.

Disadvantages - Since the pesticide is carried by an air-blast, these types of equipment must operate under calm conditions. Windy conditions interfere with the normal pattern of application of the blower. Larger models may not be able to treat hard to get at areas.

4. Low Volume Air Sprayers (mist blowers) - Mist blowers are characterized by high air velocities and somewhat lower water volumes than conventional air-blast sprayers. This type of sprayer depends on a metering device which may or may not be a conventional nozzle, that operates at low pressures and depends on the high speed air for liquid break-up.

Advantages - A considerable saving in time and labor is possible with low volume sprayers because less water is handled than with conventional air-blast sprayers.

Disadvantages - Calibration is more critical and favorable weather for spraying is more essential. Coverage with very low volumes on some crops may be less satisfactory than with normal volumes delivered by conventional air-blast sprayers.

5. Ultra-low Volume Sprayers (ULV) - Ultra-low volume spraying is accomplished by applying the chemical concentrate directly without the use of water or any other liquid carrier. Many ultra-low volume ground sprayers use a fan which delivers high speed air to help break-up and transport the spray droplets.

Advantages - The main advantage of ULV sprayers is the labor and time saved due to the elimination of water.

Disadvantages - There is an increased risk to the applicator from handling and spraying the concentrated pesticide. In addition, there are only a limited number of pesticides cleared for ULV application at the present time.

6. Aerosol Generators (foggers) - Aerosol generators and foggers break certain pesticide formulations into very small, fine droplets (aerosols). One droplet cannot be seen. But when large numbers of droplets are formed they can be seen as fog or smoke. In some foggers, heat is used to break up the pesticide. These are called thermal aerosol generators. Other foggers break the pesticide into very fine particles by such means as rapidly whirling disks, air-blast break-up, or extremely fine nozzles. Foggers are usually used to completely fill an area with a pesticidal fog, whether it be a greenhouse, warehouse, or open recreational grounds. Insects and other pests in the treated area will be controlled when they come in contact with the aerosol fog.

Advantages - The droplets produced by foggers are so fine that they do not stick to surfaces within the area. Therefore foggers using fairly safe formulations can be used in populated areas for mosquito or other insect control without leaving unsightly residues. The droplets float in the area and penetrate tiny cracks and crevices or through heavy vegetation to reach pests in hard to get places. Because they blanket an area it is difficult for pests to escape exposure.

Disadvantages - Since most of the droplets produced by foggers do not stick, little if any residual control is possible. As soon as the aerosol moves out of an area, other pests can move back in. Also, the droplets produced are so fine that they drift for long distances and may cause unwanted contamination or injury. Most aerosol generators require special pesticide formulations. When foggers are used, the weather conditions must be just right. For example, if an area is being treated for mosquitoes, rising air currents will carry the fog harmlessly over the pests and out of the area.

7. Dusters - Dusters blow fine particles of pesticide dusts onto the target surface. They may be very simply constructed. Dusters are used mostly by homegardeners, pest control operators, and truck gardeners for individual spot treatment of plants over small areas.

Advantages - Dusters are light weight, relatively cheap, and fast acting. They do not require water.

Disadvantages - Dusts are highly visible, drift easily, and are difficult to control, therefore dusters are less desirable for most crops or larger outdoor jobs.

8. Granule Spreaders - Granular equipment is designed to apply coarse, dry particles that are uniform in size to soil, water, and in some cases, foliage. Spreaders may work in several different ways including air-blast, whirling disks, multiple gravity feed outlets, and soil injectors. They may be broadcast or band spreaders such as the one shown here.

Advantages - Granular equipment like dusting equipment is light, relatively simple and no water is needed. Because granules are uniform in size, flow easily, and are relatively heavy, seeders and fertilizer spreaders can be used to apply granules, often without any modification.

Disadvantages - Because granular materials do not generally stick well to foliage, granule spreaders are not usually used on plants. Therefore, the applicator will need other machinery for controlling most leaf-feeding insects and most plant diseases.

9. Soil Injectors - Soil injection equipment is frequently used to apply fumigation materials to the soil to control nematodes and other soil borne pathogens or insects. The most common method of soil application involves the use of chisel cultivators or shanks which have a liquid or granular tube down the back of the shank permitting materials to be placed in the soil to a depth of a foot or more. With volatile materials, the shanks may be spaced at 12" or more and a continuous effective band or coverage will be obtained. Sweep type elevator shovels, with a series of nozzles on the trailing edge, may be used to apply a single band or continuous cover beneath the soil surface.

Advantages - Since soil applied materials are not as likely to cause phytotoxic damage as foliar applications and, since applications can be much more precise when injected into the soil, it has long been common practice to use undiluted technical or minimum dilution materials at very low rates of a gallon or less per acre.

Disadvantages - Because the pressure orifice or low application rates must be quite small, these are difficult to keep from plugging. This can be overcome however, through the use of gravity feed systems or the use of positive displacement type pumps which are usually driven from a ground wheel of the applicator machine.

Hand Equipment

Hand operated sprayers and dusters are most commonly used by individuals

for their own relatively small pest problems. The custom applicator, however, will often find it convenient and efficient to have hand sprayers for small jobs that do not require larger powered equipment or that require only a small amount of spray. They are also especially helpful on small jobs in hard-to-get-at areas where spray equipment must be carried in.

There are several types of hand application equipment which includes:

a. Intermittent Discharge Sprayers which spray material with each stroke of the pump.

b. Continuous Pressure sprayers which discharge spray as long as the pump is being operated.

c. Aerosol Bombs which have pressurized cans or tanks with a discharge valve and nozzle and are essentially self-contained.

d. "Knapsack" Granular Applicators that are crank operated, spinning disk types.

e. "Knapsack" Hand Sprayers that are carried on the back with a capacity up to five gallons. A hand operated piston or diaphragm pump provides the pressure.

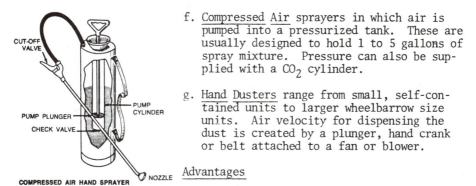

COMPRESSED AIR HAND SPRAYER

CUT-OFF VALVE

PUMP PLUNGER

CHECK VALVE

PUMP CYLINDER

NOZZLE

f. Compressed Air sprayers in which air is pumped into a pressurized tank. These are usually designed to hold 1 to 5 gallons of spray mixture. Pressure can also be supplied with a CO_2 cylinder.

g. Hand Dusters range from small, self-contained units to larger wheelbarrow size units. Air velocity for dispensing the dust is created by a plunger, hand crank or belt attached to a fan or blower.

Advantages

Hand sprayers and dusters are economical, uncomplicated, lightweight, and yet will do a surprising amount of work and adapt to many different problems. The spray and dust is easily controlled as to direction and drift because relatively little material is used and at low pressures.

Disadvantages

Hand sprayers and dusters are efficient and practical for small jobs only. Wettable powders tend to clog regular sprayer nozzles and agitation is frequently poor.

SPRAY EQUIPMENT PARTS

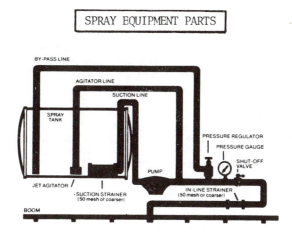

BY-PASS LINE

AGITATOR LINE

SUCTION LINE

SPRAY TANK

PRESSURE REGULATOR
PRESSURE GAUGE
SHUT-OFF VALVE

PUMP

JET AGITATOR

SUCTION STRAINER (50 mesh or coarser)

IN-LINE STRAINER (50 mesh or coarser)

BOOM

Schematic Outline of a Sprayer System

To do a proper job of applying pesticides, an operator must have the correct equipment and operate it correctly. Selection of a good quality sprayer will depend on careful attention to each of the following parts:

Tanks

Tanks should be made of rust resistant steel, fiberglass or polyethylene to avoid rust, sediment, plugging, and restriction problems. Tanks from 55-150 gallons are available on mounted sprayers and 200 gallons or larger on pull type sprayers. Only one tank per spray rig is recommended as more than one tank makes agitation difficult. Tanks need to have a large covered opening in the top with a removable strainer, to make filling, inspection, and cleaning easy. A drain plug in the bottom is necessary for complete drainage when cleaning.

Tank Agitators

The return flow from the regulator valve usually gives sufficient agitation for solutions or emulsions. Wettable powder materials usually need more agitation to keep them in suspension. Agitation can be accomplished by mechanical or hydraulic means. Mechanical agitation is more desirable for suspension sprays, but this type of agitation is difficult to secure in tractor or trailer mounted sprayers that use a tractor power take-off to operate the pumps.

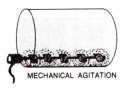
MECHANICAL AGITATION

Most field sprayers have hydraulic agitation of the by-pass or jet type. The by-pass type uses the return hose from the pressure relief valve to agitate the tank. The by-pass agitator hose must extend to the bottom of the sprayer tank. By-pass agitation is not sufficient on field sprayers with tanks larger than 55 gallons unless a centrifugal pump is used.

JET AGITATION

Piston roller and gear pumps require a "jet" agitator on 100 gallon or larger tanks. The jet must be connected to the pressure side of the sprayer and must operate at the operating pressure of the sprayer. Jet agitators must not be attached to the by-pass line. Jet agitators require about three gallons per minute of the pump output to keep material in suspension in a 55 gallon tank. Larger tanks should have three or more outlets on a manifold type that is placed horizontally about one to two inches from the bottom of the tank and will allow the jets to discharge horizontally across the bottom of the tank. The holes in the agitator manifold must be sized so that the pump can maintain the operating pressure when all the nozzles are operating. A 1/8 inch hole will allow approximately three gallons per minute to pass through at 40 psi.

The jet stream coming from a jet agitator should pass through at least a foot of liquid before striking the side or bottom of the tank. Abrasive materials under high velocity will eventually wear a hole in the tank unless it passes through sufficient solution.

If the sprayer is stopped, even briefly, it may be necessary to mechanically work the powders back in to suspension so that the jet agitator can keep them there.

Pumps

The pump is the heart of the sprayer. There are many types of pumps on the market with advantages and disadvantages specific to each pump. The most important factors to consider in selecting a pump are as follows:

- Capacity. The pumps should be of sufficient capacity to supply the boom output and provide for by-pass agitation. The boom output is calculated from the number and size of nozzles used. Hydraulic agitation, if used, should be added to the boom output. Select the pump which will provide the boom output plus about 50% more for agitation through the by-pass return. Pump capacities are given in either gallons per hour (GPA) or gallons per minute (GPM).

- Pressure. The pump must be able to produce a desired operating pressure at the capacity required for the spraying job to be done. The amount of pressure is indicated in pounds per square inch (psi). Some designated as low pressure can produce high pressure but will wear out rapidly if operated under high pressure conditions.

- Resistance to Corrosion and Wear. The pump must be able to handle the chemical spray materials without excessive corrosion or wear. Some pumps will handle abrasive materials like wettable powders with much less wear than others. Chemical reaction and corrosion effects certain materials more than others.

- Repairs. Pumps should be designed so that repairs can be made economically and quickly.

- Type of Drive. The pump should be readily adaptable to the available power source. The average farm tractor does not have a high speed power take-off unit, and therefore is not adapted to pumps designed to operate faster than 500-1000 rpm. High speed pumps will require an auxiliary power source or speed increaser. Roller pumps can be attached directly to the tractor power take off but should be secured with a chain to prevent the case from rotating.

Types of Pumps

Piston Pump

The piston pump is a positive displacement pump that can develop high pressures.  Although the most expensive type of pump, it can be used for many different jobs. Piston pumps can apply both corrosive and abrasive materials. These pumps have a stainless steel piston with a leather, rubber, or plastic packing gland. In a more expensive and durable pump, the cylinder walls may be lined with or may be entirely of ceramic materials. The less expensive piston pumps are constructed with replacable liners for easy and economical servicing.

Roller Impeller Pump

The roller impeller pump is probably the most widely used pump in agriculture today. It is adaptable in a wide range of pressures, volumes, materials, and situations and is relatively low priced.

The pumping action of the roller impeller pump is created by the roller following the eccentric housing, going out of the slots to take on liquid and in to expell it. It maintains a constant pressure and flow, making it possible for more accuracy in controlling the amount and placement of spray material.

This pump can be adapted to handle a wide range of chemicals. Nylon covered rollers usually are used with a cast nickel-iron alloy housing to handle the non-abrasives. Rubber covered rollers are used for the more abrasive type materials such as wettable powders. The roller pump is a compact unit that can be mounted easily to available power sources. The wearing parts are conveniently and economically replaced when worn.

Flexible Impeller Pump

The flexible impeller pump has flexible vanes that compress to force the liquid out of the pump. The flexible impeller pump is generally limited to pressures below 50 psi, but will handle all except highly abrasive materials.

External Gear Pump

The external gear pump is a semi-positive type pump that can develop moderate pressures. It will maintain a uniform pressure and volume, but the volume is limited. In this pump, liquid is pulled in where the rotating gears separate and is pushed out on the opposite sides as they roll together. These gears are the principal parts of the pump and take all the wear, so this pump is not adapted to abrasive materials. It is well suited to pumping oil suspension or emulsions at high volumes and pressures. The gears and housing generally are made of bronze and the gear shafts of stainless steel. It is a compact unit that is easy to mount to a power source such as the power take off of a tractor.

Internal Gear Pump

The internal gear pump is similar to the external gear pump except that one gear is running inside the other. The internal gear pump also has the capabilities of developing high pressure but because of the metal on metal gear system it may have a limited life where abrasive or corrosive materials are used. It too is a compact unit that is easy to mount to a power source such as the power take-off of a tractor.

Centrifugal Pump

A centrifugal pump handles coarse or abrasive materials effectively. Pumping action is created by a high speed impeller that literally throws the materials out of the pump. This high speed limits its use on tractors for applying agricultural chemicals. This type of pump is used most frequently as a water pump for high volumes and low pressures. Pressure and pumping capacity decreases with wear and roughness of impeller and housing. This pump requires priming unless located below the level of supply tank.

Diaphragm Pump

The pumping action of a diaphragm pump is caused by the movement of the diaphragm. This pump may be constructed with either one or two diaphragms. The most popular is a single diaphragm because of its lower cost. Liquid is drawn into the chamber by the downward stroke and is forced out by the upward stroke.

Chief limitations of this pump are pulsating pressures and the relatively small volume capacity. Stroke pulsations are evened out by small air chambers in the pump head. Two diaphragms help minimize these pulsations and give a greater volume.

The valves, usually made of neoprene, are spring loaded. The diaphragms usually are made of combined materials that will best resist the expected action of chemicals to be pumped.

This pump can handle a wide range of chemicals, including abrasives and coarse materials. It will operate for a long time if the material in the diaphragm is not deteriorated by the chemical being pumped and the bearings are of the roller type. This is a medium priced unit and servicing is easy and economical.

Vane Pump

The vane pump is similar to the roller impeller pump. The vanes are metal and are spring-loaded so that they will compress as they follow the eccentric housing, going out of the slots to take on liquid and being compressed back into the slots to force the liquid out at a reasonably high pressure. The vane pump is most successful in handling materials that have some lubricating properties.

Filter Screens (strainers)

Filter screens are used in sprayers to prevent foreign materials from entering and wearing precision parts of the sprayer. These strainers which give uniform flow, also prevent nozzle tips from clogging. Screens

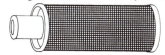

are normally placed at the entrance to the pump intake line, in the line from the pressure regulator to the boom, and in each nozzle. The mesh or size of the screens on the filters should be large enough to allow passage of wettable powders (50 mesh) and emulsifiable liquid concentrates (100 mesh). Mesh refers to the number of openings per square inch of the screen material. Usually 12 to 50 mesh screens are used in the tank strainers, 25-50 mesh screens in the in-line hose strainer, and 50-100 mesh screens in the nozzle tips. The screens in the nozzle tips should be sized according to the opening in the nozzle. Filters should be checked and cleaned often to prevent poor coverage and loss of pressure.

Pressure Regulator

The pressure regulator or relief valve maintains the required pressure in  the system. It is a spring-loaded valve that opens to prevent excessive pressure in the line and return some of the solution to the tank. Most pressure regulators are adjustable to permit changes in the working pressure if desired.

Pressure Gauge

Accurate pressure measurement is important since pressure is one factor which can be controlled and may determine the amount of liquid being sprayed. Spray nozzles are designed to operate within certain pressure limits and you should not exceed them. High pressures can cause dangerous fogging and drift while low pressure may increase droplet size to the point that improper coverage is obtained.

Check pressure gauges periodically to determine their accuracy. The pressure gauge is a delicate instrument and should be handled with care. It will indicate malfunctions by showing fluctuations in pressures. Pressure gauges should be selected to give accurate readings within the range of pressures normally used in the spraying system. For example, a gauge reading up to 500 pounds per square inch would not be satisfactory for operating at pressures of 30 to 40 pounds because it would not be sensitive enough to register significant pressure changes.

Control Valves

Quick acting control valves should be installed between the pressure regulator and the boom to control the flow of spray materials. One valve may be used to cut off the flow to the entire system, or a combination of two or three valves may be used to control flow to one or more sections of the boom. Special selector valves are available which control the flow of spray materials to any section of the boom.

These controls should be located in a place where the operator can handle them without moving from his position or taking his eyes off his driving.

Hoses and Fittings

Hoses should be of neoprene or other soil resistant materials and strong enough to take peak pressures. Hose test pressure should be twice the operating pressure. The suction hose should be made of 2-ply fabric. It should be larger than the pressure hoses because it must provide the total pump flow through the suction hose, suction strainer, and valves.

Pressures are sometimes encountered which are much higher than the average operating pressures. Hoses on the pressure side of the system should be reinforced to prevent accidental breaking, which could subject the operator to dangerous spray materials.

Fittings, including clamps should be designed for quick, easy attachment and removal.

Nozzles

The nozzle and its adaptation to the spray application requirement is important in the effective use of agricultural chemicals. The nozzle helps control the rate, the uniformity, the thoroughness, and the safety of pesticide application. All your efforts in identifying a pest, selecting a material to overcome that pest, and obtaining a proper piece of spray equipment for applying the pesticide can be a total waste if nozzles aren't properly selected, installed, and maintained. Nozzles are the prime elements in the system that can make or break you.

Nozzle performance is the key to total system performance in most sprayer systems. Factors to consider are: nozzle type; nozzle size; nozzle condition; nozzle orientation (position on boom); and nozzle spacing on the boom.

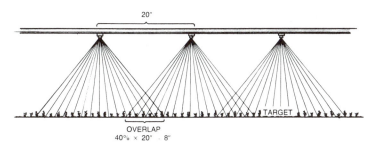

20"

OVERLAP
40% × 20" - 8"

TARGET

TYPICAL BOOM SETUP WITH PROPER PATTERN OVERLAP

Boom too low or too high will give uneven patterns

Boom should be level to sprayed surface for uniform coverage

Nozzles should be aligned parallel with the boom   Do not use 80° and 65° nozzles together

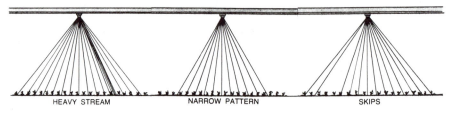

HEAVY STREAM   NARROW PATTERN   SKIPS

SYMPTOMS OF WORN NOZZLES

With several manufacturers producing a variety of nozzles specifically designed to do any of the wide range of jobs, precision application is within the grasp of any applicator who really intentionally sets out to do a good job. The problem lies in the applicator's ability to select the proper nozzle, use it correctly, and maintain the system at peak efficiency. Although nozzles exist for virtually every purpose, a few standard ones are basic to agricultural spray application. Briefly they may be described as follows:

Flat fan
This nozzle produces a nearly flat fan in several selected angles and de-
 posits a flat, eliptical pattern on the ground. The outer edges of the deposit are of reduced volume, which permits the fan patterns to be overlapped to produce uniform coverage across the boom width. These are used for broadcasts or boom spraying for such as weed control work. Drift is less than that with standard cone nozzles. Because the rate tapers at the edges, the nozzles must be overlapped approximately 50% for even distribution.

Even Flat Fan
This nozzles fills in the outer portions of the flat fan spray to produce
 even coverage across the entire width of the applied pattern. These nozzles are for band spraying and produce a uniform spray pattern throughout. They are popular and efficient for coverage of a given strip through the field over a crop row. Even flat fan spray nozzles should not be used for broadcast spraying.

Hollow Cone Spray Pattern Nozzles

These nozzles are designed for moderate to high pressures and are used where thorough coverage of crop foliage and very uniform distribution is desired. For abrasive materials (wettable powders) the disk type should be used to reduce wear.

Solid Cone Spray Pattern Nozzles

These are used for hand spraying, spot spraying, and moderate-pressure foliar application of insecticides and growth regulators. The cone nozzle contributes to better coverage of foliage and is most popular.

Flooding Nozzles

Flooding nozzles may be used for herbicides and fertilizer solutions. Flooding nozzles operate at low pressures equal to those of whirl-chamber nozzles and have even wider spray patterns (up to 160°). Flooding nozzles normally operate at low pressure with large droplets and can cover a wide area so that perhaps use of a boom is unnecessary. This is an advantage on extremely rough terrain or where many obstacles would hinder boom operation.

Off-Set Nozzles

Off-set nozzles lack the uniformity of the flat fan but can provide reasonably uniform coverage over wide areas for roadside and ditch bank weed control. These nozzles can be used on the end of a boom to increase effective boom swath but are not recommended for general use.

Whirl-Chamber Nozzles

This nozzle is available as a wide angle (120°) hollow cone nozzle that may be used in place of a flat spray nozzle. Clogging is minimized because of design. Less drift occurs due to the lower boom height and large droplet size. Uniformity of coverage remains fairly constant with changes in boom height.

Nozzle Construction

Spray nozzles are made of the following metals for the reasons indicated:

§ Brass -- most commonly used, relatively inexpensive

§ Stainless steel -- non-corrosive, relatively expensive

§ Aluminum and Monel -- resistant to moderately corrosive materials

§ Hardened Tungsten Carbide Tips -- used for highly abrasive materials

§ Plastic -- used for non-abrasive materials, corrosive resistant, inexpensive. Not recommended for high pressure spray rigs

Items to Remember

• Select a nozzle that will provide the desired droplet size, volume of flow, and pattern.

• Provide that nozzle with a material free of foreign particles, and provide it at a properly regulated pressure.

• Mount the nozzles so that their location, relative to the target is maintained constant and proper.

• Move the nozzles through the field at constant speed to ensure uniform application rates.

• After installation, properly calibrate the system to ensure that, in all respects, it is applying uniform coverage at the desired rates.

• Maintain the system in peak condition by periodic inspections and calibrations with particularly detailed attention to the performance of each nozzle.

A Word About Nozzle Tip Numbers

Nozzle tip numbers are fairly uniform among the various manufacturers. The first two numbers refer to the angle of spray discharge. The last four numbers refer to the gallons per minute of the nozzle tip at 40 psi. For example, nozzle tip number 650067 means a nozzle tip with a 65° angle of fan discharge and delivering 0.067 gpm at 40 psi. The decimal point is placed by counting three figures from the left side, i.e. a Tee Jet® 8004 nozzle tip delivers 0.4 gpm. For spraying heights where the boom is 17-19 inches from the surface to be sprayed, 80° nozzle tips are recommended, for 19-21 inches 73° nozzle tips, and for 21-23 inches 65° nozzle tips are recommended. Risk of drift is generally greater with the wider nozzle tip angle.

Spray discs are used in handguns. The number on the cap represents the diameter of the orifice (opening) in the disc in increments of 1/64 of an inch. A number 3 disc is 3/64 of an inch and a number 10 is 10/64 etc. The larger the disc number the coarser the droplet size and the more gallons per minute delivered at a fixed pressure. Refer to manufacturer's charts to purchase the proper disc for your operation and equipment.

NEW TYPES OF APPLICATION EQUIPMENT BEING PERFECTED

There are several types of equipment that have been developed in recent years that differ from the "conventional" equipment.

Rope wick applicators, sometimes known as carpet applicators, are becoming popular for use against weeds that have grown higher than the crop. The wick or wipe applicators are tractor mounted and driven through the field as the herbicide is brushed off on the unwanted weeds at a height above the crop.

Wick applicators are made with a reservoir boom made of various size PVC with upright openings on one end for easy fill-up and drain for easy cleaning. Most of the PVC booms are of 3-inch diameter.

The rope wick applicator shown above is only one of many different designs and makes. Some wick applicators use ropes interlaced between two horizontal bars while others use tufts of material other than rope. The ropes themselves are made of different materials, but work much the same way as a wick in a kerosene lamp.

The height of the front-mounted booms can be hydraulically adjusted to best fit the situation with each crop. There is no spray drift problem because the only chemical that is used is that which is wiped on the weeds by the wick. There are a number of applicators on the market.

Although several herbicides show promise with this type of application, Roundup® has been successful and is still the most popular herbicide for this type of application.

Electrostatic sprayers are non-mechanical spraying systems which produce electrically charged droplets of uniform size directly from oil-based solutions without using moving parts. The most significant part of this idea is that droplet size and movement are both controlled so that drift is virtually eliminated.

The heart of this technique is a charged "nozzle" through which the liquid flows by gravity, picking up its electric charge on the way. The uniform, charged droplets move along specific lines of the electric

field existing between the positive nozzle and the negatively charged target plant. This electric field envelopes the entire plant. Since all droplets carry the same charge and are mutually repellent, each follows a specific trajectory. The result is an even distribution of chemical over the upper and lower surfaces of all the leaves and stems.

Because droplets are physically attracted to the target at quite high velocity, drift is dramatically reduced and spraying is possible under conditions that would rule out the use of conventional equipment.

Droplet size can be controlled between a range of 40-200 microns, depending on the applied voltage. Voltages are usually high, about 20,000 volts, but currents are very low, so operator safety is not a problem.

Electrostatic sprayers have worked well in the laboratory, but there are still problems in the field and such companies as ICI state that it will be several years before their Electrodyn® system is ready for tractor-mounted field application work.

Spinning disk nozzles that revolve about 5,000 rpm are said to provide a more uniform droplet and utilize less total nozzles as they are spaced about twice the distance as conventional nozzles. Although they provide a more uniform droplet, the size of the droplet can be critical and sometimes can be subject to excessive drift.

Electrocution of weeds using wire fingers on chains that discharge high voltages from a machine into weeds is receiving new attention after being tried and discarded in the 1940's.

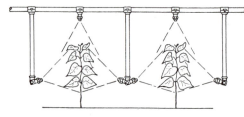

Conventional spray application equipment utilizing directed sprays and shielded sprays is not new, but is being perfected by several companies. The idea is to get better coverage with shielded sprays between rows for weed control, or to get better coverage of crops with insecticides or fungicides in a row by directing the spray from the sides as well as the top.

Recirculating ground sprayers have been developed that catch excess spray as the spray rig moves through the field. This technique is not well perfected as yet but does have the advantage of preventing excess spray from reaching the ground and the recirculation of that spray which is a savings that allows for more acreage covered with less spray and less environmental contamination.

New nozzles are constantly being experimented with. Some newer ones on the market include Raindrop® nozzles that provide larger droplets and less drift. Several types of nozzles and booms have been developed for aircraft which include microfoil booms and half a dozen different nozzle tips.

Spraying Systems Co. offers an "off-on" spray tip and crop assembly known as Quickjet® which allows for quick changing of nozzle tips, screen cleaning, etc., while keeping self-alignment of flat spray patterns. The caps are color-coded in eight different colors so that one nozzle tip size can be assigned a certain color and prevent intermingling of several sizes.

A method which minimizes, if not eliminates, human contact with concentrated pesticides during mixing and loading activities is referred to as a "CLOSED SYSTEM". Closed systems meter and transfer pesticide products from the shipping container to mixing or application tanks, and often rinse the emptied containers with a minimum of human exposure to the concentrate pesticide. In addition to reducing the risk of human exposure in metering and transferring pesticides, closed systems:

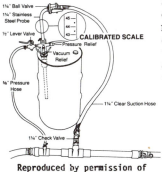

Reproduced by permission of
Mazzei Injector Corporation

-- Provide greater accuracy in measuring dosage

-- Reduce or eliminate fill site contamination

-- Reduce the possibility of water source contamination

-- Reduce the need for wearing hot, uncomfortable, protective clothing (wearing protective gloves and clean clothing remains advisable)

If provision for rinsing emptied containers is made, closed systems also:

-- Decontaminate the container by rinsing to make container easier to handle and dispose of

-- Avoid wasting pesticide remaining in unrinsed containers

-- Can reduce the time and effort required for rinsing and mixing

Closed systems are required by law in California for handling highly toxic liquid pesticides. Other states may take similar action. At least one pesticide is labeled and marketed for use only in closed systems to reduce possible health hazards.

SPRAYER MAINTENANCE AND CLEANING

Most trouble with sprayers can be traced to foreign matter that clogs screens and nozzles and sometimes wears out pumps and nozzles. Pump wear and deterioration are brought about by ordinary use, but they are also accelerated by misuse. The following suggestions will help minimize labor problems and prolong the useful life of the pump and sprayer.

1. Use clean water. Use water that looks clean enough to drink. A small amount of silt or sand particles can rapidly wear pumps and other parts of the sprayer system. Water pumped directly from a well is best. Water pumped from ponds or stock tanks should be filtered before filling the tank.

2. Keep screens in place. A sprayer system usually has screens in three places: a coarse screen on the suction hose; a medium screen between the pump and the boom; and a fine screen in the nozzle. The nozzle screen should be fine enough to filter particles which will plug the tip orifice.

3. Use chemicals that the sprayer and pump were designed to use. For example, liquid fertilizers are corrosive to copper, bronze, ordinary steel, and galvanized surfaces. If the pump is made from one of these materials it may be completely ruined by just one application of the liquid fertilizer. Stainless steel is not adversely affected by liquid fertilizers, and pumps made from this substance should be used for applying these types of fertilizers.

4. Never use a metal object to clean nozzles. To clean, remove the tips and screens and clean them in water or a detergent solution using a soft brush. The orifice in a nozzle tip is a precision machine opening. Cleaning with a pin, knife, or other metalic object can completely change the spray pattern and capacity of the tip.

CLEAN NOZZLE
TIPS WITH A
SOFT BRUSH

5. Flush sprayers before using them. New sprayers may contain large amounts of metalic chips and dirt from the manufacturing process. Sprayers which have been idle for awhile may contain bits of rust and dirt. Remove the nozzles and flush the sprayer with clean water. Clean all screens and nozzles thoroughly before trying to use sprayer.

6. Clean sprayer thoroughly after use. After each day's use thoroughly flush the sprayer with water, inside and out, to prevent corrosion and accumulation of chemicals. Be sure to discharge cleaning water where it will not contaminate water supplies, streams, crops, or other plants and where puddles will not be accessible to children, pets, livestock, or wildlife.

SPRAYER STORAGE

Be sure to prepare the spray system for off-season storage. When changing chemicals or when finished spraying for the season, clean the sprayer thoroughly both inside and out. Some chemicals, such as 2,4-D, are particularly persistent in the sprayer and must be removed completely to prevent possible crop damage of other spraying operations. For thorough cleaning between chemicals or at the end of the season, the following procedure is recommended:

1) Remove and clean all screens and tips in kerosene or a detergent solution using a soft brush.

2) Mix one box (about one half pound) of detergent with 30 gallons of water in the tank. Circulate the mixture through the by-pass for thirty minutes, then flush it out through the boom.

3) Replace the screens and nozzle tips.

4) Fill the tank about 1/3-1/2 full of water then add one quart of house hold ammonia to each 25 gallons of water. Circulate the mixture for five minutes, allowing some to go out through the nozzles. Keep the remainder of the solution in the system overnight, and then run it out through the nozzles on the following morning.

5) Flush the system with a tank full of clean water by spraying through the boom with nozzles removed.

6) When the pump is not in use, fill it with a light oil and store it in a dry place. If the pump has grease fittings, lubricate them moderately from time to time. Over lubrication can break seals and cause the pump to leak.

7) Remove nozzles and screens and place them in a light oil for storage.

8) Drain all parts to prevent freeze damage.

9) Cover openings so that insects, dirt, and other foreign material cannot get into the system.

10) Store the sprayer, hoses, and boom in a dry storage area.

W A R N I N G

Never use a pocket knife or other metal object to clean a nozzle. It will damage the precisely finished nozzle edges and ruin the nozzle performance. A round wooden toothpick is much better. Better still, remove the nozzle tip and back-flush it with air or water first before trying anything else. Don't blow through it by mouth. Chemicals are poisons!

AERIAL APPLICATION EQUIPMENT

Aerial dispersal equipment must be light weight to provide for maximum pay-
loads, strongly built and mounted to withstand the wear and the loads that
they encounter. Operating controls must be simple. All components should
be designed and installed to permit thorough cleaning between jobs, to
avoid residual contamination from previous uses.

A clean aircraft is not only good advertising, it makes daily inspections
quicker and easier. Well cared for equipment will inspire the customer's
confidence in the operator.

With the marketing of agricultural aircraft and disseminating equipment,
the conversion of general aircraft for agricultural use is disappearing.
The following general points should be considered whether you operate a
manufactured or converted aircraft. They apply to both fixed-wing and
rotary-wing aircraft.

Spray Dispersal Equipment

Spray dispersal equipment consists of the following items: tank(s), pump,
pressure regulator, line filter, flow control valve, boom and nozzles.
On fixed wing aircraft the tank is built into the fuselage. The pump,
pressure regulator, line filter and control valve are mounted under and
outside the fuselage. The boom and nozzles are usually attached to the
trailing edge of the wing.

On rotary-wing aircraft, the tanks are mounted on the sides of the frame
to keep the load in line with the main rotor shaft. The tanks are coupled
at their bases with a cross-over pipe which feeds the pump. The pump is
driven by the engine. The filter, regulator and valve are attached to
the lower frame of the fuselage. The spray boom and nozzles are mounted
to the toe of the landing skids, or to the frame under the main rotor.
The toe mounting puts the boom and nozzles where the pilot can see them.
Mounting the boom under the main rotor gives a slightly wider swath.

Spray equipment provides effective swath widths of 40 to 60 feet in the
range of one to 10 gallons per acre, when flown at heights of 5 to 8
feet above the ground or crop. Special applications such as ULV can
give much wider swaths, but they are flown at 15 to 25 feet altitude.

Tanks

Top loading tank openings should be large enough to permit pouring materials into them. All tanks should be fitted with emergency dump valves
and the valve action should be checked each time the equipment is flushed.
They should be fitted with vents to permit free flow of the materials at
emergency dump rates. Bottom-loading systems should have capacity enough
to match the loading pump. Spray tanks need agitation that reaches the
bottom to maintain suspensions and mixtures of materials. If hydraulic
agitation (pumping the material back into the tank) is used, a rule-of-
thumb flow rate is 10 gpm agitation flow for every 100 gallons capacity
of the tank. The pump outlet and the emergency dump should connect to
the bottom of the tank to permit complete emptying.

Pumps

All water-based sprays can be handled with centrifugal pumps. Centrifugal pumps are mechanically simple, light weight, flexible in their output and have a maximum pressure determined by design. Normal power sources for these pumps are:

Wind-driven fan, using the forward motion of the aircraft.

Direct coupling to the aircraft engine, with or without a clutch.

Hydraulic or electric drive from the aircraft engine to a hydraulic or electric motor coupled to the pump.

The size of centrifugal pumps is selected to provide adequate flow (gpm) to handle the maximum nozzle capacity (and agitation, if included) and the maximum pressure (psi) to give the needed boom pressure and to take care of the pressure loss (about 5 psi) in the pipes, valve and filter. Specialized applications (ULV, bi-fluid, microbial sprays, etc.) call for special pumps.

Controls

Controls should include a shut-off valve and a pressure regulator between the pump and the boom. The valve should be a quick-acting gate or ball valve. The pressure regulator should be a bypass type which returns the relief liquid to the tank(s). If the pump is powered by a wind driven fan, a brake should be installed to stop the spray pump when not in use.

Filters

Filter screens at the nozzles prevent the spray tips plugging with sediment that normally collects in a spray system. It is wise to have a line filter installed behind the pump so that sediment is not forced into the parts of the valve and the regulator, creating unnecessary wear. Screen size should be 50 or 100 mesh, depending on the size of the nozzle tips. A pressure gauge, coupled to the line beyond the filter will indicate (by low pressure) when the line filter is starting to get clogged. Clean the line filter daily during spray operations.

Plumbing

To prevent excessive pressure losses:

--For application rates over 2 gallons per acre, all main piping and fittings should be at least 1 1/2 inches inside diameter.

--For application rates of 1/2 to 2 gallons per acre, all main piping and fittings should be at least 1 inch inside diameter.

--For ULV applications, hoses to individual nozzles should be 1/8 inch I.D. Main line hoses and fittings should be at least 3/8 inch I.D.

Tees should be the size of the incoming pipe and fitted with reducers to couple the branch pipes. All joints should be made with the minimum of pockets that can trap materials. The number of bends should be kept to a minimum and all bends made with as large a radius as possible.

All spray plumbing connections using hoses should be double-clamped or clamped and safety-wired to secure the connections.

Booms

Booms need strong support. Placing the boom at the trailing edge of the wing reduces the drag resistance of the boom. They should be far enough back to avoid airstream interference with the control surfaces. Booms should be fitted with caps having openings the full size of the pipe to permit flushing and cleaning with a "bottle" brush when needed. Twin-boom or return line systems are used to eliminate air pockets at the boom tips and also to provide agitation in the boom for chemicals that settle out rapidly. Nozzle orientation to the direction of flight is important for droplet size control. If the boom cannot be rotated, use 45° elbows or swivel connectors to provide for nozzle adjustment.

Nozzles *

1) Nozzle Location

The location of the outboard nozzle on each end of the boom is critical. Since wingtip vortexes or main rotor vortexes are used to develop the width of the pattern, the end nozzles must be inboard enough to prevent the vortex trapping the fine droplets. This entrapment creates peaking in the pattern and drift. On fixed-wing aircraft, the propeller disturbance shifts the spray from the right to the left (as the pilot sees it). Nozzles need shifting to the right within the area of this disturbance to compensate for the uneven pattern. The choice of nozzles and the amount of shift cannot be determined without testing.

2) Nozzle Selection

A great deal of information is available from nozzle manufacturers for use by operators. Only general guides will be given here. Nozzles perform satisfactorily if they are operated at the manufacturer's recommendations of pressure and flow is not restricted. Departures from these values must be made with care since output and droplet size both vary with pressure. Nozzle types and operating pressures should be selected that can handle the material being applied and break it up into droplets of desired size.

Water solutions of pesticides should be pumped with nozzle pressures of 35 to 45 psi to get manufacturer's recommended breakup. Suspensions, heavy emulsions and slurries should be handled with nozzles having large openings to prevent clogging. Bi-fluid systems use special mixing chambers and tips. Foaming nozzles are marketed that have special passages for air incorporation with the spray to create foams.

---

* See additional information in Section on Aerial Equipment Calibration

Positive shut-off of nozzles is achieved by the use of:

--Diaphragm check valves
--Ball check valves
--"suck back" connection on pump, working on boom or return line.

These items will need regular attention if they are to perform properly. Aged or worn diaphragms or scarred ball seats need replacement. Any springs used to maintain pressure will require checking to see that they move freely. Seating faces should be smooth and free from cracks. Return lines need flushing to be sure that they are not clogged.

As a guide, the following suggestions are given to select the nozzle type for a treatment:

1) The atomizing or hollow cone type spray gives the finest breakup.

2) The flat spray nozzles have intermediate breakup.

3) The solid cone nozzles have the coarsest breakup.

4) Spinning nozzles tend to have a narrower range of droplet sizes than the hydraulic nozzles (1,2 and 3 above)

Ultra Low Volume Systems

ULV spraying is a recent development in aircraft spray work, using concentrate materials (no water added) in a spray made up of fine droplets. Some pesticides exhibit better action in their concentrated form. Less evaporation takes place since water is absent from the spray. The density of the droplets is slightly greater than the same sized droplets of water-base sprays, increasing the rate of fall. Application rates are reduced so more acres can be treated before reloading is needed. At present, however, only a few materials are licensed for this use.

Plans are available for the modification of aircraft dilute spray systems to handle ULV. If the aircraft is used for different spray jobs during the season, the operator can install a small ULV system, entirely separate from the dilute system. This system can be removed when the ULV applications are done, making a cleanup much easier.

One system already marketed consists of a 5-gallon stainless steel pressure tank to hold the chemical. Pressure is obtained from a liquid carbon dioxide fire extinguisher hooked up to the tank with an air pressure regulator. An electric valve controls the flow from the tank, using a 1/2 inch or 3/4 inch solenoid valve with a 12-volt DC coil. Half-inch hose leads from the valve to a tee which center-feeds the ULV boom. This boom is made of hoses and tees feeding the nozzles. The ULV boom and nozzles are clamped to the regular boom for support. The solenoid valve can be wired to the aircraft electrical system or a dry cell battery in the cockpit and to a push-button switch taped to the control stick. The 5-gallon tank and fire extinguisher can be set in the regular spray tank, where it should stand in an upright position.

The following points should be observed: ULV spraying must create fine droplets to be effective. Flat fan nozzles discharging 0.1 gpm or less are recommended operating at 40 to 55 psi. Spinning nozzles can also be used. Do not use less than four flat fan nozzles to avoid gaps in the distribution pattern. On a helicopter, a single spinning nozzle may provide sufficient output if very low rates are needed (e.g. mosquito control). Since this system produces fine droplets, the extreme outboard nozzles must be located away from the wingtips on fixed-wing aircraft. This avoids spray entrainment in the wingtip vortices. Use 2/3 wingspan as a guide for the limit. Shift the central nozzles to the right to compensate for higher than usual, 20 to 25 feet above the ground, to provide for a wider swath and greater uniformity.

When full strength chemicals are used, the solvent carrier for the chemical has to be considered. The carrier for Malathion will attack rubber and neoprene. The seals in the solenoid valve should be of teflon or viton. Hoses may have to be replaced at the end of the season if they are not resistant to the solvent. Diaphragms in the nozzle bodies should be checked frequently and replaced each season. Nozzle screens are important since plugging is easier with the smaller tips. Use 100-mesh screens.

In calculating conversions from gallons per acre to ounces per acre (liquid), use the factor: 1 gallon = 128 ounces

Granular Dispersal Equipment

Dusts, impregnated granulars, granular fertilizers, prilled fertilizers and seeds all fit into this category. For the finer materials (smaller than 60 mesh) some agitation may be needed in the hopper to prevent bridging. Since most commercial granulars contain fines, the seals between the components of the equipment (the tank, to the metering gate, to the disseminator) require frequent inspection to make sure that no leaks occur. Seal strips that look mechanically tight on the ground will often leak under the influence of air pressures developed in flight.

The size, shape and weight of the particles and the "flowability" of the material affect the swath width, the application rate and the pattern. Common effective swath widths are 35 to 40 feet when applied at 10 to 15 feet altitude. Some disseminators at low application rates achieve satisfactory swath widths of 50 feet.

Fixed-Wing

The usual disseminator for fixed-wing aircraft is the ram-air or venturi type. The throat of the venturi is attached to the fuselage under the hopper outlet. It has a broad aft section to impart lateral motion to the material. Some ram-air spreaders can deliver up to 250 lbs. per acre. Overloading the disseminator tends to peak the application pattern at the center and to reduce the swath width. Overloading should be avoided because it can lead to uneven flow making the pattern erratic.

Rotary Wing

Two systems are being used. One system uses saddle tanks mounted next to the engine, each tank having its own disseminator system. The metering devices have to be linked to prevent one tank being emptied faster than the other. The second system uses a single hopper and disseminator, hanging on a cable below the rotary-wing aircraft. A single tank and disseminator avoids the problem of balancing the output of two metering devices. Also, being on a cable, the equipment is more remote from the aircraft where corrosive granules create problems. The disseminators are either spinning plate "slingers" or some form of blower and ducting. The lower forward speed of the rotary aircraft prevents the use of the ram-air principle.

Ground Support Equipment for Aircraft

Efficient ground equipment is necessary for an efficient aerial operation. Time saved on the ground in loading aircraft safely and quickly may make the difference between profit and loss in the total operation. The supply equipment and ground crew will vary greatly from one operation to another depending on the needs of the specific aerial applicator.

Loading of Solid Materials

Dusts,granules and seeds are often put into the hopper directly from the container or sack through the large filling door of the hopper. This operation can be made easier and faster by loading a separate hopper of equal capacity on the ground before the airplane lands and hoisting this above the hopper door and dropping it into the hopper of the airplane. Other mechanical loaders can also be used for loading solid materials into aircraft, such as auger loaders.

Loading of Liquid Materials

Supply tanks for transport of the spray carrier to the field of operations on a truck or trailer may range from 200 to 5,000 gallons capacity. It is usually desirable to carry the diluent as well as the mixed chemical on the same tanker in order to save time going to and from a water source at each new mix. Smaller nurse tanks of 200 to 400 gallons are sometimes used for mixing of chemicals and supplying them directly to the aircraft in a large operation where several airplanes are to be loaded. The mixing tank may consist of two compartments, one for mixing the chemical with the carrier and the other for keeping the chemical ready for loading into the aircraft. Adequate agitation is essential for some mixtures.

Most aircraft in use are loaded through the bottom in which the spray concentrate is pumped into the aircraft tank through a bottom loading connection on the side of the aircraft, followed by the diluent. This closed system method eliminates much of the handling of toxic chemicals and makes a much safer operation for the loading crew.

| OPERATIONS |

General

When an aircraft has been calibrated, the airspeed, spraying pressure or gate setting for granulars, height of flight and the effective swath width are fixed. Applications must be made at the same settings.

Ferrying height between the airstrip and the field should be done at a minimum of 500 feet height, loaded or empty. Avoid flying over farm buildings, feedlots or residential areas both for noise and for possible leakage. Courtesy to your neighbor costs so little and pays real dividends.

Field Flight Patterns

With rectangular fields, the normal procedure is to fly back and forth across the field in parallel lines. Flight directions should be parallel to the long axis of the field (reducing the number of turns). Where cross-winds occur, treatment should start on the downwind side of the field to save the pilot flying through the previous swath as shown in the diagram below.

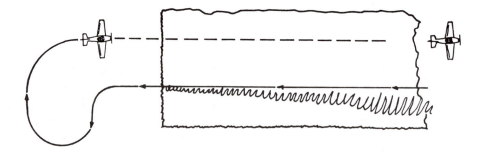

Where this fits in with crop rows or orchard rows the pilot can line up the aircraft with a crop row.

If the area is too rugged or steep for these patterns, flight lines should follow along the contours of the slopes. Where spot areas are too steep for contour work (mountainous terrain) make all treatments downslope.

Swath Marking

Swaths can be marked with flags set above the height of the crop to guide the pilot. This method is useful if the field is going to be treated several times during the season.

Automatic flagging systems are in common use now. These devices, attached to the aircraft and controlled by the pilot, release weighted streamers. These streamers give the pilot a visible mark to help him judge the next swath.

Research recently completed by Intermark, Incorporated, a research firm headquartered in Salt Lake City, Utah (reported in the December 1981 issue of Ag-Pilot International), revealed a marked difference between cost factors related to manual versus aerial flagging of agricultural crops. Two popular flagging methodologies were evaluated in the study. The methods were: the manual flagging technique (utilizing individuals to physically mark the displacement of each swath, on the ground); and the automatic flagging method (the pilot-controlled marking system that drops paper flags to mark the actual chemical swath). These two techniques were compared and analyzed with regard to the cost considerations of using each technique, as well as each system's effect on the flying efficiencies of the pilot and aircraft.

Automatic flagging device mounted on aircraft wing

Aircraft spraying and using automatic flagging technique

There were considerable variations in cost efficiencies between the two marking systems. The automatic flagging system showed a greater tendency to have fixed costs. Once the purchase of the automatic flagging equipment is made, variable costs (costs associated with usage) remain comparatively low. Maintenance of the equipment is also minimal, contributing further to the predictability of the fixed costs associated with the automatic flagman marking system.

Costs associated with the manual flagging system were found to be more variable, fluctuating directly with the amount of time required. These fluctuating time factors included pilot air time, fuel costs, aircraft depreciation, flagger salaries, fringe packages, liability insurance expenses, as well as travel and fuel costs associated with transporting flaggers to the field.

A special point of concern brought out by the study involves the growing cost of health and safety precautions required when utilizing a manual flagging system. The requirements of OSHA and DOA regulations on the legal responsibilities of the air ag contractors are considerable.

A simulation of flagging costs performed in the study indicated that direct flagging costs for a manual system ranged from 40% to 160% greater than the costs associated with automatic flagging, depending upon the flight pattern utilized. Further, it was hypothesized in the study that the larger the job, the greater the differential would become.

The efficiency of the flying pattern can directly increase, or decrease, the flagging costs, as well as the overall cost of the job. The Intermark study determined that the single factor most affecting the cost of the program was the flight time of the aircraft itself. Therefore, reducing flight time could appreciably reduce the overall cost of any particular application job.

The most commonly used flight pattern for spraying has been the "keyhole" method. But a new, seemingly more efficient pattern, has recently emerged. It's called the "racetrack" method because of its familiar shape.

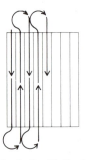
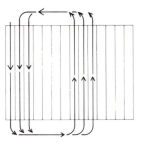

Keyhole Method   Racetrack Method

Utilizing the keyhole method allows the pilot to spray sequential strips across the field. But it requires more air time because a greater turn distance is required to realign the aircraft with the spray path. The racetrack method minimizes the turn arc to 180° and alternately sprays two halves of the field. The actual flight time saved by utilizing the new racetrack method was shown to approach 50% according to the Intermark study.

The study indicated that there were limitations in the compatibility of manual flagging with the racetrack turning method. It was shown to be unfeasible for one individual to flag a field for a racetrack flying program. A flagger would be required for each of the flagging lanes (one in each end of the field), thus doubling the direct flagging cost. The automatic flagging system, however, is controlled by the pilot so no additional costs are involved in utilizing the racetrack methodology. In fact, analysis of the racetrack method, used in conjunction with an automatic flagman system, showed a decrease in overall costs, based on reduced field costs and flight time depreciation of the fixed costs equipment. This combination consistently resulted in a 17% greater profit margin than the most efficient manual flagging system.

The conclusions drawn by the Intermark study must be necessarily tempered by each individual's own flagging needs.

Turnaround

At the end of each swath the pilot should stop the disseminator and pull up out of the field before beginning his turn. The turn should be completed before dropping into the field again. He should fly far enough beyond the field going out to turn to permit slight course corrections before dropping into the field again for the next swath as shown in the diagrams below.

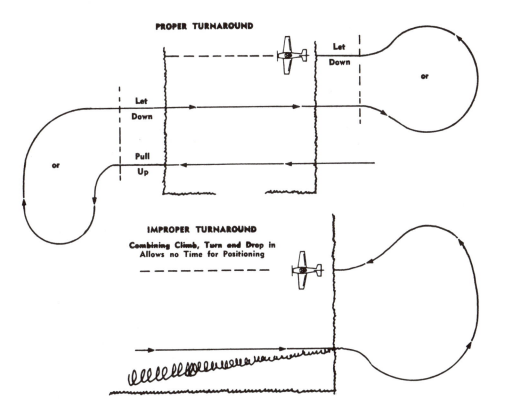

12
PESTICIDE
EQUIPMENT
CALIBRATION

Pesticide Equipment Calibration

Millions of dollars go into research to determine application rates of various pesticides on certain crops and animals. Proper rates must be applied to provide effective control. Accurate application is also essential to keep residues at acceptable levels.

The application of the right pesticide at the right time, and at the proper rate is important to prevent contamination of the environment. To get the correct rate, application equipment must be properly adjusted and operated. The correct application rate is one factor which the operator can control.

Accurate application of pesticides depends on accurate calibration of the application equipment. Calibration means to determine the output of the equipment under controlled and precise conditions.

Some of the factors that affect calibration are:

Equipment -- Set the equipment up to do the desired job according to the operator's manual or other information sources before attempting to calibrate. Select the correct orifices or nozzles for the flow rate and pressure to be used. (Refer to the section on PESTICIDE APPLICATION EQUIPMENT).

Speed -- Speed is a major variable with most application equipment. With equipment that has a constant discharge rate, the application rate is inversely proportional to the speed. If you drive twice as fast you apply 1/2 the rate. Spraying equipment must be calibrated at exactly the same speed that will be used in the field. A sprayer calibrated at 4 mph and driven at 3 mph will over spray 25% and the same sprayer driven at 5 mph will under spray 20%.

Most newer tractors are equipped with a ground speed indicator. These are generally satisfactory for determining ground speed. If the spray equipment is not equipped with a satisfactory ground speed indicator, ground speed can be determined by using the following procedure:

Step (1) -- Set two markers in the field 88 feet apart (88 ft is 1/60th of a mile).

Step (2) -- Select gear and throttle settings on your equipment.

Step (3) -- From a running start check the time in seconds required to drive the 88 feet.

Step (4) -- The speed is determined from the following formula:

$$\text{Speed (mph)} = \frac{60}{\text{Time required to travel 88 ft. (seconds)}}$$

Example: Assume the time required to travel 88 ft. was 20 seconds. The traveling speed of the equipment then is

$$\frac{60}{20} = 3 \text{ mph}$$

Pressure -- The pressure on a nozzle or orifice will vary its output rate. Most nozzle orifices have been machined to give the most effective spray pattern at 30 to 40 psi. Pressures higher than this produce a large number of fine droplets and increase the potential of having the pesticide move from the target area by drift. Low pressures can create large droplets, weak spray patterns and poor coverage. Anytime it is necessary to change the spray rate to an extent that cannot be done within the recommended pressure range, a different set of nozzles should be selected. Remember too to always use the same pressure for spraying that was used for calibration. For pressure adjustment, pressure has to be increased four times to double the spray rates.

Density -- The weight per gallon of liquid will vary the discharge rate through the nozzle or orifice at a given pressure. The heavier a liquid the slower it is discharged at the same pressure. Most nozzles are rated for water. When applying lighter materials (oils),or heavier materials (fertilizer and fumigants), adjustments must be made for the density of the material.

Viscosity -- The viscosity or thickness of a material will affect the flow rate. Sprayers are usually calibrated with water. If the viscosity of the spray material is considerably different than water, calibrate with the liquid that will be used in spraying. Most liquid pesticides have a viscosity very close to water and thus viscosity is usually not a factor.

CALIBRATION METHODS

Calibration is simply the adjustment of a machine to make it deliver the right amount of spray on a given area. There are several methods of calibration, but all involve determining how much spray is being delivered and then changing some settings to give the correct rate. A sprayer should be calibrated at the beginning of the spray season and more frequently as a larger number of acres are sprayed to adjust for possible nozzle wear.

The most accurate way to do this is to fill the tank with water only. Spray one acre, and then measure the number of gallons of water necessary to refill the tank. This is the application rate in gallons per acre. Do not apply a mixed pesticide spray while calibrating a sprayer!

Select nozzles which will deliver the calculated volume at a pressure less than 40 psi for weed control, 50-60 psi for insect control and 100 psi for plant disease control.

When a smaller acreage is used for calibration, it is advisable to measure the amount of water in quarts or pints, especially when a calibrated measuring can is not available. Accurate measurements become more important as the test area becomes smaller.

TO ADJUST THE SPRAY RATE

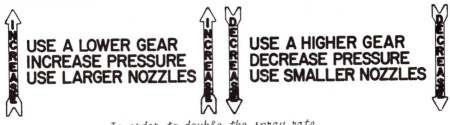

USE A LOWER GEAR
INCREASE PRESSURE
USE LARGER NOZZLES

USE A HIGHER GEAR
DECREASE PRESSURE
USE SMALLER NOZZLES

*In order to double the spray rate,
pressure has to be increased four times.*

Hand Sprayers

Spray equipment with single nozzles or sometimes short booms with three or
four nozzles are used for spot treatment and application of pesticides to
smaller areas. Single high pressure nozzles are also widely used for
rights-of-way spray operations, "ornamental" spraying, and for spot treatment of noxious weed patches. Accurate calibration of this type of equipment is very simple but seldom done. Pesticide applicators should pay
strict attention to correct calibration of hand sprayers because overappli-
cation of pesticides is a common occurance with this type of equipment.

a. Compressed Air Sprayers

Step (1) -- Mark out a square rod (16 1/2 ft X 16 1/2 ft).

Step (2) -- Determine the pressure and spray pattern needed to get good
coverage of the area being treated.

Step (3) -- Fill the sprayer and pump up the sprayer to thirty or forty
pounds pressure, if it has a pressure gauge or count the number of
strokes used to pump up sprayer.

Step (4) -- Spray the square rod, walking at the same speed you plan to
use in spraying the actual area to be treated. While spraying, time
the operation in seconds needed to spray the area.

Step (5) -- Using a suitable measuring device, catch spray from the nozzle while spraying for the same amount of time it took to cover the
one square rod area. Measure the amount of spray.

Step (6) -- Determine the amount applied per acre by using the following
formula:

Gallons/acre = amount sprayed out (cups) X 10

There are 160 square rods per acre and 16 cups/gallon. By dividing
16 into 160 we get a constant of 10 that can be used in our formula.

Example: Assume that three cups of water were applied to one square rod. The application rate is 3 X 10 = 30 gallons/acre.

b. High Pressure Hand Gun

Single nozzle, high pressure equipment is calibrated in exactly the same way as compressed air or knapsack sprayers. Accuracy is increased if the plot (1 square rod) is marked out on the species of plant to be sprayed. After determining the time required (in seconds) to spray this square rod plot, catch the spray from the nozzle for the same amount of time required to spray this square rod plot and measure. The rate of application is determined by the same formula used above.

Boom Sprayers

a. Refill Method

The refill method is probably the simplest of the many methods available for calibrating boom-type sprayers.

Step (1) -- Measure the effective spray width of the boom. This is the width covered by the spray at ground level.

Step (2) -- Divide the effective width of the boom into 43,560 (square feet/acre) to determine the distance the sprayer must travel to cover an acre. Measure this distance on the ground to be sprayed. Since the distance required to spray is usually quite large, the common practice is to reduce the course to a fraction of an acre, i.e. 1/10 or 1/16 of an acre.

Step (3) -- Fill the spray tank and adjust the pressure (30-40 psi for most uses) and the tractor or applicator speed to the speed to be used in the field.

Step (4) -- Fill the spray tank to a known reference line and spray the measured distance.

Step (5) -- Measure carefully the amount of water required to refill the tank to the reference line. It is desirable to make two to three runs to obtain more accurate calibration. Returning the sprayer to exactly the same spot each time it is refilled will also increase accuracy.

Step (6) .-- Multiply the number of gallons required to refill the tank to the previously designated reference line by the reciprocal of the fraction of an acre sprayed (1/10, 1/6, 1/4, etc.) to determine the delivery rate in gallons per acre at the speed and pressure utilized.

Example: A sprayer with a 20 foot effective spray width is calibrated on 1/10 of an acre and requires 4 gallons of water to refill the tank after the calibration run.

Step (1) -- 20 feet effective boom width

Step (2) -- $\frac{43,560}{20}$ = 2,178 linear feet necessary to cover one acre

1/10 of an acre = 218 linear feet

Step (3) -- 4 gallons of water is required to refill the tank

Step (4) -- 4 X 10 = 40 gallons per acre

b. Another Refill Method

Step (1) -- Fill the sprayer tank to the very top or to some mark.

Step (2) -- Spray exactly 660 feet (40 rods) at the same speed and pressure to be used when spraying.

Step (3) -- Return to the exact starting position and refill the tank to the starting level, measuring in gallons the amount of water used.

Step (4) -- Calculate the application rate as follows:

$$\frac{\text{gallons used X 66*}}{\text{width of spray swath in feet}} = \text{gallons per acre sprayer is applying}$$

Example: After spraying a swath 12 feet wide and 660 feet long, 8 gallons of water are required to refill the sprayer. The application rate is:

$$\frac{\text{8 gallons used X 66}}{\text{spray swath (12 feet)}} = 44 \text{ gallons per acre}$$

c. Nozzle Method

The nozzle method of calibration is a quick and accurate way to calibrate any sprayer as long as the ground speed is known and can be accurately controlled. It can be used to calibrate in the shop or in the farm yard and is valuable as a quick check for nozzle wear. By using this method it is possible to accurately predict the spray rate at any controlled speed.

The nozzle method requires checking only one nozzle on the sprayer, but assumes all nozzles are delivering the same amount. Check all nozzles at first to be sure they are delivering nearly the same amount of spray. The nozzle method is based on the formula:

---

* A constant derived by dividing 43,560 square feet/acre by 660 feet sprayed and represents the width required to spray 1 acre.

$$\text{Spray rate (gals. per acre)} = \frac{\text{one nozzle output (ounces per minute) X 46.4 *}}{\text{One nozzle coverage (inches) X speed (miles per hour)}}$$

Step (1) -- Set the pressure the same as is to be used in the field  and catch the water from one nozzle for one minute -- measure water carefully.

Step (2) -- Measure coverage of a nozzle in inches. On a boom sprayer, the coverage is the same as the nozzle spacing on the boom.

Step (3) -- Multiply the amount (ounces of water) pumped in one minute (from Step 1) by 46.4 (a constant).

Step (4) -- Multiply the forward speed of the sprayer (miles per hour) by the nozzle spacing (inches).

Step (5) -- Divide the answer obtained in Step (3) by the answer in Step 4. This is the gallons of water the sprayer is delivering per acre.

Example: A sprayer has 16 nozzles spaced 18 inches apart and the boom covers a 24 foot swath. When operated at 40 psi, one nozzle delivers 40 ounces of water in 1 minute. The sprayer is to be operated at 4 mph. What is the application rate?

Step (1) -- 40 ounces per minute (measured)

Step (2) -- Nozzle spacing = 18 inches

Step (3) -- 40 ounces X 46.4 = 1,856

Step (4) -- 4 mph X 18 inches = 72

Step (5) -- $\frac{1,856}{72}$ = 25.8 gallons per acre applied

Caution -- Be sure all nozzles are delivering at nearly the same rate when using this method of calibration.

---

* This constant applies to delivery measured in ounces, derived as follows:

$$\text{Constant} = \frac{43,560 \text{ square feet/acre X 12 inches/foot}}{88 \text{ feet/ minute (1mph) X 128 ounces/gallon}} = 46.4$$

Constants to use when the delivery is measured in other units are:

milliliters - 1.57   pints - 742.5   gallons - 5940

Band Applicators

Preemergence spraying (spraying after the crop is planted but before it comes up) or preplant spraying (application before the crop is planted) are becoming widely used practices in cultivated crops. Many of the herbicides used are expensive and one way to overcome the problem of cost is to apply the chemical as a band spray. Band spraying is the application of the herbicide in a band, usually about 1/3 to 1/2 as wide as the row spacing, immediately over the crop row. Leaving the area between the crop rows unsprayed. In this way, only 1/3 to 1/2 as much material is used per cropped acre as when full coverage spraying is used; with a resultant saving in chemical cost. The area between the rows can be cultivated clean to reduce the weed infestation.

When application rates are recommended for weed control chemicals, such as 2 pounds per acre, this much active ingredient is to be applied to the area covered by spray. With full coverage spraying the entire field would receive this amount of chemical, but with band spraying only the sprayed band receives chemical at this rate. Thus, if 14 inch bands were sprayed on 42 inch rows, for every acre of cropland treated only 1/3 of an acre will be sprayed. Therefore, if 2 pounds per acre of chemical were recommended, 2 pounds of active ingredient would be applied to each acre actually sprayed but only 2/3 pounds of chemical would be required to treat an acre of cropland.

The purpose of calibration is to determine the amount of spray applied to the band area. This figure is used to determine the amount of chemical to mix with carrier (water) in the tank. The concentration is figured exactly the same way as it would be if the spray was full coverage.

a. Calibration

As with full coverage spraying, calibration is very important. The margin of selectivity or safety of preplant or preemergence herbicides on such crops as sugar beets, fieldbeans, and corn, is sometimes narrow and accurate application is quite necessary.

Calibration can be done with various calibration jars on the market, or by using various other methods:

Refill Method

1. Measure off a known distance, such as 300 or 400 feet.

2. Fill the sprayer tank with water to a known mark. Spray the measured area at the same speed and pressure that would be used in the field.

3. Refill the tank to the known mark, measuring carefully the amount of water used.

4. Calculate the gallons per acre (gpa) sprayed by the following formula:

$$\frac{43,560 \text{ X gallons used}}{\text{Distance} \quad \text{Band Width} \quad \text{no. of}} = \text{gpa}$$

Traveled (in feet) bands
(feet)

Example: If 1/2 gallon of water were sprayed on a 300 foot strip in two 14 inch bands, the acre rate would be:

$$\frac{43,560 \text{ X } .5}{300 \text{ X } 1.2 \text{ X } 2} = 30 \text{ gpa sprayed}$$

Example: If the same amount of water was sprayed on two 6 inch bands the acre rate would be:

$$\frac{43,560 \text{ X } .5}{300 \text{ X } .5 \text{ X } 2} = 72.6 \text{ gpa sprayed}$$

Nozzle Method

The nozzle method described for a boom sprayer can also be used to calibrate a band sprayer. The formula is:

$$\frac{\text{Spray}}{\text{rate}} = \frac{\text{nozzle output (ounces per minute) X 46.4}}{\text{nozzle coverage (inches) X speed (mph)}}$$
$$\text{(gpa)} \qquad \text{(band width)}$$

Step (1) -- Adjust the pressure to the amount that will be used in the field and collect the spray from each nozzle for 1 minute and measure. (If all nozzles deliver equal amounts as they should, only one nozzle needs to be measured.)

Step (2) -- Measure the band width (coverage) in inches.

Step (3) -- Substitute the values from Steps 1 and 2 into the formula and calculate the gallons per acre sprayed on the band.

Example: If the nozzle delivers 36 ounces of water in 1 minute at 30 pounds pressure and the speed to be used is 3 mph, the spray rate on a 14 inch band will be:

$$\frac{36 \text{ X } 46.4}{14 \text{ X } 3} = 39.8 \text{ gpa.}$$

In the above example if the band width were reduced to 7 inches the spray rate on the band would be:

$$\frac{36 \text{ X } 46.6}{7 \text{ X } 3} = 79.8 \text{ gpa}$$

Changing the width of the band has a pronounced effect on the spray rate and thus on the concentration of chemical in the spray tank. It should be pointed out, however, that if the speed and row spacing remain the same in the two examples, the amount of water used per crop acre will also remain the same.

Air Blast Sprayers

1. Observe carefully the nozzle arrangement on the machine as to location and sizes of nozzles.

2. Determine the rate of nozzle discharge at the normal operating pressure by consulting the service manual that accompanies the machine, or ask for help from your sprayer dealer or distributor.

3. Check the actual output per minute of the sprayer by either timing the emptying of a portion, or all, of a tank of water. If your spray tank is not marked or graduated it may be necessary to measure known amounts of water into the tank before starting your calibration.

4. Make sure the speed of travel of the tractor and sprayer is correct. You may determine your speed in the following manner:

Determine the number of tree spaces per minute that you will pass at the speed you wish to travel using the following formulae:

$$\text{Tree spaces per minute} = \frac{\text{Miles per hour desired X 88}}{\text{Tree spacing in feet}}$$

Example: You have a tree spacing of 20 feet and you wish to go 1 1/2 miles per hour.

$$\text{Tree spaces} = \frac{1.5 \text{ X } 88}{20} \quad \text{(Ans. 6.6 tree spaces)}$$

This means that your sprayer should cover 6.6 tree spaces in one minute to be traveling 1.5 miles per hour. Check this by starting the sprayer and tractor exactly at one tree and then travel 6.6 trees and determine the time that was required. If you covered this distance in less than a minute you should slow down, if you took more than a minute you must speed up. Repeat this several times to make sure that you are accurate. Then mark the spot on your throttle and your gear setting. If the tractor has an automatic transmission (selectospeed) mark the speed setting on the selection lever.

5. Check your pump pressure and nozzles often to make sure that you are maintaining a good spray pattern.

6. Read carefully for any special instructions on handling of certain chemicals in air blast sprayers.

7. Be sure to get your mixing ratios correct.

Example: Let's assume that you are going to apply 16 pounds of compound Z per acre and you are applying 600 gallons of water per acre. You have a sprayer with a 400 gallon tank. How much material do you use per tank? 400 gallons of water will do $\frac{400}{600}$ of an acre or 2/3 of an acre

Consequently, 2/3 of 16 = 10 2/3 or about 10 1/2 to 11 pounds of compound Z per 400 gallons.

Example: Let's assume that you are going to apply 16 pounds of compound Z per acre and you are applying 90 gallons of water per acre.
Your sprayer has a 400 gallon tank. How much do you use per tank?

$$\begin{array}{r} 4.4 \text{ acres} \\ 90\overline{)\ 400} \\ \underline{360} \\ 400 \end{array}$$

400 gallons of water will do

4.4 X 16 = 70.4 pounds per 400 gallon tank

Fumigation Applicators

Fumigants are metered from the system with orifices. These orifices serve
the same purpose as spray nozzles except that they are placed in the line
going to the injector. The orifice size is selected in the same manner
as a nozzle. First determine the required flow rate in gallons per hour
formula. Then, using an orifice (spray nozzle) catalog or a flow formula,
determine the size orifice needed.

(a) Low Pressure Fumigators

The low volatile fumigator may be calibrated by collecting the material
applied over a given area.

The fumigant must remain as a liquid until it passes through the metering
orifices. With the low volatile fumigators this is a problem only within the metering pump at high speeds. Even though the low volatile fumigants are slow to vaporize they will form vapor in the metering pump at
high speeds due to the sudden vacuum created on the intake stroke. If
this happens, the rate is drastically reduced.

There are two methods of collecting the material.

Method 1

1. Measure 100 feet in the field.

2. With the applicator equipment running at the desired speed and en-
gaging the soil, collect the material applied in the 100 foot section
from one or more soil tubes. (The entire applicator output may be collected.)

3. Determine the application width being collected.

Example: If the material is collected from 6 tubes spaced 8 inches
apart, the application width is 48 inches. If the material for one
row is collected the application width is the row spacing.

4. Compare the material collected with the Row Fumigation Table on
the following page to determine the rate.

**ROW FUMIGATION OR BAND SPRAYING
RATE TABLE**

Rate Gal. per acre	Quantity per 100 ft. row									
	24-in. row		30-in. row		36-in. row		42-in. row		48-in. row	
	oz	cc	oz	cc	oz	cc	oz	cc	oz	cc
1	.6	17.4	.7	21.7	.9	26.1	1.0	30.4	1.2	34.8
3	1.7	52.1	2.2	65.0	2.6	78.2	3.1	91.2	3.5	104.3
5	2.9	86.9	3.7	108.4	4.4	130.3	5.1	152.1	5.9	173.8
7	4.1	121.6	5.1	151.7	6.2	182.5	7.2	212.9	8.2	235.1
9	5.3	156.4	6.1	195.1	7.9	234.6	9.3	273.7	10.6	312.8
12	7.1	208.5	8.8	260.1	10.6	312.8	12.3	365.0	14.1	417.1
15	8.8	260.7	11.0	325.2	13.2	391.0	15.4	456.2	17.6	521.3

Method 2

1. Measure 100 feet in the field.

2. While operating the equipment with the fumigant shut-off, record the time required to travel the 100 feet.

3. Determine the application width for the orifice or orifices to be used. For row application use the row width and collect all material for one row.

4. While the application equipment is not moving adjust orifices and/ or pressure to apply the quantity shown in the table above for the desired rate in the time required to move 100 feet through the field.

b. High Pressure Fumigators

To calibrate the high volatile applicators (methyl bromide) the container must be weighed before and after applying fumigant to a measured plot. The material cannot be collected because it would immediately volatilize.

Weigh the container of fumigant. Apply the fumigant to 100 linear feet of row for row crops or 1000 square feet of area for broadcast treatments. Re-weigh the container to determine the amount of fumigant used. Multiply the amount (lbs) used per 100 linear feet by:

145 for 30" width rows to give the lbs per acre

121 for 36" width rows to give the lbs per acre

104 for 42" width rows to give the lbs per acre

90 for 48" width rows to give the lbs per acre

For broadcast applications multiply the amount used per 1000 square feet by 43.5 to determine the rate being applied per acre.

Remember that speed, pressure, and applicator injector spacing will affect the rate.

Granular Applicators

Companies manufacturing granular applicators have rate-guide charts in their operator's manuals. These charts show the proper setting for a desired application. These settings are usually reliable, but make a field check to insure accuracy.

a. Field Calibration

Equipment needed

1. Scales (milk scales, kitchen scales or postage scales, etc. that will weigh material to 0.1 pound).

2. 100-foot tape.

3. Two stakes.

4. Bucket, plastic bag or paper sack.

Procedure

1. Measure and mark 300 feet in the field.

2. Fill one hopper with granules.

3. Disconnect the delivery tubes from the applicator.

4. Catch the material from the applicator in a suitable container (bucket, plastic bag, paper sack, etc.) while driving the tractor at a speed that will be used in the field over the 300-foot measured course.

5. Weigh in pounds the granules discharged by the applicator.

6. Calculate the square feet of the test area.

   (a) For insecticide application, multiply the row width in feet times the distance covered (300 ft.).

   (b) For herbicide banding, multiply the band width in feet times the distance covered (300 ft.).

7. Calculate the rate per acre, multiply the pounds collected (step (4) by 43,560 and divide the answer by the square feet of the area (step (6).

$$\text{Pounds per acre} = \frac{43,560 \times \text{pounds applied over test area}}{\text{area of measured course in square feet}}$$

8. Adjust applicator and repeat the process until the desired rate is obtained. After one hopper has been adjusted, the other hoppers may be calibrated by adjusting each one to discharge the amount of granules over the 300-foot course.

b. Shop Calibration

PTO driven granular applicators can be calibrated in the shop by calculating the distance that will be traveled in 1 minute, then collect
granules for 1 minute with the applicator running at field speed.

Example: The application will be run at 3 mph in the field and the
amount of granules collected in 1 minute at this speed is 1/8 pound
per tube on a three row distributor. The band width will be 12 inches.
The application rate will be:

$$\frac{43,560 \times 0.375}{3 \times (3 \times 88)}* = 20.6 \text{ lb. formulation per acre}$$

c. Calibrating Granular Insecticides

The label on some granular insecticides recommends applying so many
ounces of formulation per 1,000 feet of row. Calibration is complete
when the desired amount of formulation is collected in the prescribed
distance.

Many companies offer calibration tubes for their materials which can
be attached to the distributor tube and will give a direct reading for
that particular formulation. These tubes can be used only for the
designated chemical.

d. Small Area Calibration on a smooth, hard surface

Set the spreader at a setting you feel should be correct. Apply the
material evenly over a 100 sq. ft. area. Sweep up the material and
weigh it. This will give you the weight of material per 100 sq. ft.
Multiply the weight by 10 and you have the amount of material the
spreader will apply at that setting for 1,000 sq. ft. For example,
if you picked up one pound of material for the 100 sq. ft., then you
were applying 10 lbs. per 1,000 sq. ft. If the desired rate of application is higher than 10 lbs. per 1,000 sq. ft., increase the setting
on the spreader and repeat the procedure. If the desired rate is less
than 10 lbs. per 1,000 sq. ft., lower the setting and repeat the procedure. Once you have found the correct setting, make a note of it in
your records for future reference.

Some spreaders have attachments that collect the material so you do
not have to sweep it up. Simply weigh the contents of the container
after you have run it over the 100 sq. ft. area.

---

* 88 feet at 1 mph = feet per minute

| AERIAL EQUIPMENT CALIBRATION |

Fixed-wing aircraft and helicopters exhibit similar flight characteristics (wingtip vortex and main rotor vortex). Since the airflow patterns around and in the wake of each aircraft are sufficiently different, each type and series of aircraft needs testing. Changing the horsepower of the engine, the type of propeller or wingtip shape, will change the distribution pattern. Generalizations can be used to guide the operator on nozzle placement or granular disseminator adjustment. Pattern testing is needed to check the effect of each feature added to the aircraft.

Pattern tests should be made in calm air to avoid cross-wind distortion. If wind is unavoidable, the tests should be made in a direction parallel to the wind. Testing should be carried out in winds less than 3 mph at all times. The best time for this is in the early morning before the sun heats up the ground, creating eddies and inversions. The tests must duplicate the use for which the application is required in terms of airspeed, height of flight, nozzle pressure or gate setting for granulars, nozzle angle and placement or disseminator adjustment, etc. It is better to test with the same materials to be applied if at all possible.

Substitute materials do not always act in quite the same manner as the chemicals. This is evident with granulars, where minor changes in the surface characteristics of the granules (shape, surface finish, fineness or grind, etc.) alter the discharge rate.

Spray Testing

The nozzle type and pressure should be selected for the material being used and the atomization required for the job. The application rate (gallons per acre) will be set by the chemical being applied and crop being treated as listed in chemical recommendation handbooks or on the manufacturer's label. Since each aircraft exhibits a normal or effective swath width, this value should be used with the following tables to determine the acres per minute being treated.

Computation of Acreage and Materials

Formula:

$$\text{Acres covered} = \frac{\text{Length of Swath (miles) X Width of Swath (feet)}}{8.25}$$

The number of acres in a swath of given width and length can be determined from the chart on the following page.

Example: An aircraft with a 40-foot effective swath treats a strip 1 mile long. To find the number of acres, follow the 40-foot vertical column down until it intersects the 1-mile line. The answer to the nearest tenth is 4.8 acres. For swath widths other than those shown interpolate or use combinations of the figures shown. To determine the amount of chemical required, multiply the acres by the desired rate of application.

ACREAGE CHART

Swath Length (Miles)	30'	35'	40'	45'	50'	75'	100'	200'
1/4	.9	1.1	1.2	1.4	1.5	2.3	3.0	6.1
1/2	1.8	2.1	2.4	2.7	3.0	4.5	6.1	12.1
3/4	2.7	3.2	3.6	4.1	4.6	6.8	9.1	18.2
1	3.6	4.2	4.8	5.5	6.1	9.1	12.1	24.2
2	7.2	8.4	9.8	10.9	12.1	18.2	24.2	48.5
3	10.8	12.6	14.5	16.4	18.2	27.3	36.4	72.7
4	14.4	16.8	19.4	21.8	24.2	36.4	48.5	97.0
5	18.0	21.0	24.2	27.3	30.3	45.5	60.6	121.1

Aircraft Calibration

Formula: Acres per minute = $\dfrac{2 \times \text{Swath Width} \times \text{Miles Per Hour}}{1{,}000}$

The chart that follows shows the rate, in acres per minute, at which spray or dry material can be applied when swath width and speed of aircraft are known. For swath widths or aircraft speeds other than those shown, interpolate or use combinations of the figures shown. To find the rate of flow in gallons per minute or pounds per minute, multiply the acres per minute figure by the number of gallons or pounds per acre to be applied.

Example: A 100 mile per hour aircraft has a 40-foot effective swath. Follow the vertical 40-foot column down until the figure opposite 100 miles per hour is intersected. The aircraft would cover 8.0 acres per minute. If 1 gallon of spray is to be applied per acre, the aircraft should be calibrated to disperse liquid at the rate of 1 X 8.0 or 8.0 gallons per minute. (If 10 pounds of dry material is to be applied per acre, the aircraft should be calibrated to disperse material at the rate of 10 X 8.0 or 80 pounds per minute.)

ACRES PER MINUTE CHART

Speed (M.P.H.)	Acres per Minute Covered for a Given Swath Width							
	30'	35'	40'	45'	50'	75'	100'	200'
40	2.4	2.8	3.2	3.6	4.0	6.0	8.0	16.0
50	3.0	3.5	4.0	4.5	5.0	7.5	10.0	20.0
60	3.6	4.2	4.8	5.4	6.0	9.0	12.0	24.0
70	4.2	4.9	5.6	6.3	7.0	10.5	14.0	28.0
80	4.8	5.6	6.4	7.2	8.0	12.0	16.0	32.0
90	5.4	6.3	7.2	8.1	9.0	13.5	18.0	36.0
100	6.0	7.0	8.0	9.0	10.0	15.0	20.0	40.0
110	6.6	7.7	8.8	9.9	11.0	16.5	22.0	44.0
120	7.2	8.4	9.6	10.8	12.0	18.0	24.0	48.0

To determine gallons (or pounds) per acre discharged from the aircraft,
divide the gallons (or pounds) per minute discharged by the acres per
minute that the aircraft covers in a swath.

Knowing the gallons per minute required, the number of nozzles can be
calculated based on the manufacturer's data for that type and pressure.
The pressure and the airspeed are now fixed for the tests and the application.

Discharge Calibration

Having installed the desired type, size and number of nozzles, the output
of the system should be checked to see that the correct discharge in gallons per minute is taking place. If the pump can be run at operating
speed with the aircraft stationary, nozzle discharge can be checked with
a measuring container and a stop watch. Boom pressure must remain constant. If this stationary test cannot be done, the aircraft should be
parked and the tank(s) filled with water to a suitable mark. The aircraft is flown and the spray system is run for a timed period (30, 60,
90 or 120 seconds). The aircraft is brought back to the same point used
previously and the amount of water used is determined by reading the
tank scale(s), or refilling to the first mark using measuring devices.

Swath Pattern Tests

With the application rate now established, the swath pattern should be
checked to see that the distribution across the swath is as uniform as
possible.

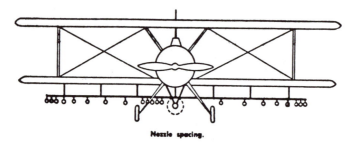

Nozzle spacing.

The best method of spray pattern testing consists of adding a tracer
(dye, fluorescent material, etc.) to water in the tank(s) of the aircraft. The aircraft is then flown at the chosen airspeed and height and
the spraying system is operated at the chosen pressure. One pass is
made over a row of target plates or cards laid out at right angles to the
direction of flight. The aircraft flies over the center of the target
line 100 to 150 feet wide. The targets are collected and the spray deposit on each target is measured by the quantity of tracer. From the results, the distribution curve of the pattern can be determined. Corrections to the nozzle location can be made and the results checked
by further testing.

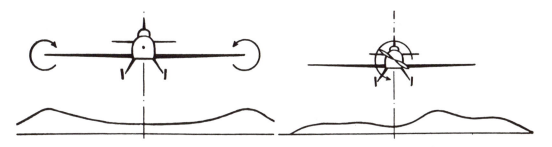

Deposit tendency:
Excessive spray in wingtip
vortices.

Deposit tendency:
Pattern distorted by pro-
peller airstream.

A less satisfactory method is to lay out a roll of paper tape (adding machine tape) and visually inspect the resultant pattern. Interpretation of the spray pattern using this method is at best, only a rough estimate of the uniformity of the deposit pattern.

<u>Droplet Size</u>

Controlling droplet size is the key to success in applying liquid chemicals from aircraft. Droplet size is affected by the following factors:

-- shape of the nozzle cone, flat fan, whirljet, flood, etc.
-- size of the tip
-- spraying pressure (20,40 or 80 psi)
-- liquid or mixture of liquids being used (water, emulsion, chemical wetting agent, evaporation control agent, etc.)
-- angle of the nozzle with respect to the direction of flight(down, forward, back).
-- airspeed of the aircraft
-- density, viscosity and surface tension of the liquid
-- evaporative conditions in the air, between the point of release from aircraft and the point of impingement on ground

All commercial spraying nozzles tend to produce a range of droplet sizes The hydraulic nozzles (flat fan, hollow cone, solid cone, floodjet, etc.) produce a broad range of sizes. The choice of these nozzles shifts the range as a whole as well as the width of the range of sizes. The spinning nozzles, using rotating discs, screens or brushes, produce a narrower range than the hydraulic nozzles. Overloading the spinning element will produce larger droplets and wider range of droplets.

Pesticide work normally requires fine droplets to be effective, depending on the mode of operation of the chemical. This provides a large number of droplets. Herbicides (and systemic insecticides) require coarse droplets so that the plant will absorb the chemical. Fertilizer slurries can use very coarse droplets where they are normal NPK mixtures to be absorbed by the roots. Foliar fertilizers, however, should be applied like herbicides.

To increase droplet size:

--lower spray pressure (not below 20 psi) and add nozzles to keep application rates up;

or --rotate the spray boom so that the nozzles discharge down and back with respect to the direction of flight;

or --change to larger tips of the same type, adjusting the number of nozzles being used;

or --use thickening agents in the spray;

or --stop spraying until cooler or calmer weather exists.

The control of drift is making the use of spray thickeners popular, reducing the number of fine droplets that create problems. These agents are often used with herbicides where sensitive crops are growing in the adjacent fields. Some sprays and spray mixtures demand fine droplets because they are phytotoxic to the crop when they are applied to the crop in coarse droplets (e.g., cover sprays in orchards).

To reduce droplet size:

--increase spray pressure (not above 60 psi)

or --rotate spray boom so that nozzles discharge down and forward with respect to the direction of flight;

or --change to smaller tips of the same type, adjusting the number of nozzles being used to keep the gallonage constant;

or --use thinning agents (wetting agents) in the spray.

It should be realized that all these changes affect the pattern and the rate of application with the exception of spray boom rotation. Rotation affects only the droplet size with little or no effect on the pattern. The other changes will interact to affect the application unless you compensate for them.

Droplet size can be varied by changing the direction of the nozzle orifice in relation to the airstream. For example, when the same orifice and pump pressure are used, the coarsest droplets can be made by aiming the orifice straight back (see Figure 1); somewhat smaller droplets can be made by aiming the orifice straight downward (see Figure 2); and the finest droplets can be made by aiming the orifice forward and downward at about 45 degrees (see Figure 3); aiming the nozzle more directly into the airstream will cause the spray to collect on the body of the nozzle or attachments and fall in large drops (see Figure 4). This wastes material.

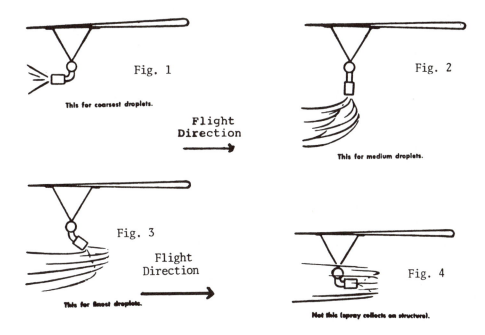

Fig. 1

This for coarsest droplets.

Fig. 2

This for medium droplets.

Flight
Direction
⟶

Fig. 3

This for finest droplets.

Fig. 4

Not this (spray collects on structural).

Flight
Direction
⟶

Granular Materials Testing

Disseminators are sensitive to adjust and the differences between granular
materials have a pronounced effect on the rate of delivery and the pattern.
Some disseminators are restricted as to quantity or type of materials being
handled. These limitations should be checked before testing.

Discharge Calibration

Several runs should be made with the disseminating equipment installed to
determine the quantity of material metered out for a given gate setting.
If the disseminating equipment can be run with the aircraft on the ground,
the material can be caught in large linen or paper bags and weighed. Ram-
air disseminators require actual flight tests to get true discharge rates
since the air currents and the engine vibration in flight affect the metering gate discharge rate. After running the disseminator for a given time
(30,60,90,120 seconds) collected material is weighed. If flight tests
are used, the quantity needed to refill the hopper is weighed. Where
flight is needed to calibrate the system, use blank granulars (the granular
carrier without the pesticide of the same type used to carry the chemical).
Test for three gate settings to determine the gate setting that will give
the required discharge in pounds/minute of granular material. Use the figures in the charts on the previous pages to convert pounds per minute discharge to pounds per acre applied.

Obstructions

Where obstructions occur (trees, power and telephone lines or buildings) at the beginning or end of the swath, it is preferable to turn the equipment on late or shut off early. Then when the field is completed, fly one or two swaths crossways (parallel to the obstruction) to finish out the field. Do not run the disseminator when dropping in or pulling out of the field --the pattern will be distorted. Obstructions inside the field should be treated in the same way. Skip the treatment as you avoid the obstruction, then at the finish, come back and spot treat the skipped part flying at right angles to the rest of the job.

Areas adjacent to buildings, residences and livestock should be treated with extra care. Try to fly parallel to the property line, leaving a border of untreated crop, to avoid possible drift onto unwanted areas. Adjust pullout and drop in paths and avoid making turns over houses. Use caution when fields include or adjacent to waterways, canals, or reservoirs. Treat fields with care if sensitive crops are planted next to them. Be certain that beekeepers are warned if they have beehives near the field to be treated and you are applying chemicals harmful to the bees.

Ferrying

Ferrying height between airstrip and worksite should be at least 500 feet, whether loaded or empty. When possible avoid flight over farm buildings feedlots or residential areas.

Speed

Calibration of the dispersal apparatus on the aircraft is actually rate of flow per minute. No device is at present available to aerial applicators to change flow rate automatically and proportionately as the speed of flight changes. Once the dispersal apparatus has been calibrated, the speed of flight should be kept at the calibrated speed as closely as possible during each swath run. Increasing the speed will result in too light a deposit; decreasing the speed will result in over application.

Height

Height of flight during application is usually governed by the form of chemical being applied. The amount of drift is increased as the height of application is increased. When deposit pattern and calibration have been set, the height of the application should be at the same height from which deposit pattern and calibration was made. Keep this height constant during each swath run to obtain uniform coverage. Whenever possible, application height should be increased up to a height equal to one wing span (where this can be done without increasing drift hazard or decreasing penetration of chemical into foliage canopy). This increases safety of application and swath width.

The following reference is highly recommended:

The Use of Aircraft in Agriculture
By:Norman B. Akesson & Wesley E. Yates
    Food and Agriculture Organization of the United Nations, Rome.1974

Can be obtained by writing the Extension Service, University of California, Berkeley, California.

13
PESTICIDE CALCULATIONS AND USEFUL FORMULAE

Pesticide Calculations & Useful Formulae

The success of a spraying operation, whether you are spraying small areas or large fields, depends upon accurate control of the application rate. After the equipment is accurately calibrated to apply the volume of spray desired, you must determine how much chemical to put in the tank to apply the correct dosage recommended.

In mixing a finished spray, it is most important to add the correct amount of pesticide to the tank. Too little may result in a poor job, while too much may result in injury to the treated surface, illegal residues on food crops, or unnecessary expense. Directions for mixing are given on the label and only very simple calculations are necessary.

To determine how much chemical to put in the tank it is necessary to know the capacity of the tank so that you can determine the number of acres that can be sprayed with one tankful of spray. This is found by dividing the capacity of the tank (in gallons) by the gallons applied per acre as determined when you calibrated the sprayer. For example, if the capacity of the tank is 200 gallons and the sprayer is applying 10 gallons per acre, you know that 200 gallons divided by 10 gallons per acre equals 20 acres that you can spray per tankful.

Most, but not all, recommendations are made on pounds of active ingredient to be applied per acre. To determine the amount of chemical to add to the tank, you multiply the acres one tank will spray by the recommended rate per acre. For example, if the tank will spray 20 acres and the recommended rate is 1 1/2 pounds of active ingredient per acre, you simply multiply 20 times 1 1/2 to arrive at 30 pounds of active ingredient per tank to spray 20 acres. The following are calculations and formulas for various pesticide mixing.

CALCULATIONS FOR MIXING

Liquid Mixing

The amount of active ingredient in liquid formulations is usually expressed as pounds of active ingredient per gallon. To determine the total material needed to treat field crops at a certain rate of active ingredient per acre as specified on the label calculate as follows:

$$\frac{\text{No. of acres(area actually treated) X rate per acre}}{\text{lbs. of active ingredient per gallon concentrate}} = \frac{\text{No. gallons of}}{\text{concentrate needed}}$$

Example: To treat 60 acres with a pesticide at a rate of 1.5 pounds per acre using a concentrate that contains 4.0 pounds per gallon will require 22.5 gallons of concentrate.

$$\frac{60 \times 1.5}{4} = \frac{22.5 \text{ gallons of the concentrate needed to treat the 60 acre}}{\text{field.}}$$

Wettable Powder Mixing

Wettable powders have active ingredients expressed as a percentage of the total weight and may vary from 25% to 80% active ingredient. To determine the total amount of material needed at a certain rate of active ingredient per acre as specified on the label calculate as follows:

$$\frac{100 \; X \; number \; of \; acres \; X \; rate \; per \; acre}{percent \; strength} = \frac{number \; of \; pounds \; of \; material}{needed}$$

Example: To treat 60 acres with a pesticide at a rate of 2.0 pounds per acre using a wettable powder that is 80 percent active ingredient will require 150 pounds of the wettable powder.

$$\frac{100 \; X \; 60 \; X \; 2.0}{80} = 150 \; pounds \; of \; the \; wettable \; powder \; to \; treat \; the \; 60-acre \; field.$$

Percentage Mixing

Sometimes you will find directions telling you to make a finished spray of a specific percentage, for instance a 1% spray. The pesticide may be formulated as a 57% emulsifiable concentrate. To make a 1% finished spray you would add 1 part of pesticide to 56 parts of water. For example, 1 fluid ounce in 56 fluid ounces (1 3/4 quarts) of water.

When mixing percentages you should remember that 1 gallon of water weighs about 8.3 pounds and 100 gallons weigh about 830 pounds. Thus, to make a 1% mix of pesticide in 100 gallons of water you must add 8.3 pounds of active ingredient of pesticide to 100 gallons of water. The formulas are as follows:

a. Formula for Wettable Powder Percentage Mixing

To figure amount of wettable powder to add to get a given percentage of active ingredient (actual pesticide) in the tank:

$$\frac{(gals. \; of \; spray \; wanted) \; X \; (\% \; active \; ingredient \; wanted) \; X \; 8.3 \; (lbs/gal)}{(\% \; active \; ingredient \; in \; pesticide \; used)}$$

Example: How many pounds of an 80% wettable powder are needed to make 50 gallons of 3.5% spray for application by mist blower?

$$\frac{50 \; (gals \; wanted) \; X \; 3.5 \; (\% \; wanted) \; X \; 8.3 \; (pounds/gallon)}{80 \; (\% \; active \; ingredient)} = \frac{1452.5}{80} = 18.1 \; lbs \; 80\% \; WP$$

b. Formula for Emulsifiable Concentrate Percentage Mixing

To figure amount of emulsifiable concentrate to add to get a given percentage of active ingredient (actual pesticide) in the tank.

$$\frac{(gals. \; of \; spray \; wanted) \; X \; (\% \; active \; ingredient \; wanted) \; X \; 8.3 \; (lbs/gal)}{(pounds \; of \; active \; ingredient \; per \; gallon \; of \; concentrate) \; X \; 100}$$

Example: How many gallons of 25% emulsifiable concentrate (2 pounds pesticide per gallon) are needed to make 100 gallons of 1% spray?

$$\frac{100 \text{ (gallons wanted) X 1 (\% wanted) X 8.3 pounds/gallon}}{2 \text{ (lbs active) X 100}} = \frac{830.0}{200}$$

$$= 200\overline{)\begin{array}{r} 4.15 \\ 830.00 \\ 800 \\ \hline 300 \\ 200 \\ \hline 1000 \end{array}}$$

= 4.15 gallons of 25% EC

c. Preparation of a Cattle Spray From a Wettable Powder

To obtain the quantity of pesticide needed to give a desired strength in the diluted spray, use the following formula:

$$\frac{\text{No. Cattle X gal. spray to be used/head X 8.3 X strength of spray desired}}{\text{Strength of wettable powder concentrate}} =$$

= lbs. wettable powder concentrate needed

Example: 150 head of cattle are to be sprayed with a .5 percent pesticide spray at the rate of 1 gallon of spray per animal. 50% pesticide wettable powder is to be used.

$$\frac{150 \text{ X } 1 \text{ X } 8.3 \text{ X } .5}{50} = \frac{12.45 \text{ lbs. of 50\% pesticide needed to prepare}}{150 \text{ gals. of 0.5\% pesticide finished spray}}$$

d. Preparation of a Spray From Wettable Powder to Treat an Orchard

Figured exactly as above except trees are substituted for cattle. The number of gallons used per tree will vary depending on size of tree.

e. Oil Sprays

To treat a certain number of trees with a certain percent oil spray during the dormant season.

NOTE: *Oil is figured on a volume basis.*

$$\frac{\text{No. trees X gals. finished spray per tree X strength of spray desired}}{\text{strength of oil concentrate}}$$

Example: An orchardist wishes to spray 200 peach trees with a 3% dormant oil spray. The oil emulsion concentrate contains 85% oil. The trees require 3 gallons of spray for good coverage.

$$\frac{200 \text{ X } 3 \text{ X } 3}{85} = \frac{21.2 \text{ gals. of 85\% oil emulsion concentrate required to}}{\text{make 600 gals. of 3 percent finished oil spray.}}$$

Dust Mixing

To figure the lbs. of insecticide needed to mix a dust containing a given percent of active ingredient:

% active ingredient wanted X lbs. of mixed dust wanted
% active ingredient in insecticide used

Example: 5 pounds of 3% dust are wanted. How much talc should be added to a 50% active ingredient insecticide powder to make the dust?

$$\frac{3 \times 5}{50} = 0.3 \text{ lb. of the 50\% insecticide}$$

Then add 4.7 lbs. talc to make the 5 lbs. 3% dust.

Granular Materials Calculation

These materials have active ingredients expressed as a percentage of the total weight and are usually manufactured in strengths of 5% to 15%. To determine how much formulation (commercial product) you need to apply to meet the recommended rate expressed in pounds of active ingredient per acre the following formula is used:

$$\frac{\text{Pounds/acre desired}}{\text{\% active ingredient}} = \text{Pounds of commercial product to use per acre}$$

Example: 40 acres are to be treated with a 15% granular pesticide at the rate of 1 lb per acre active ingredient. How many total pounds of commercial product are needed.

$$\frac{1.0}{.15} \times 40 = 6.66 \text{ lb/acre} \times 40 \text{ acres} = 266.4 \text{ pounds}$$

Square Feet Calculation and Mixing

Some labels will give mixing instructions in terms of amounts of active ingredient to be applied per 1000 square feet. This is often the case with lawn, golf course and turfgrass treatments. The calculation for 1000 square feet areas can be carried out as follows:

Example: If an insecticide contains 4 lbs. of active ingredient per gallon and the recommended rate of application is 0.5 lb. of active ingredient per acre, how much would be needed (in ozs.) to treat 20,000 square feet?

If one gallon of the insecticide contains 4 lbs. of active ingredient then one gallon can be used to treat 8 acres at the rate of 0.5 lb. active ingredient per acre.

    1 gallon = 4 lbs. of active ingredient

    1/8 gallon = 0.5 lb. of active ingredient

The next step is to convert gallons to ounces. There are 128 ozs. per gallon, so 1/8 gallon = 16 ozs. Sixteen ozs. of the insecticide would contain 0.5 lb. of active ingredient and would treat one acre.

Next, determine what percent of an acre is 20,000 square feet

$$\frac{20,000}{43,560} = 46\%$$

Since 16 ozs. of the insecticide will treat one acre and the area to be treated (20,000 sq. ft.) is 46% of an acre, then 46% of 16 ozs. will treat the area at the recommended rate.

.46(46%) x 16 ozs. = 7.36 ozs.

7.36 ozs. of the insecticide will treat 20,000 sq. ft. at the rate of 1 lb. of active ingredient per acre.

Example: The label on a liquid pesticide indicates a 25% concentration of the active ingredient. The recommended rate of application is 2 ozs. of active ingredient per 1000 square feet. How much of the pesticide will it take to treat a 5000 square feet turf area?

Since the pesticide is only 25% active this means that the rate of application for the pesticide must be 4 times ($\frac{100\%}{25\%}$ = 4) the rate for the active ingredient.

2 oz. a.i./1000 sq. ft. X 4 = 8 oz. of total pesticide per 1000 sq. ft.

Turf area is 5000 sq. ft. so 5 X 8 oz. = 40 oz. of pesticide required.

USEFUL FORMULAE AND TABLES

Figuring Volume of Ponds

Before treating a pond it is necessary to determine the pond volume in order to calculate the amount of a pesticide needed. The volume of a pond is based on the surface acreage plus the average depth of the water.

Surface acreage of ponds can be found as follows:

1) By comparing one pond with another pond of the same shape and size for which the acreage is known.

2) If a pond is rectangular in shape, the surface acreage equals the length in feet times the width in feet divided by 43,560, that is:

$$\text{Surface acres} = \frac{\text{length in feet X width in feet}}{43,560}$$

Example: If the pond were 200 feet on each side and 100 feet wide at each end, the surface acreage would be:

$$\frac{200 \times 100}{43,560} = \frac{20,000}{43,560} = 0.46$$

3) If a pond is circular in shape, measure the total distance (in feet) around the edge of the pond. Multiply this number by itself, and divide by 547,390.

$$\text{surface acres} = \frac{(\text{total feet of shoreline})^2}{547,390}$$

Example: A round pond with a total distance around the edge of 600 feet -- the surface acreage is:

$$\frac{600 \times 600}{547,390} = \frac{360,000}{547,390} = 0.66 \text{ acres}$$

Most farm ponds have uniformly sloping bottoms. Thus the average depth may be found by dividing the greatest depth of water by two. The volume of the pond can then be determined using the formula:

Volume in acre-feet = surface acreage X 1/2 maximum depth

Example: A pond has a surface area of 0.5 acres, and the greatest depth is 10 feet. The volume of water in this pond is:

$$0.5 \times 5 = 2.5 \text{ acre feet}$$

Based on the volume of a pond, the following tables can be used to determine the correct amount of a pesticide that must be added to obtain control.

POUNDS OF PESTICIDE TO USE FOR OBTAINING THE DESIRED CONCENTRATION
IN PONDS OF VARIOUS VOLUMES

Pond Volume		Concentration of Pesticide Desired, in parts per million							
Acre-feet	Gallons	1/10 ppm	1/4 ppm	1/2 ppm	3/4 ppm	1 ppm	5 ppm	10 ppm	40 ppm
0.1	32,585	0.03	0.07	0.14	0.2	0.3	1.4	2.7	10.9
0.2	65,170	0.05	0.14	0.27	0.4	0.5	2.7	5.4	21.7
0.3	97,755	0.08	0.20	0.41	0.6	0.8	4.1	8.1	32.6
0.4	130,340	0.11	0.27	0.54	0.8	1.1	5.4	10.9	43.4
0.5	162,925	0.14	0.34	0.68	1.0	1.4	6.8	13.6	54.3
0.6	195,510	0.16	0.41	0.82	1.2	1.6	8.2	16.3	65.2
0.7	228,095	0.19	0.48	0.95	1.4	1.9	9.5	19.0	76.0
0.8	260,680	0.22	0.54	1.09	1.6	2.2	10.9	21.7	86.9
0.9	293,265	0.24	0.61	1.22	1.8	2.4	12.2	24.4	97.8
1.0	325,850	0.27	0.68	1.36	2.0	2.7	13.6	27.2	108.6
2.0	651,700	0.54	1.36	2.72	4.1	5.4	27.2	54.3	217.3
3.0	977,550	0.81	2.04	4.08	6.1	8.1	40.8	81.5	325.9

DATA FOR CHANGING HUNDREDTHS OF POUNDS TO OUNCES

Hundredths of pounds	Corresponding number of ounces
0.06	1
0.13	2
0.19	3
0.25	4
0.31	5
0.37	6
0.44	7
0.50	8
0.56	9
0.62	10
0.69	11
0.75	12
0.81	13
0.88	14
0.94	15
1.00	16

CONVERSION TABLES

Linear Measure

1 inch	=	2.54 centimeters
12 inches = 1 foot	=	0.3048 meter
3 feet = 1 yard	=	0.9144 meter
5 1/2 yards or 16 1/2 feet = 1 rod	=	5.029 meters
1,760 yards or 5,280 feet = 1 (statute) mile	=	1,609.3 meters

Square Measure

1 square inch	=	6.452 square centimeters
144 square inches = 1 square foot	=	929 square centimeters
9 square feet = 1 square yard	=	0.8361 square meter
30 1/4 square yards = 1 square rod	=	25.29 square meters
160 square rods or 4,840 square yards or 43,560 square feet	= 1 acre	= 0.4047 hectare
640 acres = 1 square mile	=	259 hectares or 2.59 square kilometers
16.5 feet = 1 rod		
272 square ft. = 1 square rod		

Avoirdupois Weight

```
437.5 grains = 1 ounce           =  28.3495 grams
   16 ounces = 1 pound           = 453.59   grams
  100 pounds = 1 hundredweight   =  45.36   kilograms
2,000 pounds = 1 ton            = 907.18   kilograms
```

Liquid Measure

```
 3 teaspoonfuls = 1 tablespoonful                =  14.8   ml
 2 tablespoonfuls = 1 fluid ounce                =  29.6   ml
16 tablespoonfuls = 1 cup                        = 237     ml
 8 fluid ounces = 1 cup
         2 cups = 1 pint (16 fluid ounces)       = 473.167 ml.
        2 pints = 1 quart                        =   0.946 liters
       4 quarts = 1 gallon (8.34 pounds water)   =   3.785 liters
```

Application Factors

```
1 cup per square rod = 10 gallons per acre
1 pint per square rod = 20 gallons per acre
1 quart per square rod = 40 gallons per acre
1 gallon per square rod = 160 gallons per acre
```

FIGURING CAPACITY OF SPRAYER TANKS

The capacity of tanks of hand or power sprayers in gallons can be calculated as follows:

Cylindrical tanks(circular cross section): Multiply length in inches by square of diameter in inches, multiply the product by 0.0034.

Tanks with elliptical cross section: Multiply length in inches by short diameter in inches by long diameter in inches, multiply the product by 0.0034.

Rectangular tanks (square or oblong cross section): Multiply length by width by depth, all in inches; multiply product by 0.004329.

CONVENIENT CONVERSION FACTORS

Multiply	By	To Get
Acres	43,560.	Square feet
Acres	4,840.	Square yards
Bushels	2,150.42	Cubic inches
Bushels	4.	Pecks
Bushels	64.	Pints
Bushels	32.	Quarts
Centimeters	0.3937	Inches
Centimeters	0.01	Meters
Centimeters	10.	Millimeters
Cubic feet	1,728.	Cubic inches
Cubic feet	0.03704	Cubic yards
Cubic feet	7.4805	Gallons
Cubic feet	59.84	Pints (liquid)
Cubic feet	29.92	Quarts (liquid)
Cubic inches	16.39	Cubic centimeters
Cubic meters	1,000,000.	Cubic centimeters
Cubic meters	35.31	Cubic feet
Cubic meters	61,023.	Cubic inches
Cubic meters	1.308	Cubic yards
Cubic meters	264.2	Gallons
Cubic meters	2,113.	Pints (liquid)
Cubic meters	1,057.	Quarts (liquid)
Cubic yards	27.	Cubic feet
Cubic yards	46,656.	Cubic inches
Cubic yards	0.7646	Cubic meters
Cubic yards	202.	Gallons
Cubic yards	1,616.	Pints (liquid)
Cubic yards	807.9	Quarts (liquid)
Feet	30.48	Centimeters
Feet	12.	Inches
Feet	0.3048	Meters
Feet	1/3 or 0.33333	Yards
Feet per minute	0.01667	Feet per second
Feet per minute	0.01136	Miles per hour
Gallons	3,785.	Cubic Centimeters
Gallons	0.1337	Cubic feet
Gallons	231.	Cubic inches
Gallons	128.	Ounces (liquid)
Gallons	8.	Pints (liquid)
Gallons	4.	Quarts (liquid)
Gallons of water	8.3453	Pounds of water
Grains	0.0648	Grams
Grams	15.43	Grains
Grams	0.001	Kilograms
Grams	1,000.	Milligrams
Grams	0.0353	Ounces
Grams per liter	1,000.	Parts per million
Inches	2.54	Centimeters
Inches	0.08333	Feet
Inches	0.02778	Yards
Kilograms	1,000.	Grams
Kilograms	2.205	Pounds
Kilometers	3,281.	Feet
Kilometers	1,000.	Meters
Kilometers	0.6214	Miles
Kilometers	1,094.	Yards
Liters	1,000.	Cubic centimeters
Liters	0.0353	Cubic feet
Liters	61.02	Cubic inches
Liters	0.001	Cubic meters
Liters	0.2642	Gallons
Liters	2.113	Pints (liquid)
Liters	1.057	Quarts (liquid)
Meters	100.	Centimeters
Meters	3.281	Feet
Meters	39.37	Inches
Meters	0.001	Kilometers
Meters	1,000.	Millimeters
Meters	1.094	Yards

Multiply	By	To Get
Miles	5,280.	Feet
Miles	320.	Rods
Miles	1,760.	Yards
Miles per hour	88.	Feet per minute
Miles per hour	1.467	Feet per second
Miles per minute	88.	Feet per second
Miles per minute	60.	Miles per hour
Ounces (dry)	437.5	Grains
Ounces (dry)	28.3495	Grams
Ounces (dry)	0.0625	Pounds
Ounces (liquid)	1.805	Cubic inches
Ounces (liquid)	0.0078125	Gallons
Ounces (liquid)	29.573	Milliliters (cubic centimeters)
Ounces (liquid)	0.0625	Pints (liquid)
Ounces (liquid)	0.03125	Quarts (liquid)
Parts per million	0.0584	Grains per U. S. gallon
Parts per million	0.001	Grams per liter
Parts per million	8.345	Pounds per million gallons
Pecks	0.25	Bushels
Pecks	537.605	Cubic inches
Pecks	16.	Pints (dry)
Pecks	8.	Quarts (dry)
Pints (dry)	0.015625	Bushels
Pints (dry)	33.6003	Cubic inches
Pints (dry)	0.0625	Pecks
Pints (dry)	0.5	Quarts (dry)
Pints (liquid)	28.875	Cubic inches
Pints (liquid)	0.125	Gallons
Pints (liquid)	0.4732	Liters
Pints (liquid)	16.	Ounces (liquid)
Pints (liquid)	0.5	Quarts (liquid)
Pounds	7,000.	Grains
Pounds	453.5924	Grams
Pounds	16.	Ounces
Pounds	0.0005	Tons
Pounds of water	0.01602	Cubic feet
Pounds of water	27.68	Cubic inches
Pounds of water	0.1198	Gallons
Quarts (dry)	0.03125	Bushels
Quarts (dry)	67.20	Cubic inches
Quarts (dry)	0.125	Pecks
Quarts (dry)	2.	Pints (dry)
Quarts (liquid)	57.75	Cubic inches
Quarts (liquid)	0.25	Gallons
Quarts (liquid)	0.9463	Liters
Quarts (liquid)	32.	Ounces (liquid)
Quarts (liquid)	2.	Pints (liquid)
Rods	16.5	Feet
Square feet	144.	Square inches
Square feet	0.11111	Square yards
Square inches	0.00694	Square feet
Square miles	640.	Acres
Square miles	28,878,400.	Square feet
Square miles	3,097,600.	Square yards
Square yards	0.0002066	Acres
Square yards	9.	Square feet
Square yards	1,296.	Square inches
Temperature ($^{\circ}$C.) +17.98	1.8	Temperature $^{\circ}$F.
Temperature ($^{\circ}$F.) - 32	5/9 or 0.5555	Temperature. $^{\circ}$C.
Ton	907.1849	Kilograms
Ton	32,000.	Ounces
Ton	2,000.	Pounds
Yards	3.	Feet
Yards	36.	Inches
Yards	0.9144	Meters
Yards	0.000568	Miles

PESTICIDE DILUTION TABLE
(Amount of pesticide formulation for each one gallon of water)

Pesticide Formulation	Percentage of Actual Chemical Wanted								
	0.0313%	0.0625%	0.125%	0.25%	0.5%	1.0%	2.0%	3.0	5.0%
Wettable Powder (WP)									
15% WP	2½ t.	5 t.	10 t.	7 T.	1 C	2 C	4 C	6 C	10 C
25% WP	1½ t.	3 t.	6 t.	12 t.	8 C	1 C	2 C	3 C	5 C
40% WP	1 t.	2 t.	4 t.	8 t.	5 T	10 T	1½ C	2 C	3½ C
50% WP	3/4 t.	1½ t.	3 t.	6 t.	4 T	8 T	1 C	1½ C	2½ C
75% WP	1/2 t.	1 t.	2 t.	4 t.	8 t.	5 T	10 T	1 C	2 C
Emulsifiable Concentrate (EC)									
10%-12% EC 1 lb actual/gal	2 t.	4 t.	8 t.	15 t.	10 t.	2/3 pt.	1 1/3 pt.	1 qt.	3½ pt.
15%-20% EC 1.5 lb actual/gal	1½ t.	3 t.	6 t.	12 t.	7½ T	½ pt.	1 pt.	1½ pt.	2' pt.
25% EC 2.5 lb actual/gal	1 t	2 t	4 t	8 t	5 T	10T	2/3 pt	1 pt	1 3/4 pt
33%-35% EC 3 lb actual/gal	3/4 t	1¼ t	3 t	6 t	4 T	8 T	½ pt	3/4 pt	1 1/3 pt
40%-50% EC 4 lb actual/gal	1/2 t	1 t	2 t	4 t	8 t	5 T	10 T	½ pt	4/5 pt
57% EC 5 lb actual/gal	7/16 t	7/8 t	1 3/4 t	3½ t	7 t	4½T	9 T	14 T	1½ C
60%-65% EC 6 lb actual/gal	3/8 t	3/4 t	1/2 T	1 T	2 T	4 T	8 T	12 T	1½ C
70%-75% EC 8 lb actual/gal	1/4 t	1/2 t	1 t	2 t	4 t	8 t	5 T	7½ T	13 T

gal = gallon   1 lb = pound   pt = pint   C = cup   T = tablespoon   t = teaspoon
3 level teaspoonfuls = 1 level tablespoonful   2 tablespoonfuls = 1 fluid ounce
8 fluid ounces or 16 tablespoons = 1 cupful   4 quarts = 1 gallon or 128 fluid ounces
2 cupfuls = 1 pint   2 pints = 1 quart or 32 fluid ounces

DILUTION TABLE FOR PESTICIDES GIVEN IN POUNDS PER ACRE

Amount of Actual Chemical Recommended Per Acre

Amount of Formulation Needed to Obtain the Above Amounts of Actual Chemical

Formulation	1/8 lb.	1/4 lb.	1/2 lb.	3/4 lb.	1 lb.	1 1/2 lbs.	2 lbs.	2 1/2 lbs.	3 lbs.
10% - 12% Emulsion Concentrate (contains one lb. chemical per gal.)	1 pt.	1 qt.	2 qts.	3 qts.	1 gal.	1 1/2 gals.	2 gals.	2 1/2 gals.	3 gals.
15% - 20% Emulsion Concentrate (contains 1 1/4 lbs. chemical per gal.)	1/3 qt.	2/3 qt.	1 1/3 qts.	2 qts.	2 2/3 qts.	1 gal.	1 1/3 gals.	1 2/3 gals.	2 gals.
25% Emulsion Concentrate (contains 2 lbs. chemical per gal.)	1/2 pt.	1 pt.	1 qt.	3 pts.	2 qts.	3 qts.	1 gal.	5 qts.	1 1/2 gals.
40% - 50% Emulsion Concentrate (contains 4 lbs. chemical per gal.)	1/4 pt.	1/2 pt.	1 pt.	1 1/2 pts.	1 qt.	3 pts.	2 qts.	5 pts.	3 qts.
60% - 65% Emulsion Concentrate (contains 6 lbs. chemical per gal.)	1/6 pt.	1/3 pt.	2/3 pt.	1 pt.	1 1/3 pts.	1 qt.	2 2/3 pts.	3 1/3 pts.	2 qts.
70% - 75% Emulsion Concentrate (contains 8 lbs. chemical per gal.)	1/8 pt.	1/4 pt.	1/2 pt.	3/4 pt.	1 pt.	1 1/2 pts.	1 qt.	2 1/2 pts.	3 pts.
25% Wettable Powder	1/2 lb.	1 lb.	2 lbs.	3 lbs.	4 lbs.	6 lbs.	8 lbs.	10 lbs.	12 lbs.
40% Wettable Powder	5 ozs.	10 ozs.	1 1/4 lbs.	1 7/8 lbs.	2 1/2 lbs.	3 3/4 lbs.	5 lbs.	6 1/4 lbs.	7 1/2 lbs.
50% Wettable Powder	1/4 lb.	1/2 lb.	1 lb.	1 1/2 lbs.	2 lbs.	3 lbs.	4 lbs.	5 lbs.	6 lbs.
75% Wettable Powder	1/6 lb.	1/3 lb.	2/3 lb.	1 lb.	1 1/3 lbs.	2 lbs.	2 2/3 lbs.	3 1/3 lbs.	4 lbs.
80% Wettable Powder	2 1/2 ozs.	5 ozs.	5/8 lb	15/16 lb.	1 1/4 lbs.	1 7/8 lbs.	2 1/2 lbs.	3 1/8 lbs.	3 3/4 lbs.
1% Dust	12 1/2 lbs.	25 lbs.	50 lbs.	75 lbs.	100 lbs.	150 lbs.	200 lbs.	250 lbs.	300 lbs.
5% Dust	2 1/2 lbs.	5 lbs.	10 lbs.	15 lbs.	20 lbs.	30 lbs.	40 lbs.	50 lbs.	60 lbs.
10% Dust	1 1/4 lbs.	2 1/2 lbs.	5 lbs.	7 1/2 lbs.	10 lbs.	15 lbs.	20 lbs.	25 lbs.	30 lbs.

CONVERSION TABLES FOR LIQUID FORMULATIONS

Per 1000 square feet

Concentration of Active Ingredient in Formulation,

(cc or tablespoon (tbsp) of formulation per 1000 square feet)

Rate Desired lbs/A	1		2		2.5		3		4		5		6	
	cc	tbsp.	cc	tbsp.	cc	tbsp.	cc	tbsp.	cc	tbsp.	cc	tbsp.	cc	tbsp.
1	87	(5 3/4)	43	(3)	35	(2 1/3)	29	(2)	22	(1 1/2)	17	(1 1/4)	14	(1)
2	173	(11 1/2)	87	(5 3/4)	69	(4 2/3)	58	(3 3/4)	43	(3)	35	(2 1/3)	29	(2)
3	260	(18 1/3)	130	(8 2/3)	104	(7)	87	(5 3/4)	65	(4 1/3)	52	(3 1/2)	43	(3)
4	348	(23 1/4)	174	(11 2/3)	139	(9 1/4)	116	(7 3/4)	87	(5 3/4)	70	(4 2/3)	58	(3 3/4)
5	434	(29)	217	(14 1/2)	174	(11 2/3)	145	(9 2/3)	109*	(7 1/4)	87	(5 3/4)	72	(4 3/4)
6	521	(34 3/4)	260	(17 1/3)	208	(13 3/4)	174	(11 2/3)	130	(8 2/3)	104	(7)	87	(5 3/4)
7	608	(40 1/2)	304	(20 1/4)	243	(16 1/4)	203	(13 1/2)	152	(10)	122	(8)	101	(6 3/4)
8	694	(46 1/4)	347	(23)	278	(18 1/2)	231	(15 1/2)	174	(11 2/3)	139	(9 1/4)	116	(7 3/4)
9	781	(52)	390	(26)	312	(20 3/4)	260	(17 1/3)	195	(13)	156	(10 1/2)	130	(8 2/3)
10	867	(57 3/4)	433	(28 3/4)	347	(23)	289	(19 1/4)	217	(14 1/2)	173	(11 1/2)	144	(9 2/3)

* Example: To spray a 1000 sq ft area at the rate of 5 lb/A active ingredient using a
formulation containing 4 lb/gal active ingredient, use 109 cc or 7¼ tablespoons
of the 4 lb/gal formulation in the amount of carrier your application equipment is applying per unit area (1000 sq. ft.).

CONVERSION TABLES FOR DRY FORMULATIONS

Per 1000 square feet

Rate Desired lb/A	Concentration of Active Ingredient in Formulation												
	100%	90%	80%	75%	70%	60%	50%	40%	30%	25%	20%	10%	5%
	(Grams of formulation per 1000 square feet)												
1	10	12	13	14	15	17	21	26	35	42	52	104	208
2	21	23	26	28	30	35	42	52	69	83	104	208	417
3	31	35	39	42	45	52	63	78	104	125	156	312	625
4	42	46	52	56	60	69	83*	104	139	167	208	417	833
5	52	58	65	69	74	87	104	130	174	208	260	521	1040
6	63	69	78	83	89	104	125	156	208	250	312	625	1250
7	73	81	91	97	104	121	146	182	243	292	364	729	1460
8	83	93	104	111	119	139	167	208	278	333	417	833	1670
9	94	104	117	125	134	156	187	234	312	375	469	937	1870
10	104	116	130	139	149	174	208	260	347	417	521	1040	2080

* Example: To treat a 1000 sq ft area at the rate of 4 lb/A active ingredient
using a formulation containing 50% active ingredient, use 83 grams
of the 50% formulation in the amount of carrier your application
equipment is applying per unit area (1000 sq ft).

• SPRAY CONCENTRATION CONVERSION CHART

Ounces Per 100 Gallons	PPM	% Solution	Grams Per 100 Liters
⅔	50	.005	5
1	75	.0075	7.5
	100	.01	10
2 (⅛ lb.)	150	.015	15
2⅔	200	.02	20
3⅓	250	.025	25
4 (¼ lb.)	300	.03	30
5⅓	400	.04	40
6⅔	500	.05	50
8 (½ lb.)	600	.06	60
9⅓	700	.07	70
10⅔	800	.08	80
12 (¾ lb.)	900	.09	90
13⅓	1000	0.10	100
16 (1 lb.)	1200	0.12	120
20 (1¼ lb.)	1500	0.15	150
24 (1½ lb.)	1800	0.18	180

ADDITIONAL USEFUL TABLES

A DIPSTICK GAUGE FOR 55-GALLON DRUMS*

You will need a 3-foot, flat stick. A length of 1 x 2, or an old yardstick, works fine. Mark the correct gallonage figure directly on the stick for each inch. Half inches can be estimated. Use the appropriate listing below, depending on whether your drum is mounted vertically or sideways.

Horizonal Drum	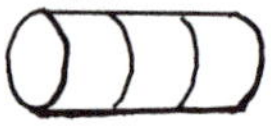	Upright Drum
Inches	Gallons	Inches
22	55	33
21	54	32½
20¼	53	31¾
19¾	52	31¼
19¼	51	30½
18¾	50	30
18¼	49	29½
17¾	48	28¾
17½	47	28¼
17	46	27½
16¾	45	27
16½	44	26½
16	43	25¾
15¾	42	25¼
15¼	41	24½
15	40	24
14¾	39	23½
14¼	38	22¾
14	37	22¼
13¾	36	21½
13¼	35	21
13	34	20½
12¾	33	19¾
12¼	32	19¼
12	31	18½
11¾	30	18
11½	29	17½
11 1/8	28	16¾
10 7/8	27	16¼
10½	26	15½
10¼	25	15
10	24	14½
9¾	23	13¾
9¼	22	13¼
9	21	12½
8¾	20	12
8¼	19	11½
8	18	10¾
7¾	17	10¼
7¼	16	9½
7	15	9
6¾	14	8½
6¼	13	7¾
6	12	7¼
5¾	11	6½
5½	10	6
5	9	5½
4½	8	4¾
4	7	4¼
3¾	6	3½
3¼	5	3
3	4	2½
2½	3	1¾
1¾	2	1¼
1¼	1	½
0	0	0

*Most drums are 22 inches in diameter, 33 inches tall and weigh, when full (55 gallons), 500 pounds.

CONVERSION TABLE FOR GRANULAR RATES

If % of active ingredients in granules is	and recommended rate of active ingredients is	then apply this amount of granules
5 %	½ lb/acre	10 lbs./acre
5 %	1 lb/acre	20 lbs./acre
5 %	1 ½ lbs./acre	30 lbs./acre
5 %	2 lbs./acre	40 lbs./acre
5 %	3 lbs./acre	60 lbs./acre
5 %	4 lbs./acre	80 lbs./acre
10 %	½ lb./acre	5 lbs./acre
10 %	1 lb./acre	10 lbs./acre
10 %	1 ½ lbs./acre	15 lbs./acre
10 %	2 lbs./acre	20 lbs./acre
10 %	3 lbs./acre	30 lbs./acre
10 %	4 lbs./acre	40 lbs./acre
20 %	½ lb./acre	2 ½ lbs./acre
20 %	1 lb./acre	5 lbs./acre
20 %	1 ½ lbs./acre	7 ½ lbs./acre
20 %	2 lbs./acre	10 lbs./acre
20 %	3 lbs./acre	15 lbs./acre
20 %	4 lbs./acre	20 lbs./acre
25 %	½ lb./acre	2 lbs./acre
25 %	1 lb./acre	4 lbs./acre
25 %	1 ½ lbs./acre	6 lbs./acre
25 %	2 lbs./acre	8 lbs./acre
25 %	3 lbs./acre	12 lbs./acre
25 %	4 lbs./acre	16 lbs./acre

WIDTH OF AREA COVERED TO ACRES PER MILE TRAVELED

Width of Strip (feet)	Acres/mile
6	.72
10	1.21
12	1.45
16	1.93
18	2.18
20	2.42
25	3.02
30	3.63
50	6.04
75	9.06
100	12.1
150	18.14
200	24.2
300	36.3

AIDS IN CONVERTING DOSAGE, VOLUME, RATES AND AMOUNTS (APPROX.)

WETTABLE POWDERS

RATES		APPROXIMATE AMOUNT FOR LESS THAN ONE ACRE			
Amt. per Acre	Area treated per oz.	⅓ acre	¼ acre	4,000 sq. ft.	1,000 sq. ft.
1 lb.	2722 sq. ft.	5⅓ oz.	4 oz.	1.4 oz.	.37 oz.
2 lb.	1360 sq. ft.	11 oz.	½ lb.	2.9 oz.	¾ oz.
3 lb.	907 sq. ft.	1 lb.	¾ lb.	4⅓ oz.	1.1 oz.
4 lb.	681 sq. ft.	1⅓ lb.	1 lb.	5¾ oz.	1½ oz.
5 lb.	545 sq. ft.	1⅔ lb.	1¼ lb.	7⅕ oz.	1⅘ oz.
6 lb.	454 sq. ft.	2 lb.	1½ lb.	8⅖ oz.	2⅕ oz.
7 lb.	380 sq. ft.	2⅓ lb.	1¾ lb.	10 oz.	2½ oz.
8 lb.	316 sq. ft.	2⅔ lb.	2 lb.	12⅖ oz.	3⅕ oz.
9 lb.	303 sq. ft.	3 lb.	2¼ lb.	12.9 oz.	3⅓ oz.
10 lb.	272 sq. ft.	3⅓ lb.	2½ lb.	14⅖ oz.	3¾ oz.
11 lb.	250 sq. ft.	3⅔ lb.	2¾ lb.	15⅘ oz.	4 oz.
12 lb.	222 sq. ft.	4 lb.	3 lb.	1⅛ lb.	4½ oz.
13 lb.	209 sq. ft.	4⅓ lb.	3¼ lb.	1⅕ lb.	4⅘ oz.
14 lb.	194 sq. ft.	4⅔ lb.	3½ lb.	1¼ lb.	5⅕ oz.
15 lb.	180 sq. ft.	5 lb.	3¾ lb.	1⅓ lb.	5⅗ oz.
16 lb.	170 sq. ft.	5⅓ lb.	4 lb.	1 lb. 7 oz.	6 oz.

LIQUIDS

RATES		APPROXIMATE AMOUNT FOR LESS THAN ONE ACRE			
Amt. per Acre	Area treated per gal.	⅓ acre	¼ acre	4,000 sq. ft.	1,000 sq. ft.
⅛ pt. (12 tsp.)	64 acres	4 tsp.	3 tsp.	1 tsp.	⅛ tsp.
⅙ pt. (16 tsp.)	48 acres	5⅓ tsp.	4 tsp.	1.5 tsp.	⅜ tsp.
¼ pt. (8 TBS.)	32 acres	8 tsp.	2 TBS.	2.3 tsp.	½ tsp.
⅓ pt. (32 tsp.)	24 acres	11 tsp.	8 tsp.	2⅘ tsp.	¾ tsp.
⅖ pt. (38 tsp.)	20 acres	13 tsp.	9½ tsp.	3⅓ tsp.	1 tsp.
½ pt. (16 TBS.)	16 acres	16 tsp.	4 TBS.	4⅓ tsp.	1⅛ tsp.
¾ pt. (24 TBS.)	11 acres	8 TBS.	6 TBS.	6½ tsp.	1¾ tsp.
1 pt. (32 TBS.)	8 acres	⅔ C.	8 TBS.	2⅘ tsp.	2¼ tsp.
1⅓ pt. (43 TBS.)	6 acres	14 TBS.	11 TBS.	3⅘ tsp.	1 TBS.
⅕ gal. (51 TBS.)	5 acres	17 TBS.	13 TBS.	4½ TBS.	4 tsp.
1 qt. (64 TBS.)	4 acres	⅔ pt.	½ pt.	5¾ TBS.	1½ TBS.
⅓ gal. (85 TBS.)	3 acres	⅞ pt.	¾ pt.	⅛ pt.	2 TBS.
2 qt. (128 TBS.)	2 acres	⅔ qt.	1 pt.	⅖ pt.	3 TBS.
3 qt. (190 TBS.)	1⅓ acres	1 qt.	¾ qt.	½ pt.	4½ TBS.
1 gal. (256 TBS.)	1 acre	1⅓ qt.	1 qt.	⅘ pt.	6 TBS.

DETERMINING SMALL QUANTITIES OF SPRAY MATERIALS

LIQUIDS			POWDER		
AMT. PER 100 GAL.	AMT. PER 1 GAL.	APPROX. DILUTIONS	AMT. PER 100 GAL.	AMT. PER 1 GAL.	APPROX. DILUTIONS
¼ pt.	¼ tsp.	1-3200	½ lb.	¾ tsp.	1-1600
1 pt.	1 tsp.	1-800	⅝ lb.	1 tsp.	1-1400
1½ pt.	1½ tsp.	1-550	1 lb.	1½ tsp.	1-800
1 qt.	2 tsp.	1-400	1¼ lb.	2 tsp.	1-700
3 pt.	1 Tbs.	1-233	1½ lb.	2½ tsp.	1-600
2 qt.	4 tsp.	1-200	2 lb.	1 Tbs.	1-400
3 qt.	2 Tbs.	1-150	2½ lb.	4 tsp.	1-350
1 gal.	8 tsp.	1-100	3 lb.	4½ tsp.	1-266
2 gal.	5 Tbs.	1-50	4 lb.	2 Tbs.	1-200
3 gal.	½ cup	1-33	5 lb.	8 tsp.	1-160
4 gal.	⅔ cup	1-25	8 lb.	4 Tbs.	1-100
5 gal.	1 cup	1-20	10 lb.	5 Tbs.	1-80
11 gal.	⅞ pt.	1-9	16 lb.	8 Tbs.	1-50

APPLICATION EQUIVALENTS

1 oz. per sq. ft. = 2722.5 lbs./acre
1 oz. per sq. yd. = 302.5 lbs./acre
1 oz. per 100 sq. ft. = 27.2 lbs./acre
1 lb. per 100 sq. ft. = 435 lbs./acre
1 lb. per 1000 sq. ft. = 43 lbs./acre
1 lb. per acre = ⅓ oz./1000 sq. ft.
5 gal. per acre = 1 pt./1000 sq. ft
100 gal. per acre = 2.5 gals./1000 sq. ft.

AREA AND CUBIC MEASURES

1 acre = 43,560 sq. ft. = 160 sq. rds.
1 acre = 1 mile 8 ft. wide or 2 miles 4 ft. wide
1 square rod = 272 sq. ft. = 30¼ sq. yds.
3⅔ square rods = 1000 sq. ft.
1 square yard = 1,296 sq. in. = 9 sq. ft.
1 square foot = 144 square inches
1 cubic foot = 1,728 cubic inches = 7.48 gal.
1 cubic yard = 27 cubic feet = 202 gal.

LIQUID MEASURES

1 teaspoon (tsp.) = 5 cubic centimeters (cc) = 60 drops
1 Tablespoon (Tbs.) = 3 tsp. = ½ fl. oz. = 15 ml.
2 tablespoons = 1 fluid ounce (fl. oz.)
1 cup = ½ pt. = 16 Tbs. = 48 tsp.
1 pt. = 32 Tbs. = 16 fl. oz. = 28.8 cubic inches
1 gal. = 4 qts. = 231 cu. in. = 8.33 lb. water

WEIGHTS

1 ounce = 2 Tbs. liquid = 28⅓ grams
1 pound = 16 oz. = 454 grams = .453 kilograms
1 gram (gm) = .03527 oz.
1 milligram (mg) = 1/1000 gram
1 kilogram (kg or kilo) = 2.2046 lbs.
1 gallon water = 8.345 lbs.

FIGURING TREES PER ACRE AND TREES PASSED PER MINUTE AT VARIOUS GROUND SPEEDS FOR SEVERAL TREE SPACINGS

Tree Spacing Ft.	10	12	16	20	25	30	35	40
Trees/acre	435.2	302.5	170	108.9	69.7	48.4	35.6	27.2
Trees/Minute at								
1 mph	8.8	7.3	5.5	4.4	3.5	2.9	2.5	2.2
1 1/2 mph ...	13.2	11.0	8.2	6.6	5.3	4.4	3.8	3.3
2 mph	17.6	14.6	11.0	8.8	7.0	5.9	5.0	4.4
2 1/2 mph ...	22.0	18.3	13.7	11.0	8.8	7.3	6.3	5.5
3 mph	26.4	22.0	16.5	13.2	10.6	8.8	7.5	6.6

FIGURING THE COST OF CHEMICALS

Using the following information, which pesticide is a "better buy"?

Example:

Pesticide A	Pesticide B
$1.45/lb	$8.90/lb.
Rate: 8 oz./1000 sq. ft.	Rate: 1 oz./1000 sq. ft.

Since there are 16 oz. in a lb., Pesticide A costs $\frac{\$1.45}{16} = \$.09/oz$

and Pesticide B costs $\frac{\$8.90}{16} = \$.56/oz$

Pesticide A is cheaper per ounce but the rate of application is eight times that of Pesticide B. It will therefore cost 8 X $.09 or $.72 per 1000 sq. ft. to use Pesticide A compared to $.56 per 1000 sq. ft. to use Pesticide B.

Pesticide B is the "Better Buy".

Example:

Pesticide A	Pesticide B
$8.00/gal	$10.60/gal
(4 lb/gal material)	(4 lb/gal material)
Rate: 2 lb/acre	Rate: 3 lb/acre

Pesticide A would cost $4.00 per acre for application since each pound costs $2.00 ($\frac{8}{2}$) and the rate is 2 pounds per acre so 2 X 2 = $4.00.

Pesticide B would cost $7.95 per acre for application since each pound costs $2.56 ($\frac{10.60}{4}$) and the rate is 3 pounds per acre, so 3 X $2.65 = $7.95.

Pesticide A is the "Better Buy".

14
PESTICIDE
TRANSPORTATION
STORAGE
DECONTAMINATION
AND
DISPOSAL

Pesticide Transportation, Storage, Decontamination & Disposal

SAFE TRANSPORT OF PESTICIDES

When transporting toxic chemicals, it is possible for accidents to happen. It is the responsibility of the person transporting the material to take preventive measures to reduce the likelihood of anyone being hurt by ac-

cidental contact with a poison. Ignorance of the nature of the materials being transported should not be a contributing factor to an "accident".

Several precautions should be taken to assure the safe transportation of pesticides and the safety and comfort of those doing the transporting. Pesticides and other chemicals of a volatile nature and those which give off noxious or poisonous fumes should never be transported in a closed vehicle with the passengers or driver. The following suggestions should be helpful in insuring safe handling and transport of pesticides:

1. Transporting Vehicles

    a. The best type of vehicle is the open type such as pick-up truck. Wettable powders and dust formulations in paper containers should be protected from rain. Avoid puncturing or tearing paper containers when handling them during loading, unloading, or storing.

    b. If pesticides are transported in a station wagon, the windows should be left open.

    c. When a sedan is used to transport these kinds of chemicals, and the materials must be carried in the trunk, provisions for adequate ventilation of both the car and the trunk must be made.

    d. Under no circumstances should pesticides be transported inside the truck cab or car. Spills may be impossible to remove from seat covers. Fumes from spilled poisons are hazardous.

2. Containers

    a. Containers of liquid formulations of pesticides and chemicals must

be tightly closed to prevent spillage and should always be in their original containers.

    b. Be sure the outside of the container is not contaminated with the pesticide.

    c. Glass containers should be avoided whenever possible, but if necessary to transport them, extra precautions must be taken to avoid breaking.

c.(continued) Form fitting, plastic foam shipping packages are excellent protection when transporting gallon jugs and other containers. These are often used by chemical companies to ship material.

d. Load the pesticide container so that it does not roll or slide from place to place. Such movement could break or puncture the container. Containers may be transported in a plastic bag, heavy cardboard box, or something similar as a further precaution.

3. Temperatures

a. Pesticides and other chemicals are affected by high and low temperatures. In warm weather the temperature inside the trunk, far exceeds the outside temperature. Materials transported in the trunk should not be allowed to remain in the vehicle after arriving at the destination.

b. Adequate protection against freezing should also be taken in the event of extremely cold weather.

4. Additional Precautions

a. Don't transport weed killing chemicals with other pesticides or fertilizer. A spill may lead to cross contamination.

b. Don't carry pesticides near your groceries. You could poison your food.

c. Don't carry pesticides with feed or livestock food. A spill could contaminate the feed or food.

d. Don't allow children to ride near the pesticides. A spill could result in injury or even death.

e. If there is a spill,clean it up immediately (refer to section on PESTICIDE DECONTAMINATION).

PESTICIDE STORAGE

Proper storage of pesticides will protect the health and well being of people, help protect against environmental contamination, and protect the chemical shelf life. There are a number of fairly rigid conditions that are required for storage of pesticides intended for agricultural and industrial uses. The following are the major items with which everyone should be completely familiar:

- Storage in a separate building is preferable. A separate, isolated storage room can be used in some instances.

- The storage area should be well ventilated and have a source of heat for pesticides that cannot tolerate cold temperatures. When chemicals are subjected to high temperatures, they may expand, causing drum heads to bulge and leak. High temperatures have also been noted to reduce the effectiveness of emulsifiers, speed up container erosion, and, in some instances, cause the pesticide to deteriorate. Low temperatures also cause problems in pesticide storage. Some compounds will freeze and upon expansion rupture metal or glass containers. Other compounds under low temperatures may settle or crystalize out of solution.

- Store all pesticides in their original, labeled containers; never store in a food, feed, or beverage container.

- Keep lids of containers tight and tops of bags closed when containers are not being used.

- Liquid containers should be stored on pallets to avoid rusting of metal containers. Dusts and powders will tend to cake when wet or subjected to extremely high humidity. They too should be stored on pallets to prevent moisture absorption.

- Check all containers frequently for leaks or tears to avoid contamination.

- Try to avoid storing unnecessarily large quantities of pesticides by keeping good records of previous requirements and making good estimates of future needs. When large amounts of pesticides are being stored, records should be kept on the kind and quantity. Make an inventory of all pesticides available in storage and MARK CONTAINERS WITH DATE OF PURCHASE.

Dispose of insecticides showing the following signs of deterioration.

FORMULATION	GENERAL SIGNS OF DETERIORATION
Emulsifiable concentrates	When milky coloration does not occur with the addition of water and when sludge is present or any separation of components is evident.

FORMULATION	GENERAL SIGNS OF DETERIORATION
Oil Sprays	When milky coloration does not occur by adding water.
Wettable Powders	Lumping occurs and the powder will not suspend in water.
Dusts	Excessive lumping.
Granulars	Excessive lumping.
Aerosols	Generally effective until the opening of the aerosol dispenser becomes obstructed.

* Do not store pesticides in the same room with food, feed, or water.

* Do not store herbicides, especially the hormone types, in the same room with other pesticides or fertilizers. Cross contamination may occur.

* Post Warning signs on the outside of all walls, doors, and windows of one pesticide storage facility where they will be readily seen by anyone attempting to enter. Keep the facility locked when not in use.

  * The storage facility should be constructed of fire resistant material and should include portable fire extinguishers and a sprinkler system if possible.

  * Do not store clothes, respirators, lunches, cigarettes, or drinks with pesticides. They may pick up poisonous fumes or dusts or soak up spilled poison.

* Plenty of soap and water should be available in the storage area. Seconds count when washing poisons from your skin.

* If possible, try to avoid storing pesticides from one year to the next. Storage life of various pesticides is considerably different. Some pesticides belonging to the chlorinated hydrocarbon group can be stored for a number of years with little or no chemical change. Other pesticides, however, such as organic phosphorous compounds tend to have a rather short storage life. Climatic conditions such as high temperatures, high humidity, and sunlight may cause a chemical break down or degradation of certain pesticides, especially when they are formulated as wettable powders or dusts. Compounds that are subject to degradation will have labels that provide instructions to prevent it. Sometimes when pesticides have been stored for long periods there is doubt as to their effectiveness. In this case, try small amounts of the pesticide according to label directions. If the treatment is satisfactory, use the pesticide up as promptly as possible.

Minimizing Pesticide Fire and Explosion Hazards

Pesticides with low flash points (140°F or less) are dangerous in storage. Pesticide formulations that have flash points at or below 20°F must have on the label "danger-extremely flammable-keep away from sparks and heated surfaces". Those products with flash points above 20°F but not over 80°F, must be labeled with the statement "warning-flammable!; keep away from heat and open flame". Pesticides with a flash point above 80°F but not exceeding 140°F must bear the statement "Do Not Use or Store Near Heat or Open Flame".

When pesticides are purchased check the label for flash point warnings and make sure that they are used and stored according to the directions to prevent the hazard of fires and explosions.

Pesticides containing oils or aeromatic petroleum distillates are the ones most likely to have such warnings on the label. Certain dry formulations also present fire and explosion hazards. Sodium chlorate is well known for its potential to ignite when in contact with organic matter, sulfur, sulfides, phosphorous, powdered metals, strong acids, or ammonium salts. Whenever a container of sodium chlorate is opened, the entire contents should be used. A container partially full should never be stored. Certain dusts or powders, particularly those that are very fine, such as sulfur, may ignite as easily as gases or vapors. The following are suggestions that should help to reduce fire hazards:

1) Keep the storage area locked at all times when not in actual use to prevent the possibility of fires being accidentally set.

2) Do not store glass containers in sunlight where they might concentrate heat rays and start fires.

3) Store combustible materials away from steam lines and heating devices.

4) Sheet rock or other fire proof materials should be used to line the storage area.

5) Large storage areas should have fire sprinkler systems. Be familiar with fire alarms, fire exits, and fire fighting equipment.

6) Storage areas should be located as far as possible away from other buildings and populated areas.

7) Local firemen should be notified of the contents of the storage facility. It may save their lives and the lives of others in the event of a fire. An emergency fire disaster plan should be developed in cooperation with the local fire department and local physicians for the protection of personnel and property. The following are suggestions that should be helpful to firemen in the event of a fire in a pesticide storage area:

⊕ If a fire occurs in a pesticide storage facility, someone thoroughly familiar with the hazards of resulting smoke, fumes, splashes, or other possible types of contamination should be on hand to warn firemen or anyone in the vicinity.

⊕ Firemen should wear the proper protective clothing and respirators.

⊕ Firemen should attempt to stay upwind of the fire while fighting it.

⊕ If necessary, residents downwind of the fire, should be evacuated.

⊕ Firemen should avoid dragging hoses through pesticide contaminated water.

⊕ Firemen should assume that all equipment used to fight such fires is contaminated and hazardous until decontaminated.

⊕ Firemen should avoid using heavy hose streams if possible, as the force of the stream spreads contamination and will cause dusts to become airborne and present a possible explosion hazard as well as a toxic hazard.

⊕ Firemen should be aware that overheated containers may erupt at any time and should try to keep a safe distance from the fire.

⊕ The runoff of water from a fire involving pesticides should be diked to prevent it from entering sewers or streams.

⊕ Firemen should wash and change clothing immediately after the fire.

⊕ All clothing, boots and other equipment should be thoroughly washed.

⊕ Firemen should prevent curious people from entering burned-out areas by erecting "Toxic Chemical" signs or barriers until clean-up is completed.

⊕ Firemen should check rubble and surrounding areas for evidence of contamination.

PESTICIDE DECONTAMINATION

The subject of decontamination concerns application and protective equipment, personnel, and areas involved in spills, fires, and highway accidents.

Decontamination of Application Equipment

Very little information is available on the best methods for decontamination of application equipment, probably because of the great variety of application equipment in use today. Common sense plus knowledge of the pesticides involved is needed most in dealing with decontamination problems. For example, we know that some chemicals such as the hormone-type herbicides even in minute amounts are injurious to many plants. Thus it is essential that equipment used for application of herbicides be thoroughly decontaminated as far as possible.

Complete decontamination is generally considered impossible for hormone-type herbicides; however, if the equipment must be used for application of other chemicals, the following methods for "cleaning" are offered:

For water soluable herbicide formulations

Method 1. --Add ½ Cup of ammonia(household) to 2 gallons of water. Flush some of the solution through the spray system and out through the booms or nozzle. Let the rest stand overnight in the spray tanks. Empty and thoroughly rinse the tank, hose, booms and/or nozzle.

Method 2. --Add 1½ ounces of sal soda(sodium carbonate or a washing soda) to 2 gallons of water. Allow to stand at least two hours. Discharge through the spray booms or nozzle. Rinse the tank well; refill twice with water and empty through the spray system each time.

For oil soluable formulations

Add 1½ cups of kerosene with a little detergent to sal soda solution (see above) and proceed as in method 2.

For either water or oil soluable formulations

Add 10 ounces of lye to 2 gallons of water and proceed as for method 2. Add activated charcoal (1 ounce, powdered) and 1 ounce of detergent to 2 gallons of water in the sprayer. Agitate for several minutes and then discharge through the spray system.

REMEMBER: Rinse water can kill and damage plants; clean sprayers away from all desired vegetation.

Both mammalian toxicity and persistence of the chemical that is contaminating the equipment must be considered very carefully; knowledge of the pesticide is very important.

One of the simplest ways to reduce the hazards of contaminated equipment is to dilute the chemical with lots of water, being very careful where the water drains.

Most pesticides are subject to chemical degradation in the presence of an alkaline or acid medium. Phosphate pesticides are more readily degradable than chlorinated hydrocarbons or carbamate-type pesticides. Equipment contaminated with aldrin, dieldrin, or endrin cannot be detoxified thoroughly by using water and detergent or lye. Check with the chemical supplier for specific directions regarding the best procedure to be followed for chemically degrading and decontaminating equipment used for spraying chlorinated hydrocarbon material.

Some equipment may be subject to corrosion or deterioration if exposed to strong alkali or acid. If these are used, rinse them off thoroughly as soon as possible and run lots of clean water through the lines and nozzles. Use caution when handling lye; it can cause severe eye and skin burns.

Several companies offer preparations that are designed for tank and equipment cleaning. These are reported to: clean tanks and equipment inside and out; remove rust and deposit buildups; neutralize acids including 2,4-D; and provide a protective film. These cleansing solutions are generally available from pesticide dealers.

Decontaminating Protective Equipment

Decontaminating respiratory devices is accomplished simply by discarding the filter pads and cartridges and washing the respirator with soap and water after each use. After washing, rinse the face piece to remove all traces of soap. Dry the respirator with a clean cloth and place the face piece in a well ventilated area to dry. When you are ready to use the respirator, insert new filters and cartridges.

Rubber boots and gloves should also be washed inside and out with soap and water and rinsed thoroughly daily or more often if contamination has occurred or is suspected.

Rubber or plastic protective pants, coats and headgear should also be washed, rinsed and dried in a method similar to that for respirators.

Decontamination of Clothing

Clothing worn by pesticide applicators should be laundered daily. The problem of how to launder pesticide contaminated clothing is sometimes difficult and is determined on the basis of the toxicity of the chemical and the type of formulation. Recent studies at the University of Nebraska (Environmental Perspectives, 3:4, 1982) show clothing is easily contaminated by pesticides. Once contaminated, it is difficult to remove all the pesticide through home-laundering procedures.

Clothing contaminated with highly toxic and concentrated pesticides must be handled most carefully as these pesticides are easily absorbed through the skin. If the clothes have been completely saturated with concentrated pesticides, they should be discarded. Clothing contaminated by moderately toxic pesticides do not warrant such drastic measures. Hazards are less pronounced in handling clothing exposed to low toxicity pesticides. However, the ease of pesticide removal through laundering does not depend on toxicity level, it depends on the formulation of the pesticide. For example, 2,4-D amine is easily removed through laundering because it is soluable in water while 2,4-D ester is much more difficult to remove through laundering.

Disposable clothing helps limit contamination of clothes because the disposable garments add an extra layer of protection. This is especially important when you are in direct contact with pesticides, such as when mixing and loading pesticides for application.

Laundering Recommendations

Wash contaminated clothing separately from the family wash. Research has shown that pesticide residues are transferred from contaminated clothing to other clothing when they are laundered together. It is vital that the person doing the wash know when pesticides have been used so all clothing can be properly laundered.

Pre-rinsing contaminated clothing before washing will help remove pesticide particles from the fabric. Pre-rinsing can be done by:

1. Pre-soaking in a suitable container prior to washing.
2. Pre-rinsing with agitation in an automatic washing machine.
3. Spraying/hosing garments outdoors.

Pre-rinsing is especially effective in dislodging the particles from clothing when a wettable powder pesticide formulation has been used.

 Clothing worn while using slightly toxic pesticides may be effectively laundered in one machine washing. It is strongly recommended that multiple washings be used on clothing contaminated with concentrated pesticides to draw out excess residues. Always wear rubber gloves or use a stick when handling highly contaminated clothing to prevent pesticide absorption into the body.

Washing in hot water removes more pesticide from the clothing than washing in other water temperatures. Remember--the hotter, the better. Avoid cold water washing. Although cold water washing might save energy, cold water temperatures are relatively ineffective in removing pesticides from clothing.

Laundry detergents, whether phosphate, carbonate, or heavy duty liquids are similarly effective in removing pesticides from fabric. However, research has shown that heavy duty liquid detergents are more effective than other detergents in removing emulsifiable concentrate pesticide formulations. Emulsifiable concentrate formulations are oil-based and heavy duty liquid detergents are known for oil-removing ability.

Laundry additives such as bleach or ammonia, do not contribute to removing pesticide residues. Either of these additives may be used, if desired, but caution must be used. Bleach should never be added to or mixed with ammonia because they react together to form a fatal chlorine gas. Be careful--do not mix ammonia and bleach.

If several garments have been contaminated, wash only one or two garments in a single load. Wash garments contaminated by the same pesticide together. Launder, using a full water level to allow the water to thoroughly flush the fabric.

During seasons when pesticides are being used daily, clothing exposed to pesticides should be laundered daily. This is especially true with highly toxic or concentrated pesticides. It is much easier to remove pesticides from clothing by daily laundering than attempting to remove residues that have accumulated over a period of time. In fact, clothing heavily contaminated by highly toxic pesticide concentrates should be destroyed by burning or burial in a safe place.

Pesticide carry-over to subsequent laundry loads is possible because the washing machine is likely to retain residues which are then released in following laundry loads. It is important to rinse the washing machine with an "empty load" using hot water and the same detergent, machine settings and cycles used for laundering the contaminated clothing.

Line drying is recommended for these items. Although heat from an automatic dryer might create additional chemical breakdown of pesticide residues, many pesticides break down when exposed to sunlight. This also eliminates the possibility of residues collecting in the dryer.

The foregoing information was adapted from University of Nebraska NEB Guide HEG 81-152, "Laundering Pesticide Contaminated Clothing," November, 1981.

Decontamination of Spray Personnel

Decontamination of personnel, especially after a splash or spill of a concentrated chemical, whether on clothing, skin, or in the eyes, should be

handled quickly and efficiently. Speed is particularly important when highly toxic chemicals are involved, but it is only slightly less urgent with the less toxic compounds. Keep in mind the factors that contribute to poisoning: the inherent toxicity of the chemical; its concentration; its physical form; and the length of time exposed. If, for example, you have spilled a highly toxic, concentrated liquid formulation on your skin or clothing, you must get the clothing off immediately and wash thoroughly with soap and water. Speed is essential. Don't forget to wash your hair and under your fingernails.

Alcohol is an excellent decontaminating agent, particularly if the area contaminated is limited. It has been found that thirty minutes after a test application of parathion to the skin, vigorous scrubbing with soap and water will remove 80% or more of the material, and alcohol will remove most of the remainder. After five hours, however, 40% cannot be washed off with soap and water and 10% will remain even after scrubbing with alcohol.

It is strongly advised that personnel who handle toxic pesticides always have readily available clean water, soap, and a clean change of clothing for emergency use. However, if these items are not available, the nearest irrigation water, pond, or practically any source of water that is not contaminated with pesticides will serve the purpose in an emergency.

Following emergency decontamination, call your physician and tell him of the accident. He may want to check for possible after effects. (See SAFETY Section).

Decontamination of Pesticides Spilled During Transportation

Two widely-known emergency services --CHEMTREC, (Chemical Transportation Emergency Center), and the Pesticide Safety Team Network(PSTN) -- have combined their communications links to provide a more unified system to deal with chemical-related accidents. CHEMTREC, which is sponsored by the Manufacturing Chemists Association, is now the contact for both services and provides a direct round-the-clock link to ten PSTN Area Coordinators across the country.

The Pesticide Safety Team Network was organized in March, 1970, to provide emergency assistance in the event of a pesticide spill or accident during transportation. The service is a cooperative voluntary program operated as a public service by the National Agricultural Chemicals Association and fourteen member companies. In the event of a major accident or spill during transportation or warehousing, the 24 hour answering service provided by CHEMTREC will relay requests for assistance to the coordinator in whose area the help is needed. (See Map on following page). The person requesting assistance will be contacted by the PSTN Coordinator in a matter of minutes. At that time he will be instructed in the immediate measures to take. The Coordinator will also determine whether to dispatch a Safety Team to the scene.

The CHEMTREC center operates 24 hours a day with a toll free nation-wide telephone number. The CHEMTREC emergency number is: 800-424-9300. This number is not for dissemination to the general public, but for official use in transportation and warehousing.

CHEMTREC is not a contact for general chemical industry information, but is a source for assistance to emergency crews and carriers involved in a chemical transportation incident. It is designed to provide immediate data on how to handle these emergencies for those trained to do so.

AREA COORDINATORS

PESTICIDE SAFETY TEAM NETWORK

CHEMTREC

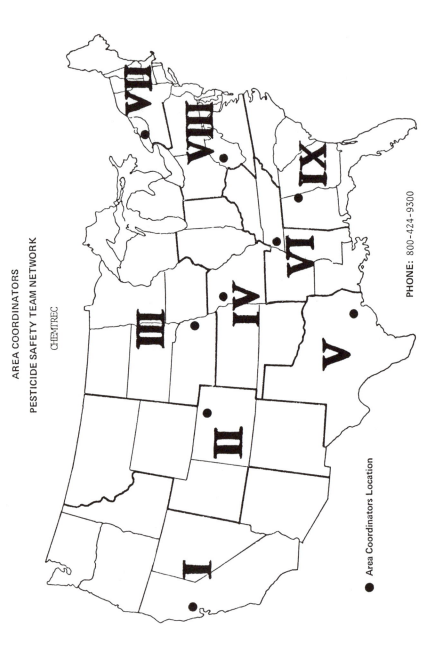

PHONE: 800-424-9300

● Area Coordinators Location

| DECONTAMINATING AGRICULTURAL PESTICIDES |

These suggestions are guides towards reducing the concentration and hazards
of agricultural pesticides in three areas: at the loading area on the farm,
where planes are loaded and when accidental spillage occurs on roads or
highways.

These suggested treatments will not decontaminate all agricultural pesticides
in these areas. But these practices should reduce the hazards for those
who handle and load pesticides into spray machines or spill them accidentally.

PESTICIDE LOADING
AREA ON THE FARMS

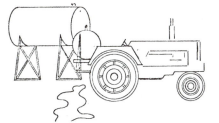

When To Decontaminate

1. If concentrate is spilled on loading
   area, treat immediately.

2. At least each 2 weeks during the spray
   season, treat the loading area.

Materials Needed

1. Eight to ten 50 pound sacks of hydrated
   lime.

2. Five one gallon jugs of sodium hypochlorite such as clorox or purex.

3. Sprinkler can or bucket.

How to Decontaminate Loading Area

- Hydrated lime and sodium hypochlorite(clorox, purex) are readily available.

- Sprinkle or spray the loading area or where spillage occurs with a mixture
  of 1 gallon of water added to 1 gallon of a hypochlorite.

- Then spread hydrated lime liberally over the area and let stand for at
  least 1 hour.

- Remove 1 to 2 inches of top soil where the accidental spillage occurs.
  Either pile the topsoil outside the area or put into an old drum and
  put fresh soil into the area.

- Keep all used containers stacked together and dispose of properly when
  convenient.

PESTICIDE LOADING AREA
ON PLANE LANDING STRIP

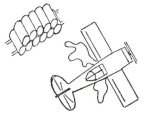

When To Decontaminate

1. When a can or drum of concentrate accidentally spills on the loading area.

2. Each week to reduce concentration on or
   in the soil where loading personnel work.

Material Needed

1. Ten to twenty 50 pound bags of hydrated
   lime.

2. Ten 1 gallon jugs of a sodium hypochlorite such as clorox or purex.

3. Sprinkling can or bucket.

How To Decontaminate Loading Area

■ Mix the sodium hypochlorite (clorox, purex) at the rate of 1 gallon to 1 gallon of water.

1. Spray or sprinkle the area with the solution.

2. Apply a liberal application of hydrated lime over the loading area and let stand for at least an hour.

■ Where convenient, you might scrape the topsoil (about 1 to 2 inches) and pile to one side or put in old drums on edge of loading area. Haul in fresh dirt to replace the removed dirt.

■ Keep all used containers stacked together and dispose of properly when convenient.

ACCIDENTAL SPILLAGE ON HIGHWAY OR ROAD

Suggested Actions

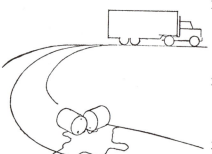

1. Do not walk in spilled pesticide. Prevent vehicles from driving over spilled chemical. Use precaution in handling leaking containers and when entering truck if spillage occurred in truck.

2. Two commonly available materials, clorox or purex and hydrated lime, will neutralize many of the agriculture pesticides.

3. Sprinkle or spray the area where spillage occurred with a mixture of the hypochlorite (clorox, purex) and water, adding 1 gallon of water to each 1 gallon of the hypochlorite.

Spread hydrated lime over the entire area and let stay at least one hour. If it is convenient, then wash the area down, especially that on the concrete or asphalt highways.

IMPORTANT: Hydrated lime and the hypochlorites, such as clorox and purex, will not neutralize all pesticides that are being used agriculturally, but they are suggested at this time as a practical approach to help reduce the hazards of pesticides in the loading area.

REMEMBER: Contact the CHEMTREC Center as well as the highway patrol, sheriff and/or police immediately after a pesticide transportation accident. The above suggestions are only emergency measures and should not take the place of action by the authorities.

PESTICIDE AND CONTAINER DISPOSAL

Proper pesticide waste and container disposal is extremely important. It is the pesticide users responsibility to see that unused chemicals and empty containers are disposed of properly. Improper disposal of pesticide wastes and containers over the past few years has resulted in incidents involving animal poisonings and environmental contamination.

Unfortunately, there are no easy or perfect means to dispose of excess pesticides and empty containers. Pesticide wastes may range in type from materials left over from excess spray mixtures, to accidental spillage, to pesticides left after a warehouse fire. Pesticide containers vary from aerosol cans or paper bags to 1 and 5 gallon cans on up to the 55 gallon drums which are sometimes hard to dispose of.

Supposedly empty pesticide containers are never really empty as several ounces of pesticides can generally be found in discarded containers. Many times the concentrations of the active ingredient in liquid formulations will tend to increase in toxicity as the solvents evaporate from the discarded container.

The National Agricultural Chemicals Association has suggested a procedure to use when rinsing containers prior to disposal that reduces their hazard potential. The procedure involves the triple rinse method as described in the following steps:

1. When the container is being emptied allow it to drain in a vertical position for thirty seconds.

2. Best results are obtained when the container is rinsed three times allowing thirty seconds for draining after each rinse.

3. Water or other diluting materials being used in a spray program can be used to rinse the container as follows: use one quart for each rinse of a one gallon can or jug; a gallon for each five gallon can; and five gallons for either thirty or fifty-five gallon drums.

4. Each rinse should be drained into the spray tank before filling it to the desired level.

Follow this Rinse and Drain Procedure

for Pesticide Containers

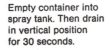
Empty container into spray tank. Then drain in vertical position for 30 seconds.

Add a measured amount of rinse water (or other dilutent) so container is 1/4 to 1/5 full. For example, one quart in a one-gallon container.

Rinse container thoroughly, pour into tank, and drain 30 sec. Repeat three times. Add enough fluid to bring tank up to level.

Crush pesticide container immediately. Sell as scrap for recycling or bury. Do not reuse.

It makes good sense to empty and triple rinse pesticide containers because that will ensure that you are getting all of the "goodie" out of the container, as well as ensuring that there will be no residue left to contaminate the environment or injure someone. The following chart attests to the fact that you are saving money when you use the triple rinse system.

EMPTY *and* RINSE

Pesticide Containers
thoroughly

"Empty" Containers
thoroughly

Laboratory tests show that even a good effort to empty a drum leaves about 6 ounces of pesticide in a 5-gallon container and 32 ounces in a 55-gallon drum.

Rinse with several quarts of water —once, twice, or THREE times and pour the rinse water into a sprayer. You save money with each of the rinses.

Your in-the-can loss with pesticides costing $20 and $30 per gallon is shown below.

The value of rinsing a 5-gallon "empty" container is illustrated below.

Empty to save money

AMOUNT OF RESIDUE	$ LOSS AT:	
	$20/GAL.	$30/Gal.
6½ OZ.	$1.00	$1.50
32 OZ.	$5.00	$7.50

AN EASY WAY TO CUT COSTS!

Rinse to save money

(Bar chart: OUNCES OF FORMULATION IN DRUMS vs. NUMBER OF RINSES, showing bars at 0, 1, 2, 3 rinses decreasing from about 6 ounces to near 0.)

Surplus or unwanted pesticide disposal can be kept at a minimum with some advanced planning. The safest means of disposal is to use the pesticide exactly as the label directs for the purpose for which it is intended. To do so requires some advanced planning to determine the kinds and amounts needed for the job and plan to use all of the spray, dust, or granule prepared for the application.

WARNING -- Never pour pesticides down the drain or flush them down the toilet!

Present methods of pesticide container and waste disposal include burial, incineration, degradation, soil injection, and in some instances permanent

storage. All pesticides and empty containers should be kept in storage until used or disposed of properly. The several methods proposed above for pesticide waste and container disposal are feasible under certain conditions. It is the users responsibility to select the most feasible method. Every possible alternative should be considered in relation to the problem, giving consideration to the environment.

The choice or selection of the disposal method will depend on several factors such as pesticide type; container type; facilities available for disposal; the nearness to communities, streams and crops; and any other geological or environmental considerations. The failure to consider any of these factors may result in more complex problems in the future.

Pesticide applicators having small quantities of unwanted pesticides or empty pesticide containers can usually dispose of them on their own premises. Most farmers or nurserymen have small enough quantities to enable them to dispose of their unwanted materials in this manner.

The Resource Conservation Recovery Act (RCRA) governs larger accumulations of pesticides, rinsings collected from aircraft washing aprons, disposal of empty containers, etc.

Many states also have hazardous waste disposal laws which the pesticide applicator must know about and comply with.

Details of some of the state laws as well as interpretations of RCRA are complicated and difficult to understand in some cases, but it is the responsibility of the pesticide applicator to make sure he knows what the law is and to obey all provisions as fully as possible.

Because of the complexity of pesticide waste and container disposal, the following rules and regulations pertaining to the disposal and storage of pesticides and their containers as published by the Environmental Protection Agency in the Federal Register, Volume 39, No. 85, May 1, 1974, are reproduced for your information and consideration.

PESTICIDES AND CONTAINERS
(Subpart C)

Procedures Not Recommended (§165.7)

No person should dispose of or store(or receive for disposal or storage) any pesticide or dispose of or store any pesticide container or pesticide container residue:

(a) In a manner inconsistent with its label or labeling.

(b) So as to cause or allow open dumping of pesticides or pesticide containers.

(c) So as to cause or allow open burning of pesticides or pesticide containers; except, the open burning by the user of small quantities of combustible containers formerly containing organic or metallo-organic pesticides, except organic mercury, lead, cadmium, or arsenic compounds, is acceptable when allowed by State and local regulations.

(d) So as to cause or allow water dumping or ocean dumping, except in conformance with regulations developed pursuant to the National Marine Protection, Research and Sanctuaries Act of 1972 (Pub. L 92-352), and to Sections 304,307, and 311 of the Federal Water Pollution Control Act as Amended (Pub. L-92-500).

(e) So as to violate any applicable Federal or State pollution control standard.

(f) So as to violate any applicable provisions of the Act.

Recommended Procedures For the Disposal of Pesticides (§165.8)

Recommended procedures for the disposal of pesticides are given below:

(a) _Organic pesticides_, (except organic mercury, lead, cadmium, and arsenic compounds which are discussed in paragraph (c) of this section) should be disposed of according to the following procedures:

    (1) Incinerate in a pesticide incinerator at the specified temperature/dwell time combination, or at such other lower temperature and related dwell time that will cause complete destruction of the pesticide. As a minimum it should be verified that all emissions meet the requirements of the Clean Air Act of 1970 (42 U.S.C. 1857 et seq.) relating to

(a)(1) gaseous emissions: specifically any performance regulations and standards promulgated under sections 111 and 112 should be adhered to. Any liquids, sludges, or solid residues generated should be disposed of in accordance with all applicable Federal, State, and local pollution control requirements. Municipal solid waste incinerators may be used to incinerate excess pesticides or pesticide containers provided they meet the criteria of a pesticide incinerator and precautions are taken to ensure proper operation.

(2) If appropriate incineration facilities are not available, organic pesticides may be disposed of by burial in a specially designated landfill. Records to locate such buried pesticides within the landfill site should be maintained.

(3) The environmental impact of the soil injection method of pesticide disposal has not been clearly defined nationally, and therefore this disposal method should be undertaken only with specific guidance. It is recommended that advice be requested from the Regional Administrator in the region where the material will be disposed of prior to undertaking such disposal by this method.

(4) There are chemical methods and procedures which will degrade some pesticides to forms which are not hazardous to the environment. However, practical methods are not available for all groups of pesticides. Until a list of such methods is available, it is recommended that advice be requested from the Regional Administrator in the region where the material will be disposed of prior to undertaking disposal by such method.

(5) If adequate incineration facilities, specially designated landfill facilities, or other approved procedures are not available, temporary storage of pesticides for disposal should be undertaken. Storage facilities, management procedures, safety precautions and fire and explosion control procedures should conform to those set forth in §165.10

(6) The effects of subsurface emplacement of liquid by well injection and the fate of injection materials are uncertain with available knowledge, and could result in serious environmental damage requiring complex and costly solutions on a long-term basis. Well injection should not be considered for pesticide disposal unless all reasonable alternative measures have been explored and found less satisfactory in terms of environmental protection. As noted in the Administrator's Decision Statement No. 5, dated February 6, 1973, the Agency's policy is to oppose well injection of fluid pesticides "without strict controls and a clean demonstration that such emplacement will not interfere with present or potential use of the subsurface environment,

(a)(6) contaminate ground water resources or otherwise damage the environment." Adequate pre-injection tests, provisions for monitoring the operation and the environmental effects, contingency plans to cope with well failures, and provisions for plugging injection wells when abandoned should be made. The Regional Administrator should be advised of each operation.

(b) Metallo-organic pesticides (except organic mercury, lead, cadmium, or arsenic compounds which are discussed in paragraph (c) of this section), should be disposed of according to the following procedures:

(1) After first subjecting such compounds to an appropriate chemical or physical treatment to recover the heavy metals from the hydrocarbon structure, incinerate in a pesticide incinerator as described in paragraph (a) (1) of this section.

(2) If appropriate treatment and incineration are not available bury in a specially designated landfill as noted in paragraph (a) (2) of this section.

(3) Disposal by soil injection of metallo-organic pesticides should be undertaken only in accordance with the procedure set forth in paragraph (a) (3) of this section.

(4) Chemical degradation methods and procedures that can be demonstrated to provide safety to public health and the environment should be undertaken only as noted in paragraph (a) (4) of this section.

(5) If adequate disposal methods as listed above in this section are not available, the pesticides should be stored according to the procedures in §165.10 until disposal facilities become available.

(6) Well injection of metallo-organic pesticides should be undertaken only in accordance with the procedures set forth in paragraph §165.8 (a) (6) of this section.

(c) Organic mercury, lead, cadmium, arsenic, and all inorganic pesticides should be disposed of according to the following procedures:

(1) Chemically deactivate the pesticides by conversion to nonhazardous compounds, and recover the heavy metal resources. Methods that are appropriate will be described and classified according to their applicability to the different groups of pesticides. Until a list of practical methods is available, however, each use of such procedures should be undertaken only as noted in paragraph §165.8 (a) (4) of this section.

(2) If chemical deactivation facilities are not available, such pesticides should be encapsulated and buried in a specially designated landfill. Records sufficient to permit location for retrieval should be maintained.

(3) If none of the above options is available, place in suitable containers (if necessary) and provide temporary storage until such time as adequate disposal facilities or procedures are available. The general criteria for acceptable storage are noted in §165.10.

Recommended Procedures for the Disposal of Pesticide Containers and Residues (§165.9)

(a) Group I: Containers. Combustible containers which formerly contained organic or metallo-organic pesticides, except organic mercury, lead, cadmium, or arsenic compounds, should be disposed of in a pesticide incinerator, or buried in a specially designated landfill, as noted in §165.8(a); except that small quantities of such containers may be burned in open fields by the user of the pesticide when such open burning is permitted by State and local regulations, or buried singly by the user in open fields with due regard for protection of surface and sub-surface water.

(b) Group II: Containers. Non-combustible containers which formerly contained organic or metallo-organic pesticides, except organic mercury, lead, cadmium, or arsenic compounds, should first be triple-rinsed. Containers in good condition may then be returned to the pesticide manufacturer or formulator, or drum reconditioner for re-use with the same chemical class of pesticide previously contained providing such reuse is legal under currently applicable U.S. Department of Transportation regulations including those set forth in 49 CFR 173.28. Other rinsed metal containers should be punctured to facilitate drainage prior to transport to a facility for recycling as scrap metal or for disposal. All rinsed containers may be crushed and disposed of by burial in a sanitary landfill, in conformance with State and local standards or buried in the field by the user of the pesticide. Unrinsed containers should be disposed of in a specially designated landfill, or subjected to incineration in a pesticide incinerator.

(c) Group III: Containers. Containers (both combustible and non-combustible) which formerly contained organic mercury, lead, cadmium, or arsenic or inorganic pesticides and which have been triple-rinsed and punctured to facilitate drainage, may be disposed of in a sanitary landfill. Such containers which are not rinsed should be encapsulated and buried in a specially designated landfill.

(d) Residue disposal. Residues and rinse liquids should be added to spray mixtures in the field. If not, they should be disposed of in the manner prescribed for each specific type of pesticide as set forth in §165.8.

Recommended Procedures and Criteria for Storage of Pesticides and Pesticide Containers (§165.10)

(a) General.

(1) Pesticides and excess pesticides and their containers whose uncontrolled release into the environment would cause reasonable

(a)(1)adverse effects on the environment should be stored only in facilities where due regard has been given to the hazardous nature of the pesticide, site selection, protective enclosures, and operating procedures, and where adequate measures are taken to assure personal safety, accident prevention, and detection of potential environmental damages. These storage procedures and criteria should be observed at sites and facilities where pesticides and excess pesticides (and their containers) that are classed as highly toxic or moderately toxic and are required to bear the signal words DANGER, POISON, or WARNING, or the skull and crossbones symbol on the label are stored. These procedures and criteria are not necessary at facilities where most pesticides registered for use in the home and garden, or pesticides classed as slightly toxic (word CAUTION on the label) are stored. All facilities where pesticides which are or may in the future be covered by an experimental use permit or other special permit are stored should be in conformance with these procedures and criteria.

(2) Temporary storage of highly toxic or moderately toxic pesticides for the period immediately prior to, and of the quantity required for a single application, may be undertaken by the user at isolated sites and facilities where flooding is unlikely, where provisions are made to prevent unauthorized entry, and where separation from water systems and buildings is sufficient to prevent contamination by runoff, percolation or wind-blown particles or vapors.

(b)Storage sites. Storage sites should be selected with due regard to the amount, toxicity, and environmental hazard of pesticides, and the number of sizes of containers to be handled. When practicable, sites should be located where flooding is unlikely and where soil texture/structure and geologic/hydrologic characteristics will prevent the contamination of any water system by runoff or percolation. Where warranted, drainage from the site should be contained (by natural or artificial barriers or dikes), monitored, and if contaminated, disposed of as an excess pesticide as discussed in §165.8. Consideration should also be given to containing windblown pesticide dusts or particles.

(c) Storage facilities. Pesticides should be stored in a dry, well ventilated, separate room, building or covered area where fire protection is provided. Where relevant and practicable, the following precautions should be taken:

(1) The entire storage facility should be secured by a climb-proof fence, and doors and gates should be kept locked to prevent unauthorized entry.

(2) Identification signs should be placed on rooms, buildings, and fences to advise of contents and warn of their hazardous nature, in accordance with suggestions given in paragraph (g) (1) (i) of this section.

(3) All items of movable equipment used for handling pesticides at the storage site which might be used for other purposes should be labeled "contaminated with pesticides" and should not be removed from the site unless thoroughly decontaminated.

(4) Provision should be made for decontamination of personnel and equipment such as delivery trucks, tarpaulin covers, etc. Where feasible, a wash basin, and shower with a delayed-closing pull chain valve should be provided. All contaminated water should be disposed of as an excess pesticide. Where required, decontamination area should be paved or lined with impervious materials, and should include gutters. Contaminated runoff should be collected, and treated as an excess pesticide.

(d) Operational procedures. Pesticide containers should be stored with the label plainly visible. If containers are not in good condition when received, the contents should be placed in a suitable container and properly relabeled. If dry excess pesticides are received in paper bags that are damaged, the bag and the contents should be placed in a sound container that can be sealed. Metal or rigid plastic containers should be checked carefully to insure that the lids and bungs are tight. Where relevant and practicable, the following provisions should be considered:

(1) Classification and separation.

  (i) Each pesticide formulation should be segregated and stored under a sign containing the name of the formulation. Rigid containers should be stored in an upright position and all containers should be stored off the ground, in an orderly way, so as to permit ready access and inspection. They should be accumulated in rows or units so that all labels are visible, and with lanes to provide effective access. A complete inventory should be maintained indicating the number and identity of containers in each storage unit.

  (ii) Excess pesticides and containers should be further segregated according to the method of disposal to ensure that entire shipments of the same class of pesticides are disposed of properly, and that accidental mixing of containers of different categories does not occur during the removal operation.

(2) Container inspection and maintenance. Containers should be checked regularly for corrosion and leaks. If such is found, the container should be transferred to a sound, suitable, larger container and be properly labeled. Materials such as absorptive clay, hydrated lime, and sodium hypochlorite should be kept on hand for use as appropriate for the emergency treatment or detoxification of spills or leaks. (Specific information relating to other spill treatment procedures and materials will be published as it is confirmed).

(e) <u>Safety precautions</u>. In addition to precautions specified on the label and in the labeling, rules for personal safety and accident prevention similar to those listed below should be available in areas where personnel congregate:

    (1) <u>Accident prevention measures</u>.

        (i) Inspect all containers of pesticides for leaks before handling them.

        (ii) Do not mishandle containers and thereby create emergencies by carelessness.

        (iii) Do not permit unauthorized persons in the storage area.

        (iv) Do not store pesticides next to food or feed or other articles intended for consumption by humans or animals.

        (v) Inspect all vehicles prior to departure, and treat those found to be contaminated.

    (2) <u>Safety measures</u>.

        (i) Do not store food, beverages, tobacco, eating utensils, or smoking equipment in the storage or loading areas.

        (ii) Do not drink, eat food, smoke, or use tobacco in areas where pesticides are present.

        (iii) Wear rubber gloves while handling containers of pesticides.

        (iv) Do not put fingers in mouth or rub eyes while working.

        (v) Wash hands before eating, smoking, or using toilet and immediately after loading, or transferring pesticides.

        (vi) Persons working regularly with organophosphate and N-alkyl carbamate pesticides should have periodic physical examinations, including cholinesterase tests.

(f) <u>Protective clothing and respirators</u>.

    (1) When handling pesticides which are in concentrated form, protective clothing should be worn. Contaminated garments should be removed immediately, and extra sets of clean clothing should be maintained nearby.

    (2) Particular care should be taken when handling certain pesticides to protect against absorption through skin, and inhalation of fumes. Respirators or gas masks with proper canisters approved for the particular type of exposure noted in the label directions, should be used when such pesticides are handled.

(g) Fire control.

    (1) Where large quantities of pesticides are stored, or where conditions may otherwise warrant, the owner of stored pesticides should inform the local fire department, hospitals, public health officials, and police department in writing of the hazards that such pesticides may present in the event of a fire. A floor plan of the storage area indicating where different pesticide classifications are regularly stored should be provided to the fire department. The fire chief should be furnished with the home telephone numbers of:

        (i) the person(s) responsible for the pesticide storage facility,

        (ii) the appropriate Regional Administrator, who can summon the appropriate Agency emergency response team,

        (iii) the U. S. Coast Guard, and

        (iv) the Pesticide Safety Team Network of the National Agricultural Chemicals Association.

    (2) Suggestions for Fire Hazard Abatement.

        (i) Where applicable, plainly label the outside of each storage area with "DANGER", "POISON", "PESTICIDE STORAGE" signs. Consult with the local fire department regarding the use of the current hazard signal system of the National Fire Protection Association.

        (ii) Post a list on the outside of the storage area of the types of chemicals stored therein. The list should be updated to reflect changes in types stored.

    (3) Suggested Fire Fighting Precautions.

        (i) Wear air-supplied breathing apparatus and rubber clothing.

        (ii) Avoid breathing or otherwise contacting toxic smoke and fumes.

        (iii) Wash completely as soon as possible after encountering smoke and fumes.

        (iv) Contain the water used in fire fighting within the storage site drainage system.

        (v) Firemen should take cholinesterase tests after fighting a fire involving organophosphate or N-alkayl carbamate pesticides, if they have been heavily exposed to the smoke. Baseline cholinesterase tests should be part of the regular physical examination for such firemen.

(3)(vi) Evacuate persons near such fires who may come in contact with smoke or fumes or contaminated surfaces.

(h) <u>Monitoring</u>. An environmental monitoring system should be considered in the vicinity of storage facilities. Samples from the surrounding ground and surface water, wildlife, and plant environment, as appropriate, should be tested in a regular program to assure minimal environmental insult. Analyses should be performed according to "Official Methods of the Association of Official Analytical Chemists (AOAC)," and such other methods and procedures as may be suitable.

PESTICIDE-RELATED WASTES

(Subpart D)

Procedures for Disposal and Storage of Pesticide-Related Wastes (§165.11)

(a) In general all pesticide-related wastes should be disposed of as excess pesticides in accordance with the procedures set forth in §165.7 and § 165.8. Such wastes should not be disposed of by addition to an industrial effluent stream if not ordinarily a part of or contained within such industrial effluent stream, except as regulated by and in compliance with effluent standards established pursuant to sections 304 and 307 of the Federal Water Pollution Control Act as amended.

(b) Pesticide-related wastes which are to be stored should be managed in accordance with the provisions of §165.10.

15
PESTICIDE
RECORD
KEEPING

Pesticide Record Keeping

Records of pesticide usage are important to help protect yourself and your investments. In many cases the pesticide usage history of a piece of property is as important as the cropping history of the land. Often the type of crop to be planted is determined by the chemicals applied previously. This is especially important with certain weed killers and certain chlorinated hydrocarbons. Good records can sometimes be the difference between the winning or losing of a law suit when you have been accused of using pesticides wrongly.

The record will help you to:

- Improve pest control practices and efficiency.
- Avoid pesticide misuse.
- Compare applications made with results obtained.
- Purchase only amounts of pesticides needed.
- Reduce inventory carry-over.
- Establish proper use in case of residue questions.
- Establish where the error was, if any was made.
- Establish proof of use of recommended procedures in case of law suits.
- Plan cropping procedures for next year.
- Plan pesticide needs for next year.

How to Keep Records

There are a number of different types of forms that have been devised for keeping pesticide application records. Some of the forms are for use in agricultural field applications and others are for use in keeping records of pesticide applications in vegetable and fruit crops. Still other forms are needed for keeping records of pesticide applications on livestock, poultry and buildings.

Carry a pocket notebook with you and write down information as it happens-- don't trust your memory. This information can be transferred to a permanent record that can be kept in your home or office.

What Information Should Be Kept?

- Crop, animal or building treated.
- Crop variety or animal species treated.
- Pest(s) treated.
- Location and number of acres or animals treated.

- Time of day, date and year of application
- Type of equipment used.
- Pesticide used, including the name and percent of active ingredient, type of formulation, trade name, manufacturer, and lot number.
- Amount used per acre or per 100 gallons of water.
- Amount of active ingredient applied per acre or per animal.
- Stage of crop or animal development.
- Pest situation; i.e. severity of infestation and presence of beneficial species.
- Weather conditions -- temperature, wind, rainfall.
- Harvest date.
- Results of application.

Miscellaneous Information

* Including damage to crop from spray application or other damages caused by storms or other situations. Information such as estimated yield losses due to storm damage would also be valuable.

* A sketch of fields treated showing locations of ditches and surrounding crops could also be valuable.

The examples of record keeping forms on the following pages are suggested for your consideration. You may wish to devise your own record keeping forms and include other information in addition to those items already mentioned.

GROWER'S PERMANENT RECORD OF CHEMICAL PESTICIDE APPLICATIONS

INCLUDE ALL TREATMENTS: SOIL, DORMANT, PRE-PLANT, GROWING SEASON AND POST-HARVEST.

CROP

VARIETY

GROWER'S NAME & FIELD IDENTIFICATION

TOTAL ACRES IN FIELD OR ORCHARD

IDENTITY OF TREATED AREA	DATES OF TREATMENT START	DATES OF TREATMENT FINISH	NUMBER ACRES TREATED	AIR OR GRND	PESTICIDE COMMON NAME OR TRADE NAME	LIQUID WET. PWDR. DUST GRANULES	ACTIVE INGREDIENTS, % OR LBS. PER GALLON	AMOUNT OF MATERIAL PER 100 GALS.	AMOUNT OF MATERIAL OR PER ACRE	GALLONS OF MIXED SPRAY, LBS. DUST OR LBS. GRANULES PER ACRE	PEST TO BE CONTROL'D

- COPY THIS INFORMATION FROM THE PESTICIDE LABEL -

SPRAYS ONLY:

USE SPACE BELOW FOR SKETCH TO DESCRIBE TREATED AREA, IF NECESSARY.

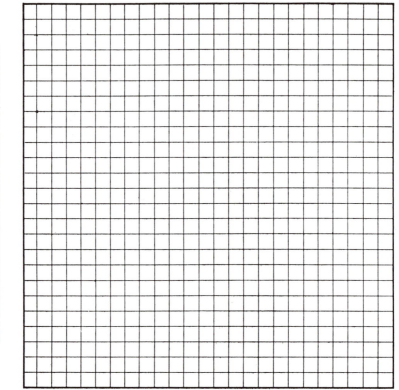

CAUTION...

BE CAREFUL WITH PESTICIDES
READ THE LABEL CAREFULLY AND COMPLETELY.
FOLLOW THE LABEL INSTRUCTIONS.
STORE PESTICIDES IN A SAFE PLACE.

CHEMICAL APPLICATION RECORD

Farm: _____   Fertilization: _____

Field: _____   Acres: _____

Soil Type: _____   Date Harvested: _____

Crop: _____ Variety: _____   Yield: _____

Crop Last Year: _____   Notes: _____

RECORD EACH APPLICATION OF EACH CHEMICAL TO THIS FIELD	APPLICATION NO. ( )	APPLICATION NO. ( )	APPLICATION NO. ( )	APPLICATION NO. ( )
DATE (DAY, MONTH, YEAR)				
ACRES TREATED				
CHEMICAL USED				
FORMULATION OF CHEMICAL				
TOTAL AMOUNT APPLIED				
STAGE OF CROP GROWTH				
PURPOSE OF APPLICATION (NAME OF WEED, INSECT, DISEASE OR OTHER REASON)				
STAGE OF DEVELOPMENT OF WEEDS, INSECTS, DISEASE				
METHOD OF APPLICATION				
SOIL CONDITIONS (WET, DRY, CLODDY)				
TEMPERATURE				
WIND (DIRECTION AND SPEED)				
CLOUD COVER				
EFFECTIVENESS OF TREATMENT				

COMMENTS

VEGETABLE CROP AND PESTICIDE RECORD

CROP & VARIETY	Grower	Block	Acres	Planted
Seed Source & Lot No.	Field Foreman	or Area		Date Year

CROP TREATMENT AND DEVELOPMENT HISTORY: Include all pre-plant, plant bed, soil, seed, growing, and post-harvest treatments, with date and hour.

Dates of Treatment		Acres Treated	Method- Spray, Dust, Granule, Etc.	Type of Equipment & Who Applied Aircraft, Sprayer, Dusters, Etc.	Pesticide- Name & Amount Active Ingredient	Amount Used Per 100 Gal. Water	Amount Applied Per Acre	Lbs. Active Per Acre	Stage of Development of Crop	Remarks- Weather, Rainfall, Pest Infestation
Start	Finish									
EXAMPLE:										
1-16	1-16-61	25	Spray	400 gallons - Boom Sprayer	Parathion 4E	1 pint	1½ pint/ 150 gal. water	.75	Tomatoes- first bloom	Leafminers & worms, Bad- Dry, warm, clear, no wind- No disease

ESTIMATED DATE OF FIRST HARVEST _____. ACTUAL HARVEST DATES: Start _____. Finish _____

FORAGE, FRUIT & VEGETABLE SPRAY RECORDS

CROP_____ VARIETY_____

PREVIOUS
CROP_____

SPRAY
TANK SIZE_____

PLANTING
DATES (Veg.)_____

HARVEST
DATE_____

List all insecticides, fungicides, miticides, herbicides and nematocide applications including seed and soil treatments.

FIELD OR BLOCK NO., SIZE & DATE	STAGE OF GROWTH	MATERIAL, FORMULATION & RATE PER A OR /100 GAL.	TOTAL ADDED /TANK	NO. OF TANKS	COMMENTS (WEATHER, METHOD OF APPLICATION, DRIVER, ETC.)

GROWER SIGNATURE_____ DATE_____

Address...........................

Producer's Name...

Year..................

CHEMICAL USE RECORD
Dairy Animals, Livestock, Poultry, and Animal Housing

Use Pesticides Safely
FOLLOW THE LABEL

This form provides a simple method for recording the use of pesticides, feed additives, or drugs in livestock or animal housing. This information is necessary in order to establish the proper time interval between the last treatment and the earliest date that meat, milk, or eggs can be marketed. You will also find this information useful for evaluating the results of treatment.

Species of Animal or Poultry	Market Class (beef, layer broiler, dairy, etc.)	Material Used and Rate of Application or Dosage	Method of Treatment (feed, drench, spray, injection, etc.)	Date of First Treatment	Date of Final Treatment	Earliest Date that Meat, Milk or Eggs Can Be Marketed	Special Precautions

READ THE LABEL

USE ONLY AS DIRECTED

LIVESTOCK TREATMENT RECORD

STORE CHEMICALS IN A SAFE PLACE

DISPOSE OF EMPTY CONTAINERS PROPERLY

Date	Number of Animals Treated	Breed and Average Weight	Pests Controlled	Pesticide Used (Kind & Amount)	Method of Treatment	Weather Conditions	Other Operations Performed (Dehorning, Castrating, Weaning, etc.)

GLOSSARY FOR PESTICIDE USERS

DID YOU KNOW

Glossary For Pesticide Users

A

ABSCISSION - The formation of a layer of cells which results in fruit, leaf or stem drop from a plant.

ABRASION - The process of wearing away by rubbing or grinding. Also, a scrape, scratch, or sore that breaks the skin.

ABRASIVE - Something that grinds down or wears away an object. For example, wettable powders are abrasive to pump and nozzles.

ABSORPTION - Movement of a pesticide from the surface into a plant, animal, or the soil. For instance, in animals, absorption may take place through the skin, breathing organs, stomach, or intestines, in plants, through leaves, stems, or roots.

ABSORPTIVE CLAY - A special type of clay powder which can take up chemicals and hold them. It is sometimes used to clean up pesticide spills.

ACARICIDE - A pesticide used to control mites and ticks. Same as miticide.

ACCELERATE - Increase in rate of speed.

ACCEPTED METHOD - A commonly used way of doing something.

ACCIDENT - Means of an unexpected, undesirable event, caused by the use or presence of a pesticide, that adversely affects man or the environment.

ACCORDANCE - Agreement. Example: To follow directions on the pesticide label.

ACCUMULATE - To build up, add to, store, or pile up.

ACCUMULATIVE PESTICIDES - Those chemicals which tend to build up in animals or the environment.

ACID - A compound whose hydrogen atom is replaceable by positive ions or radicals to form salts.

ACID EQUIVALENT - The theoretical yield of parent acid from an ester or salt such as esters of 2,4-D or the amine salt of 2,4-D.

ACRE - 43,560 square feet. An area of land about 210 feet long by 210 feet wide.

ACTIVATED CHARCOAL - Very finely ground, high quality charcoal which absorbs liquids and gases very easily.

ACTIVATOR - A material added to a pesticide to increase, either directly or indirectly, its toxicity.

ACTIVE INGREDIENT - The chemical or chemicals in a product responsible for the desired effects, which are capable, in themselves, of preventing, destroying, repelling or mitigating insects, fungi, rodents, weeds, or other pests.

ACTUAL DOSAGE - The amount of active ingredient (not formulated product) which is applied to an area or other target.

ACUTE DERMAL TOXICITY - Poisoning from a single dose of a chemical absorbed through the skin.

ACUTE INHALATION TOXICITY - How poisonous a single dose (or exposure) of a pesticide is when breathed into the lungs.

ACUTE ORAL TOXICITY - Poisoning from a single dose of a chemical taken by mouth.

ACUTE POISONING - Poisoning which occurs after a single dose (or exposure) of a pesticide.

ACUTE TOXICITY - How poisonous a pesticide is to an animal or person after a single dose (or exposure).

ADDITIVE - Any material that is added to a pesticide (not necessarily a wetting agent or a surfactant).

ADHERENCE - The property of a substance to adhere or stick to a given surface.

ADHESIVE - A substance that will cause a spray material to stick to the sprayed surface, often referred to as a sticking agent.

ADJACENT - Next to. Neighboring.

ADJUVANT - A chemical or agent added to a pesticide mixture which helps the active ingredient do a better job. Examples: Wetting agent, spreader, adhesive, emulsifying agent, penetrant.

ADSORPTION - The process by which materials are held or bound to the surface in such a manner that the chemical is only slowly available. Clay and high organic soils tend to adsorb pesticides in many instances.

ADULT - A full-grown, sexually mature insect, mite, nematode, or other animal. This stage is often the one to migrate and begin new colonies or infestations.

ADULTERATED - When strength or purity of a pesticide falls below standard claimed or if any substance has been added, or any part substracted.

ADULTICIDE - An insecticide which is toxic to the adult stage of insects.

AERATE - To bring in contact with more air. To loosen or stir up the soil mechanically or use additives such as peat. To let in or pull in fresh air after fumigation.

AERIAL APPLICATION - Treatment applied with the use of an airplane or helicopter.

AEROBIC - Living or functioning in air or free oxygen. The opposite of anaerobic.

AEROSOL SPRAY - A fine spray produced by pressurized gas that leaves very small droplets of pesticide suspended in the air.

AGITATE - To keep a pesticide chemical mixed up. To keep it from separating or settling in the spray tank.

AGITATOR - A paddle, air or hydraulic action to keep a pesticide chemical mixed in the spray or loading tank.

AGRICULTURAL COMMODITY - Any plant or part of a plant, animal, or animal product that is to be bought or sold.

AIR BLAST SPRAYER - A machine which can deliver high and low volumes of spray. It is used for orchards, shade trees, vegetables, and fly control.

ALCOHOL - Although in ordinary conversation ethyl alcohol is generally referred to merely as 'alcohol' that term applied to a long series of hydroxy organic compounds beginning with the one-carbon compound methanol, or methyl alcohol. Both methyl alcohol and ethyl alcohol are common solvents, frequently used in formulating pesticidal mixtures.

ALGAE - Nonvascular chlorophyll containing plants, usually aquatic.

ALGICIDE - A chemical intended for the control of algae, especially in water that is stored or is being used industrially.

ALIPHATIC - Chemically, those compounds that possess open-chain molecular structures. Generally are considered less toxic to plants than aromatic compounds.

ALKALI - The opposite of an acid. It is usually dangerous in concentrated form.

ALKALOID - Naturally occurring nitrogenous materials appearing in some plants which are used in preparing the botanically-derived insecticides.

ALTERNATIVE - One of two or more choices. Another way to do something.

AMIDE - A compound derived from carboxylic acids by replacing the hydroxyl of the COOH by the amino group NH_2.

AMINE - An organic compound containing hydrogen derived from ammonia by replacing one or more hydrogen atoms by as many hydrocarbon radicals.

AMPHIBIANS - Animals of the class Vertebrata that are intermediate between fish and reptiles. They are cold-blooded and have moist skin without scales, feathers or hair. Examples are frogs, toads, and salamanders.

AMPHOTERIC - An amphoteric compound has the capacity of behaving either as an acid or base. An amphoteric surface active compound is capable of anionic or cationic behavior depending on whether it is in an acidid or basic system.

ANAEROBIC - Living or functioning in the absence of air or free oxygen.

ANALOG - A compound that is very similar in both structure and formula to another compound.

ANIMAL - All vertebrate and invertebrate species.

ANIMAL SIGN - Evidence of an animal's presence in an area.

ANIONIC - An ion having a negative charge is an anion. When the surface active portion of a surfactant molecule possesses a negative charge it is termed an anionic surface active agent. Contrast with cationic.

ANIONIC SURFACTANT - A surface-active additive to a pesticide having a negative surface charge. The anionics perform better in cold, soft water. Most wetting agents are of this class.

ANNUAL - A plant that grows from a seed, produces flowers, fruit or seed the same year then dies.

ANTAGONISM - The phenomenon which results in a depression of compound activity when two or more are in close proximity or mixed together.

ANTIBIOTICS - Substances that are "against life." They are chemical substances produced by certain living cells such as bacteria, yeasts and molds. They are antagonistic or damaging to other living cells such as disease-producing bacteria. Antibiotics may kill living cells or prevent them from growing and multiplying. Penicillin is an example of an antibiotic that damages certain bacteria that cause disease in man.

ANTICOAGULANT - A chemical used in a bait to destroy rodents. It destroys the walls of the small blood vessels, and keeps the blood from clotting. As a result, the animal bleeds to death.

ANTIDOTE - A practical immediate treatment including first aid in case of poisoning; a remedy used to counteract the effects of a poison.

ANTI-DRIFT AGENT - A chemical used to reduce spray drift during the actual spraying operation by various physical factors.

ANTI-FEEDING COMPOUND - A compound which will prevent the feeding of pests on a treated material without necessarily killing or repelling them. It is not a repellent.

ANTIOXIDANT - A substance capable of chemically protecting other substances against oxidation or spoilage.

ANTI-SIPHONING DEVICE - A device attached to a filling hose to prevent water in the spray tank from draining into the water source.

ANTI-TRANSPIRANT - A chemical applied directly to a plant which reduces the rate of transpiration or water loss by the plant.

APHICIDE - A compound used to control aphids or plant lice.

APIARY - A place where colonies of bees are intentionally kept for the benefit of their honey-producing and pollinating activities.

APICULTURE - Pertaining to the care and culture of bees.

APPLICATION - The placing of a pesticide on a plant, animal, building, or soil; or its release into the air or water to prevent damage, or destroy pests.

APPLICATOR - A person or piece of equipment which applies pesticides to destroy pests or prevent damage by them.

APPLY UNIFORMLY - To spread or distribute a pesticide evenly.

AQUATIC PLANTS OR WEEDS - Plants or weeds that grow in water. The plants may float on the surface, grow up from the bottom of the body of water (emergent), or grow under the surface of the water (submergent).

AQUEOUS - Indicating the presence of water in a solution. A solution of a chemical in water.

AROMATICS - Oil or oil-like materials which kill plant and animal tissue similar to burning. Chemically, a compound having a closed ring structure.

ARS - Agricultural Research Service of the U.S. Department of Agriculture.

ARSENICALS - One of the most important groups of early insecticides, comprised principally of the arsenates and arsenites. The killing power of these materials is directly related to the percentage of metallic arsenic contained and in addition the metal component of the material (i.e., lead in lead arsenate) may also have some toxicity.

ARSONATE - An arsenical compound with a hydrocarbon group connected to the arsenic atom ; hence, an organic arsenical.

ARTIFICIAL RESPIRATION - First aid given to a person who has stopped breathing.

ASEPTIC - Free of disease causing organisms.

ASEXUAL - Reproduction not involving union of two nucleii.

ASSOCIATION OF AMERICAN PESTICIDE CONTROL OFFICIALS, INC. - This association (AAPCO) is composed of officials charged by law with the active execution of the laws regulating sale of economic poisons.

ATOMIZE - To break up a liquid into fine droplets by passing it through an apparatus under pressure.

ATROPINE SULFATE OR ATROPINE - An antidote used by doctors to treat people or animals poisoned by organic phosphate and carbamate pesticides in an attempt to save their lives.

ATTRACTANTS - Substances or devices capable of attracting insects or other pests to areas where they can be trapped or killed.

ATTRACTIVE NUISANCE - A legal term for any object which might attract children or other persons to it and then might injure or hurt them as a result. Examples are sprayers, empty pesticide containers, a bottle or food container filled with pesticides.

AUXIN - A generic term for compounds characterized by their capacity to induce elongation in shoot cells. They resemble indole-3-acetic acid in physiological action.

AVICIDE - A substance to control pest birds. Generally not designed to kill but to repel or to so affect a few individuals that others are frightened away.

AVOID PROLONGED CONTACT - Do not get exposed to a pesticide or breathe in the vapor (gas) for very long.

B

BACK-SIPHONING - Fluid or spray material siphoning from sprayer back to original source.

BACTERIA - Organisms (germs), some of which cause diseases in plants or animals. They are too small to be seen without a microscope.

BACTERICIDE - Any chemical used to kill bacteria.

BAIT - An edible material that is attractive to the pest, which normally contains a pesticide unless used as a prebait.

BAIT SHYNESS - The tendency for rodents, birds, or other pests to avoid a poisoned bait.

BALANCE OF NATURE - The developing, evolving and diversified life and state of adjustment (balance) all organisms have reached in relation to the environment. Nature is in a constant state of change and this constant evolutionary process is nature's way of trying to create a balance which is never reached.

BAND APPLICATION - An application to a continuous restricted area such as in or along a crop row rather than over the entire field area.

BARRIER APPLICATION - The use of pesticide or another agent to stop pests from entering a container, area, field or building.

BASAL TREATMENT - A treatment applied to the stems or trunks of plants at and just above the ground line.

BED - (1) A narrow flat-topped ridge on which crops are grown with a furrow on each side for drainage of excess water. (2) An area in which seedlings or sprouts are grown before transplanting.

BED-UP - To build up beds or ridges with a tillage implement.

BEETLE - A hard-shelled insect. Examples: Lady beetles, June bugs.

BENEFICIAL - Useful or helpful to man. A lady beetle is beneficial because it feeds on other insects that damage plants.

BENZENOID CHEMICAL - One having the "benzene ring" form of molecular structure (see also AROMATICS).

BIENNIAL - A plant that completes its life cycle in two years. The first year it produces leaves and stores food. The second year, it produces fruits and seeds.

BIOACTIVITY - Pertains to the property of affecting life.

BIOASSAY - The qualitative or quantitative determination of a substance by the systematic measurement of the response of living organisms as compared to measurement of the response to a standard or standard series.

BIOCIDE - A chemical which has a wide range of toxic properties, usually to members of both the plant and animal kingdoms.

BIOCONCENTRATION - The process of a pesticide becoming concentrated in plants or animals. When a chemical increases in concentration at each succeeding link in the food chain.

BIOLOGICAL CONTROL - Control of pests by means of predators, parasites and disease producing organisms.

BIOLOGICAL INSECTICIDE - A biological agent such as *Bacillus thuringiensis*, which kills insects like a chemical insecticide, and then rapidly dissipates in the environment.

BIOTIC - Relating to life.

BIPYRIDYLIUMS - A group of synthetic organic pesticides which includes the herbicide Paraquat.

BIRD REPELLENT - A substance which drives away birds or discourages them from roosting.

BLIGHT - A general term that may include spotting, discoloration, sudden wilting, or death of leaves, fruit, flowers, stems, or the entire plant.

BOOM - A section of pipe (or tubing) which connects several nozzles so that a pesticide can be applied over a wider area.

BOOT STAGE - When the seed head of a grass begins to emerge from the sheath.

BOTANICAL NAME - A scientific name made up of the genus and species. It is universal and more reliable than common names.

BOTANICAL PESTICIDE - A pesticide produced by and extracted from plants. Examples are nicotine, pyrethrum, strychnine and rotenone.

BRAND - The name, number, trade-mark or designation of a pesticide or device made by the manufacturer, distributor, importer or vendor. Each pesticide differing in the ingredient statement, analysis, manufacturer or distributor name, number or trade-mark is considered as a distinct and separate brand.

BROADCAST APPLICATION - An application over an entire area.

BROADLEAF SPECIES - Botanically, those plants classified as dicotyledoneae; morphologically - those having broad, rounded or flattened leaves as opposed to the narrow blade-like leaves of the grasses, sedges, rushes and onions.

BROAD-SPECTRUM PESTICIDE - One that controls a wide range of pests when applied correctly.

BROOD - The offspring, all of approximately the same age, rising from a single species of organism.

BRUSH - Economically useless woody plants, including a range in size from small shrubs to large trees; woody weeds.

BRUSH CONTROL - Control of woody plants.

BUCKET PUMP - The simplest hydraulic sprayer is the bucket pump. It is a plunger pump adapted by a clamp or otherwise for use in an open pail and delivering the liquid through a spray nozzle at the end of a hose.

BUD SPRAY - A pesticide application to trees any time between the time the first color appears in the ends until the buds begin to open.

BUG BOMB - An aeresol can containing insecticides.

BUILD UP - Accumulation of a pesticide in soil, animals, or in the food chain.

C

CALCULATE - To figure out by mathematical process by working with numbers.

CALIBRATE - To measure or figure out how much pesticide will be applied by the equipment to the target in a given amount of time.

CALIBRATED - Checked, measured. Knowing how much of a pesticide chemical is being applied by each nozzle or opening of a sprayer, duster or granular applicator to a given area, plant or animal.

CANCELLED - A pesticide use that is no longer registered as a legal use by the EPA. (Note: a cancellation order is less severe than a suspension order.)

CANISTER - A metal or plastic container filled with absorbent material which filters fumes and vapors from the air before they are breathed in by an applicator.

CANKER - A definite, localized, dead, often sunken or cracked area on a stem, twig, limb or trunk, surrounded by living tissues. Cankers may girdle affected parts resulting in a die-back starting in the tip.

CARBAMATES - A group of chemicals which are salts or esters of carbonic acid. Includes insecticides, herbicides, and fungicides.

CARBON DIOXIDE - Heavy colorless gas that does not support combustion.

CARCINOGEN - A substance or agent capable of producing cancer.

CARCINOGENIC - The term used to describe the cancer producing property of a substance or agent.

CARRIER - The liquid or solid material added to a chemical compound to facilitate its field application. An inert material which when used with a toxic compound improves the physical dispersion of the toxicant.

CARTRIDGE - The part of a respirator which adsorbs fumes and vapors from the air before the applicator breathes them in.

CATALYST - A substance that speeds up the rate of a chemical reaction but is not itself used up in the reaction.

CATERPILLAR - The worm-like stage of moths and butterflies that usually feeds on plants.

CATIONIC - An ion having a positive charge is a cation. When the surface active portion of a surfactant molecule possesses a positive charge it is termed a cationic surfactant.

CAUSAL ORGANISM - The organism that produces a given disease.

CAUSTIC - A chemical that will burn if it gets on the skin.

CAUTION - A warning to the user of pesticide chemicals. Used on labels of pesticide containers having slightly toxic pesticides in toxicity Category III as defined by the Federal Insecticide, Fungicide and Rodenticide Act.

CELL - The structural and functional unit of all plant and animal life. The living organism may have from one cell (bacteria) to billions (a large tree).

CENTIGRADE (C.) - A thermometer scale in which water freezes at 0 degrees C. and boils at 100 degrees C. To change to degrees Fahrenheit, multiply degrees centigrade by nine-fifths and add 32. Also, Celsius (C.), the preferred form approved by the Ninth General Conference on Weights and Measures in 1948.

CERTIFICATION - Means the recognition by a certifying agency that a person is competent and thus authorized to use or supervise the use of restricted use pesticides. (FIFRA amended)

CERTIFIED APPLICATOR - Means any individual who is certified to use or supervise the use of any restricted use pesticide covered by his certification.

cfs - Cubic feet per second of flow as water in a stream.

CHEMICAL - Often used here to mean "pesticide" chemical.

CHEMICALLY INACTIVE - Will not easily react with any other chemical or object.

CHEMICAL NAME - One that indicates the chemical composition and/or chemical structure of the compound being discussed.

CHEMICAL REACTION - When two or more substances are combined and as a result, undergo a complete change to make new substances or materials.

CHEMOSTERILANT - A chemical compound capable of producing reproductive sterilization.

CHEMOTHERAPY - The treatment of a desired plant or animal with chemicals to destroy or control a pathogen without seriously harming the plant.

CHLORATES - Herbicides and defoliants. They act as contact poisons, are translocated and may be adsorbed from the soil to kill both plant roots and tops. Chlorates cause chlorosis of leaves and a starch depletion in stems and roots when applied in less than lethal doses.

CHLORINATED HYDROCARBON - A chemical compound containing chlorine, carbon and hydrogen. DDT is a chlorinated hydrocarbon.

CHLOROSIS - The yellowing of a plant's normally green tissue because of a partial failure of the chlorophyll to develop.

CHOLINESTERASE - A body enzyme necessary for proper nerve function that is destroyed or damaged by organic phosphates or carbamates taken into the body by any path of entry.

CHOLINESTERASE INHIBITOR - Any organo-phosphate, carbamate, or other pesticide chemical that can interrupt the action of enzymes which inactivate the the acetylcholine associated with the nervous system.

CHRONIC EFFECT - A slow and long continued effect.

CHRONIC POISONING - Resulting from long periods of exposure to a chemical.

CHRONIC TOXICITY - Ability to cause injury or death from prolonged exposure.

CIDE - Suffix meaning "to kill".

CIRCULATE - To move completely through something in a path that returns to the starting point.

CLASSIFICATION - The process of assigning pesticides into groups according to common characteristics.

CLIMATIC - Having to do with weather conditions, sunshine, temperature, rainfall, moisture in the air, etc.

CLODS - Hard lumps of dirt formed by plowing or cultivating soil.

CLUSTER BREAK - The separation of pear or apple flower buds from each other in the cluster before blooming.

CO-DISTILLATION - In the process of distillation, vapors are driven off by heat and then condensed. In co-distillation, an additional material is carried off with the vapor.

COLORATION - Certain white or colorless pesticides must be treated with a coloring agent which will produce a uniformly colored product not subject to change in color beyond the minimum requirements during ordinary conditions of marketing, storage and use. Seed treatment materials are examples of pesticides which have been colored to indicate upon visual inspection what seed has been treated.

COMBUSTIBLE - Able to catch fire and burn.

COMMERCIAL APPLICATOR - Means a certified applicator (whether or not he is a private applicator with respect to some uses) who uses or supervises the use of any pesticide which is classified for restricted use for any purpose or on any property other than as provided by the definition of "private applicator".

COMMON EXPOSURE ROUTE - Means a likely way (oral, dermal, respiratory) by which a pesticide may reach and/or enter an organism.

COMMON NAME - (1) Official common names of pesticides such as atrazine, carbaryl, captan, and warfarin, that when present must appear in the active ingredients list on container labels. Other unofficial common names given to a plant or animal. It is possible for two living things to have the same common name in different places. In addition, the same living thing may have several common names.

COMPATIBLE - Two compounds are said to be compatible when they can be mixed without affecting each other's properties.

COMPETENT - Means properly qualified to perform functions associated with pesticide application, the degree of capability required being directly related to the nature of the activity and the associated responsibility.

COMPLETE METAMORPHOSIS - The process of insect development involving egg, larva, pupa and adult stages.

COMPRESSED AIR SPRAYER - Sprayer usually 1 to 3 gallon capacity with extension rod, equipped with air pump to develop pressure, often with shoulder strap for carrying. Not suitable for spraying at heights over 6 to 8 feet.

CONCENTRATE - Opposite of dilute. A liquid or dry formulation containing a high percentage of toxicant to save shipping and storage charges and yet be of convenient strength and composition for dilution.

CONCENTRATION - The amount of a substance contained in a unit volume, i.e. 4 pounds per gallon; or the percentage of the substance, as in a 50% wettable powder.

CONDEMNATION - The removal of a crop or product which does not meet the legal standards for tolerance on foods and is therefore not to be sold.

CONDITIONS FAVORABLE TO INFECTION - The situation a disease organism must have to attack and grow on or in a plant or animal. Disease, injury or destruction may follow. Some of the factors that can be important are temperature, humidity, and light intensity.

CONFINED AREAS - Rooms, buildings, and greenhouses with limited or inadequate ventilation.

CONIDIUM - (plural - conidia) A spore formed asexually, usually at the tip or side of a hypha (e.g., the conidia of the apple scab fungus; causes secondary infection.

CONIFERS - Trees and shrubs with needle-like leaves. Examples: Pine, cedar, larch, spruce, and hemlock.

CONTACT HERBICIDE - A compound that kills primarily by contact with plant tissue rather than as a result of translocation. Only that portion of a plant contacted is directly affected.

CONTACT POISON - A pesticide which kills when it touches or is touched by a pest.

CONTAMINATE - To alter or render a material unfit for a specified use, by the introduction of a foreign substance (a chemical).

CONTRACT - An agreement with someone to do a job or perform a service for him.

CONTROL - To prevent from doing damage. To reduce or keep down the number of pests so that little disease, damage, or injury occurs to a crop or property.

CONVENTIONAL - Usual. Everyday. Customary.

COPPER - Compounds of copper form one of the most useful groups of fungicides, and various forms are applied to plants, cordage, fabric and leather, and used as algaecides, seed disinfectants and wood preservatives. Copper materials have also been used as insecticides and repellents for certain insects.

CORROSIVE - Having the power to eat away slowly. Some pesticides eat or wear away rubber hoses, nozzles, and other parts of spray machinery.

COTYLEDON LEAVES - The first leaf, or pair of leaves, of the embryo of seed plants.

COUPLING AGENT - A solvent that has the ability to solubilize or to increase the amount of solubility of one material in another is referred to as a coupling agent.

COVERAGE - The amount of spread of a pesticide over a surface.

CREEPING PERENNIAL - A perennial plant that reproduces by both seed and asexual means (rhizomes, stolons, etc.).

CRITICAL AREAS - Places where pests are most likely to be troublesome.

CROOK STAGE - That stage of plant growth as it emerges from the soil, i.e., bean seedlings which have broken through the soil but before the stem becomes erect.

CROP - A plant growing where it is desired.

CROP TOLERANCE - The degree or the ability of the crop to be treated with a chemical but not injured.

CROSS-CONTAMINATION - When one pesticide gets into or mixes with another pesticide accidently. This usually occurs in a pesticide container or in a poorly cleaned sprayer.

CROSS-RESISTANT - When a pest population which has become resistant to one pesticide also becomes resistant to other chemically related pesticides.

CROWN - The point where stem and root join in a seed plant. This term is also used to describe the foliage and branches of trees.

CRUCIFERS - Plants belonging to the mustard family including mustard, cabbage, turnips, radish, etc.

CUBE - The root of a tropical plant (Lonchocarpus spp.) valued as a source of rotenone. Obtained commercially mostly from Peru. Pronounced koo-bay.

CUCURBITS - Plants belonging to the gourd family including pumpkins, cucumbers, squash, etc.

CULM - The jointed stem of a grass which is usually hollow except at the nodes or joints.

CULTURAL CONTROL - Control measures including modifications of the planting, growing, cultivating, and harvesting of crops aimed at prevention of pest damage rather than destruction of an existing infestation.

CUMULATIVE EFFECT - The result of some poisons which build up or are stored in the body so that small amounts eaten or contacted over a period of time can sicken or kill an animal or person. Examples of chemicals which accumulate are: anticoagulant rodenticides, mercury compounds, thallium sulfate.

CUMULATIVE PESTICIDES - Those chemicals which tend to accumulate or build up in the tissues of animals or in the environment (soil, water).

CURATIVE - Having the power to heal or cure.

CURATIVE PESTICIDE - A pesticide which can inhibit or eradicate a disease-causing organism after it has become established in the plant or animal.

CURRENT - Now in use.

CUSTOM APPLICATOR - Any person, who for hire, by contract or otherwise, applies by aerial, ground, or any hand or mechanical equipment, pesticides, to any waters, lands, plants, farm structures or animals.

CUTICLE - A non-living outer layer covering all or part of an organism.

CUTIN - Waxy, fatty material that forms the cuticle, or waxy layer, covering plant surfaces such as leaves.

CUT-SURFACE APPLICATION - Treatments made to frills or girdles that have been made through the bark into the wood of a tree.

CYCLODIENES - Compounds with a ring structure such as aldrin, chlordane, heptachlor, etc.

D

DAMAGING - Harmful. Injurious.

DAMPING OFF - Rotting of seeds in the soil, or the sudden wilting and death of seedlings anytime after germination. Usually caused by fungi that attack the seed or stem of the plant.

DANGER - Risk, Hazard.

DAYS TO HARVEST - The minimum number of days permitted by law between the final pesticide application and the harvest of crops. (Same as pre-harvest interval.)

DAYS TO SLAUGHTER - The minimum number of days established by law, between the final pesticide application and the date an animal is to be slaughtered.

DEBRIS - Trash or unwanted plant parts or remains.

DECIDUOUS PLANTS - Those plants that are perennial in habit but lose their leaves during the winter.

DECONTAMINATE - To make safe by removing any pesticide from equipment or other surfaces as directed on a pesticide label or by an agricultural authority.

DEFLOCCULATING AGENT - A substance that prevents rapid precipitation (settling out) of solids in the liquid in the spray tank.

DEFOLIANT - A preparation intended for causing leaves to drop from crop plants such as cotton, soybeans or tomatoes, usually to facilitate harvest.

DEFOLIATE - To lose or become stripped of leaves.

DEGRADABILITY - The ability of a chemical to decompose or break down into less complex compounds or elements.

DEGRADATION - Breakdown of a complex chemical by the action of microbes, water, air, sunlight, or other agents.

DEGRADE - Decompose, or breakdown.

DEGREE OF EXPOSURE - The amount or extent to which a person has been in contact with a toxic pesticide.

DELAYED ACTION - With some herbicidal chemicals delayed response is expected. Considerable time may elapse before maximum effects can be observed. Usually treated plants stop developing soon after treatment, then gradually die.

DELAYED DORMANT SPRAY - A spray applied to deciduous trees during the period from the first swelling of the buds until the first color starts to show.

DELETERIOUS - Harmful. Injurious.

DENSITY - The size of a population within a given unit of space.

DEOXYGENATION - Depletion of oxygen.

DEPLETED - Exhausted. Reduced.

DEPOSIT - The amount of pesticide laid down immediately following an application.

DERMAL - Of or pertaining to the skin.

DERMAL TOXICITY - Dermal Toxicity is the passage of pesticides into the body through the unbroken skin. Most pesticides pass through unbroken skin to some extent. Many are not adsorbed readily unless in certain solvents. However, many including most organophosphates, will pass through unbroken skin as technical material and in all formulations. Dermal exposure results from spillage onto skin and clothing (including gloves and other protective equipment), from drift or from damaged or improperly maintained equipment. It is important to be aware of hazards resulting from the dermal toxicity of the materials you use. This is the greatest hazard to people handling pesticides.

DERRIS SPECIES - Formerly the most important plant sources of rotenone-containing roots. Grown principally in Malaya and East Indies. U.S. industry now depends on *Lonchocarpus* species from South America, mostly peru.

DESICCANT - A compound that promotes drying or removal of moisture from plant tissues.(See defoliant for difference in these two terms).

DESICCATION - Dehydration or the removal of tissue moisture by chemical or physical action. Drying chemicals that promote desiccation are desiccants. They are used primarily for preharvest drying of actively growing plant tissues when seed or other plant parts are developed but only partially mature: or for drying of plants which normally do not shed their leaves, such as rice, corn, small grains and cereals.

DETERGENT - A chemical (not soap) having the ability to remove soil and grime. Detergents can be used as surfactants in some pesticide sprays.

DETERIORATE - To break down or wear away. To decay or grow worse.

DETOXIFY - To make harmless.

DIAGNOSIS - Identification of the nature and cause of a problem.

DICOT (DICOTYLEDON) - A plant that has two seed leaves or cotyledons; the broadleaf plants.

DIEBACK - Death of branches or shoots of plants from the tips back toward the trunk or stem.

DILUENT - A material, liquid or solid, serving to dilute the technical toxicant to field strength for adequate plant coverage, maximum effectiveness and economy.

DILUTE - To make less concentrated. To mix a pesticide chemical with water, oil, or other material before it can be safely and correctly used.

DILUTION RATE - The amount of diluent that must be added to a unit of a pesticide to obtain the desired dosage.

DINITROS - A common designation for dinitro-phenol contact pesticides.

DIP TANK - A large tank used for dunking animals in a pesticide to protect against or destroy ectoparasites.

DIP TREATMENT - The application of a liquid chemical to a plant by momentarily immersing it, wholly or partially under the surface of the liquid, so as to coat the plant with the chemical.

DIRECT SUPERVISION - Unless otherwise prescribed by its labelling, a restricted use pesticide must be applied by a certified applicator or under a certified applicator's direct supervision.

DIRECTED APPLICATION - An application to a restricted area such as a row, bed, or at the base of plants.

DISCING - Breaking up the top few inches of the soil with a disc. This destroys many weeds and levels the soil for planting or treating with a fumigant or other pesticide.

DISEASE - A condition in which any part of a living organism is abnormal as the result of an infectious or non-infectious agent.

DISEASED - Unhealthy or abnormal. A plant or animal suffering from an infection by a fungus, bacteria, nematode, virus, other pest or poor growing conditions which disturb normal activity. As opposed to an insect injuring a plant or animal by chewing or sucking.

DISINFECTANT - Similar to bactericide. Chemicals which kill or inhibit growth of bacteria.

DISINFESTANT - An agent that kills or inactivates organisms present on the surface of the plant or plant part in the immediate environment. In the case of seeds (seed disinfestant) or soils (soil disinfestant).

DISPERSE - To spread out or scatter. To separate and move apart.

DISPERSING AGENT - A material that reduces the cohesive attraction between like particles. Dispersing and suspending agents are added during the preparation of wettable powders to facilitate wetting and suspension of the active ingredient.

DISPOSAL - The process of discarding or throwing away unused spray material, surplus pesticides, and pesticide containers.

DISSIPATE - To get rid of by scattering or spreading out. Gases and vapors dissipate through the air.

DISSOLVE - Usually refers to getting solids into suspension and/or solution which is necessary for uniform application results.

DISTRIBUTION - The amount and way in which a pesticide chemical is spread when applied to a plant, animal or other surface. Also, part of the United States or the world in which an insect, fungus, or other pest is found.

DOMESTIC ANIMAL - Tame animal used for man's benefit. Example: cow, sheep, horses.

DORMANT - Inactive. Not growing. In the case of plants, it is after the leaves fall or growth stops and before the buds open in the spring. Also, the period when seeds fail to sprout due to internal controls.

DORMANT SPRAY - A spray applied when plants are in a dormant condition. The temperature should be as high as 40° to 45°F. for application.

DOSAGE - A dose is a measured quantity, as of medicine, taken at one time or in one period of time. Dosage, therefore, is the amount of medicine in a dose. Used in connection with pesticides it refers in general to rate of application.

DOSE RATIO - Ratio between successively increasing doses.

DOUSE - Drench. Soak.

DOWNWIND - Direction toward which the wind is blowing.

DRENCH - Saturation (thorough soaking) of the soil with a pesticide, or an oral treatment of an animal.

DRENCH TREATMENT - The application of a liquid chemical to an area until the area is completely soaked.

DRIFT - Movement of spray or dust material by wind or air currents outside the intended area, usually as fine droplets, during or shortly after application.

DUCKFOOT CULTIVATOR - A field cultivator equipped with small sweep shovels.

DUST - A dry mixture consisting of the pesticide and some inert carrier such as clays, talc, attapulgites, walnut shell, calcium carbonate and others as carriers or diluents to facilitate application.

E

EARLY COVER SPRAY - A pesticide chemical applied to fruit soon after the petals fall.

ECOLOGY - The science concerned with the interrelationships of organisms and their environments.

ECONOMIC POISON - Any substance or mixture of substances intended for preventing, destroying, repelling or mitigating any insects, rodents, nematodes, fungi or weeds or any other form of life declared to be a pest and any substance or mixture of substances intended for use as a plant regulator, defoliant or desiccant.

ECONOMIC THRESHHOLD - That point of pest infestation where application of a control measure would return more money than the cost of the control procedure.

ECOSYSTEM - A community of life and its environment functioning as a unit in nature.

ECTOPARASITES - Plants and animals that live and feed on the outside of an animal or plant. Most are annoying, cause injury, or carry disease organisms. Examples: lice, fleas, some fungi.

EDIBLE - Safe to eat. A food.

EELWORMS - See NEMATODE.

EFFICACY - Capacity for serving to produce effects; effectiveness.

EMERGENCE - The action of a young plant breaking through the surface of the soil, or of an insect coming out of an egg or pupa.

EMERSED AQUATIC - A plant rooted below the water surface. However, the main plant parts are above the surface.

EMETIC - A substance used to make humans or animals vomit.

EMULSIFIABLE CONCENTRATE - Produced by dissolving the toxicant and an emulsifying agent in an organic solvent. A solvent substantially insoluble in water is usually selected since water-miscible solvents have not in general proved satisfactory. Strength usually stated in pounds of toxicant per gallon of concentrate.

EMULSIFY - To make into an emulsion. When small drops of one liquid are finely dispersed (distributed) in another liquid, an emulsion is formed. The drops are held in suspension by an emulsifying agent, which surrounds each drop and makes a coating around it.

EMULSIFYING AGENT - A material which facilitates the suspending of one liquid in another; for example, oil dispersed in water.

EMULSION - A mixture in which one liquid is suspended as minute globules in another liquid; e.g., oil in water.

ENCAPSULATION - A method of formulating pesticides, in which the active ingredient in encased in a material (often poly vinyl) resulting in sustained pesticidal release and decreased hazard. Also, a method of disposal of pesticides and pesticide containers by sealing them in sturdy, waterproof, chemical-proof containers which are then sealed in thick plastic, steel, or concrete to resist damage of breakage so the contents cannot get out.

ENCEPHALITIS - Inflammation of the brain. Virus-caused encephalitides can affect man and horses. These viruses are transmitted by mosquitoes.

ENDEMIC - Referring to a disease that is regularly present in a given region or area but does not necessarily cause significant losses. Such a disease may become EPIDEMIC.

ENDEMIC DISEASE - A disease peculiar to a particular locality.

ENDOTOXIN - Any of a group of toxic substances found in certain disease-producing bacteria and liberated by the disintegration of the bacterial cell.

ENTOMOLOGY - A branch of a science that studies insects.

ENVIRONMENT - Means water, air, land, all plants, man and other animals living therein, and the interrelationships which exist among them.

ENVIRONMENTAL MANIPULATION - Skillfull or artful management or use of the environment in such a way as to assist man in pest control.

ENZYME - A natural substance which regulates the rate of a reaction but which itself remains chemically unchanged.

EPA - Environmental Protection Agency. Responsible for the protection of the environment in the United States.

EPA ESTABLISHMENT NUMBER - A number assigned to each pesticide production plant by EPA. The number indicates the plant at which the pesticide product was produced and must appear on all labels of that plant.

EPA REGISTRATION NUMBER - A number assigned to a pesticide product by EPA when the product is registered by the manufacturer or his designated agent. The number must appear on all labels for a particular product.

EPIDEMIC - A sudden widespread increase in the incidence of a disease or organism.

EPIDERMIS - The outer cellular tissue of an animal or plant.

EPINASTY - Increased growth on upper surface of a plant organ or part (especially of leaves) causing it to bend downward.

EPIPHYTE - A plant that grows upon another plant (or on a building or telegraph wire), which it uses as a mechanical support but not as a food source. Some (such as Spanish moss) may harm the plants they grow on, however, by excluding light or smothering them. Most orchids are epiphytes.

ERADICANT - A chemical used to eliminate a pest from a plant or a place in the environment.

ERADICATION - The complete elimination of either weed, insects, disease organisms or other pests from an area.

ERODE - To wear away.

ESCAPE - A plant in a treated area that either missed treatment or failed to respond to treatment in the same manner as other treated plants.

ESSENTIAL - Necessary. A must.

ESTER - A compound formed by the union of an organic acid and an organic base (an alcohol). An example is 2,4-D and isooctyl alcohol to form the isooctyl ester of 2,4-D.

ETHANE - A colorless gas present in the gases flowing from gas wells and in refinery off-gases. One of the major raw materials for producing ethylene. (Ethane, propane and butane comprise LPG).

ETHERS - Organic compounds in which two hydrocarbon radicals are joined through an atom of oxygen.

ETHYLENE - An olefin; a basic chemical. The most widely used petro-chemical building block in the world. A colorless, sweet-smelling gas made principally from ethane and propane in the United States - and from naptha in other countries.

ETHYLENE OXIDE - Used for making chemical derivatives such as glycol ethers, ethanolamines, and synthetic detergents.

EVAPORATE - To form a gas and disappear into the air; to vaporize.

EVAPORATION - The process of a solid or liquid turning into a gas.

EVERGREEN - Plants that retain their functional leaves throughout the year.

EXCLUSION - Control of disease by preventing its introduction (e.g., by quarantines) into disease-free areas.

EXEMPTION - An exception to a policy, rule, regulation, law, or standard.

EXOSKELETON - The segmented, external skeleton of an insect; the insect's "skin".

EXPOSED SURFACES - Surfaces that can be successfully attacked by a pest. Surfaces that need to be protected by a pesticide chemical or other agent.

EXPOSURE - When contact occurs with a pesticide through skin (dermal), mouth (oral), lungs (inhalation/respiratory), or eyes.

EXPOSURE PERIOD - The length of time something has been under attack by a pest. Also, the length of time a pest is in contact with a pesticide chemical.

EXUDATE - Matter diffused from within a cell.

F

FACE SHIELD - A transparent piece of protective equipment used by a pesticide applicator to protect his face from exposure to pesticides.

FACILITATE - Make it easier to do or to understand something.

FACULTATIVE - Incidental, not necessarily compelled to live under one type of environment.

FAHRENHEIT (F.) - A thermometer scale that marks the freezing point of water at 32 degrees F. and the boiling point at 212 degrees F.

FATAL - Deadly. Able to cause death.

FDA - Food and drug Administration, U.S. Department of Health, Education and Welfare.

FEEDER ROOTS - The hair-like roots of a plant which take up water and most of the food materials needed by a plant.

FEPCA - Federal Environmental Pesticide Control Act of 1972. Greatly revised FIFRA and mandated the newly created U.S. Environmental Protection Agency (EPA) to direct its efforts toward protecting the environment from the unreasonable adverse effects of pesticides.

FIELD MARGIN - Edge of a field

FIFRA - Federal Insecticide, Fungicide, and Rodenticide Act (Amended).

FILLER - A diluent in a powdered form.

FILTER - To screen out the unwanted material; clean by straining out the undesirable parts, or a piece of equipment for doing this.

FINAL TREATMENT - Last pesticide chemical application before harvesting a crop or slaughtering an animal.

FINITE TOLERANCE - The maximum amount of pesticide which can legally remain on a food or feed crop at harvest after the pesticide has been directly applied to the crop.

FIRST AID - The first effort to help a victim while medical help is on the way.

FLACCID - Plant tissues becoming limp.

FLAG STAGE - Stage of growth in cereals and other grasses at which the sheath and leaf have been produced from which the head will emerge.

FLEXIBLE - Easy to bend.

FLOWABLE PESTICIDE - Very finely ground solid material which is suspended in a liquid and usually contains a high concentration or large amount of the active ingredient and must be mixed with water when applied.

FLUID - In a liquid form.

FLUORINE COMPOUNDS - This group includes a number of fluorine salts, many of which are highly toxic to warm-blooded animals. Sodium fluoride was the first to be used as an insecticide being applied to cockroaches, poultry lice, and ants.

FLY - An insect, usually with one pair of wings.

FOAMING AGENT - A material which causes a pesticide mixture to form a thick foam. It is used to reduce drift.

FOG - Particles between .1 and 50μ (microns) in diameter which make a fine mist.

FOG TREATMENT - The application of a pesticide as a fine mist for the control of pests.

FOGGER - An aeresol generator. A piece of pesticide equipment that breaks some pesticides into very fine droplets (aeresols or smokes) and blows or drifts the "fog" onto the target area.

FOLIAGE - The leaves, needles, and blades of plants.

FOLIAR SPRAYS - Droplets of a pesticide applied to leaves, needles and blades of plants.

FOOD CHAIN - phrase that describes how all living organisms are linked together and depend on each other for food; i.e., plant eaters, plant and meat eaters, or meat eaters.

FORMULA - A brief way of writing a complicated idea by using abbreviations and symbols.

FORMULATION - The pesticide product containing the active ingredient, the carrier and other additives required to make it ready for sale.

FRILL - One or a series of overlapping cuts made through bark into the sapwood of unwanted trees or brush into which herbicides are applied.

FRUIT SET - The number of apples or other fruit that begin to grow after the petals fall.

FUEL OILS - Petroleum fractions used for burning are used also as solvents. Distillate fuel oil has been recommended for application by itself against mosquito larvae.

FULL COVERAGE SPRAY - This term on a label signifies that the total volume of spray to be applied will thoroughly cover the crop being treated to the point of runoff or drip.

FULLER'S EARTH - A silicate mineral used as an adsorbent base or dust diluent. It is flowable and possesses excellent grindability acting as a grinding aid.

FUMES - A smoke, vapor, or gas.

FUMIGANT - The AAPCO has adopted this definition: "A substance of mixture of substances which produce gas, vapor, fume or smoke intended to destroy insects, bacteria or rodents." Fumigants may be volatile liquids and solids as well as substances already gaseous. They may be used to disinfest the interiors of buildings; objects and materials that can be enclosed so as to retain the fumigant; and the soil where crops are valuable enough to warrant the treatment.

FUNGI - Small, often microscopic plants without chlorophyll (green coloring). Fungi produce tiny threadlike growths. They grow from seed-like spores. Some fungi can infect and cause disease in plants or animals; other fungi can attack and destroy non-living things.

FUNGICIDE - A chemical used to kill fungi; a compound used to destroy or inhibit fungi (usually plant diseases).

FUNGISTAT - A material to prevent growth or multiplication of fungi without necessarily destroying the organisms after the latter have gained a foothold.

G

GALL - A growth, lump or swelling of plant tissue caused by mites, insects, bacteria, nematodes, viruses, fungi, or chemicals.

GAMMA IRRADIATION - The use of rays from a radioactive source.

GAS MASK - A device which filters out chemicals in the spray, dust or gas form from air breathed by the wearer. A full-face gas mask must be worn to protect from gases; it should be equipped with adequate canisters of absorbent materials (or with oxygen supply). Simple respirators protect from spray and dust without covering the eyes, but not from poisonous gases.

GERMICIDE - A chemical or agent that kills microorganisms such as bacteria or prevents them from causing disease.

GERMINATION - Process of germinating or beginning of vegetative growth. Often refers to the beginning of growth from a seed.

GENERAL USE PESTICIDE - A pesticide which can be purchased and used by the general public without undue hazard to the applicator and the environment as long as the instructions on the label are followed carefully (See RESTRICTED USE PESTICIDE).

GIRDLING - Practice of completely removing a band of bark all the way around a woody stem.

GPA (G.P.A.) - Gallons per acre.

gpm - Gallons per minute. A measure of liquid moved by a pump.

GRAM - A metric weight measurement equal to 1/1000th of a kilogram; approximately 28.5 grams equal 1 ounce.

GRANARY - A storage area for threshed grain.

GRANULAR PESTICIDE - A pesticide chemical mixed with or coating small pellets or a sand-like material. They are applied with seeders, spreaders, or special equipment. Granular pesticides are often used to control or destroy soil pests.

GRANULE - A type of formulation in which the active ingredient is mixed with, adsorbed, absorbed, or pressed on an inert carrier forming a small pellet.

GRASSY WEEDS - Weeds belonging to the grass family, characterized by round jointed stems, sheathing leaves which are parallel veined, and flowers borne in spikelets.

GROWING SEASON - In general the number of days from the last killing frost in the spring until the first killing frost in the fall. This varies according to the resistance of each crop to freezing temperatures.

GROWTH REGULATOR - An organic substance effective in minute amounts for controlling or modifying plant processes; organic compounds, other than nutrients, which in small amounts promote, inhibit, or otherwise modify any physiological process in plants.

GROWTH STAGES OF CEREALS - (1) Tillering stage - when a plant produces additional shoots from a single crown. (2) Jointing stage - when the internodes of the stem are elongating and the nodes can be felt by pinching the stem. (3) Boot stage - when leaf sheath swells due to the growth of developing spike or panicle. (4) Heading stage - when seed head is emerging from the sheath.

GRUBS - The larvae of certain beetles, wasps, bees and ants.

H

HABITAT - Physical place where an organism lives.

HAND DUSTER - In a hand duster a plunger expels a blast of dust-laden air. The dust chamber may be at the end of the plunger tube itself, or an enlargement at the end, or it may be located below the plunger tube.

HAND SPRAYER - Small, portable, pesticide sprayers that can be carried and operated by a man.

HARD WATER - Water with minerals such as calcium, iron and magnesium dissolved in it. Some pesticides added to hard water will curdle or settle out.

HARVEST AID - A material used to remove the leaves from cotton plants, kill potato vines, and in any other way, facilitate machine harvesting of a crop. (See DEFOLIANT, DESICCANT).

HARVEST DATE - A day a crop is removed from its site of growth, as from a tree, bush or vine, or cut as in the case of alfalfa. (Note: "Removed from" does not refer to when a crop is removed from the field, but rather to when the crop is picked, cut, dug up, etc.).

HAZARD - The probability that injury will result from use of a substance in a proposed quantity and manner. The sum of the toxicity plus the exposure to a pesticide.

HAZARDOUS - Dangerous, risky. Pesticide chemicals that may cause injury or death if not used as directed on the label.

HECTARE - In the metric system, a land measure equal to 100 ares, or 10,000 square meters. One hectare is equivalent to 2.471 acres.

HERBACEOUS PLANT - A vascular plant that remains soft or succulent and does not develop woody tissue.

HERBICIDE - A pesticide used for killing or inhibiting plant growth. A weed or grass killer.

HIGH PRESSURE SPRAYER - Same as HYDRAULIC SPRAYER.

HIGHLY TOXIC - (1) substances are considered highly toxic by law if the LD_{50} of a single oral dose is 50 milligrams or less per kilogram of body weight (LD_{50}); (2) if LC_{50} of toxicity by inhalation is 2,000 mcg or less of dust or mist per liter of air or 200 ppm or less by volume of a gas or vapor when administered by continuous inhalation for one hour to both male and female rats or to other rodent or non-rodent species if it is reasonably forseeable that such concentrations will be encountered by man, or: (3) if LD_{50} of toxicity by skin absorption is 200 milligrams or less per kilogram of body weight when administered by continuous contact for 24 hours with the bare skin of rabbits or other rodent or non-rodent species as specified.

HIGHLY VOLATILE - A liquid that quickly forms a gas or vapor (evaporates) at room temperature.

HIGH VOLUME SPRAYS - Spray applications of more than 50 gallons per acre.

HOMOLOG - A series of compounds which express similarities of structure but regular differences of formulae.

HORMONE - A naturally occurring substance in plants or animals that controls growth or other physiological processes. It is used with reference to certain man-made or synthetic chemicals that regulate or affect growth activity.

h.p. - Horsepower.

HOST - The term "host" means any plant or animal on or in which another lives for nourishment, development, or protection.

HUMIDITY - Refers to the dampness or amount of moisture in the air.

HYDRATED - The addition of water. (Chemically combined).

HYDRAULIC - Pertaining to water or other liquids.

HYDRAULIC AGITATOR - A device which keeps the tank mix from settling out by means of water flow under pressure.

HYDRAULIC SPRAYER - A machine which applies pesticides by using water at high pressure and volume to deliver the pesticide to the target. Same as high pressure sprayer.

HYDROCARBON - A chemical whose molecules contain only carbon and hydrogen atoms. The simplest hydrocarbon is methane (natural gas), each of whose molecules contains one atom of carbon and four atoms of hydrogen. Crude oil is a mixture largely of hydrocarbons.

HYDROGEN-ION CONCENTRATION - A measure of the acidity. The hydrogen-ion concentration is expressed in terms of the pH of the solution. For example, a pH of 7 is neutral, from 1 to 7 is acid, and from 7 to 14 is alkaline.

HYDROLYSIS - The splitting of a substance into the smaller units of which it is composed by the addition of water elements.

HYDROPHIL - A substance or system which attracts, or is attracted to, water is hydrophillic in nature.

HYDROPHOBE - A substance or system which repels, or is repelled by, water is hydrophobic in nature.

HYGROSCOPIC - Substances capable of absorbing water from the atmosphere under normal conditions of temperature, pressure and humidity, are called hygroscopic substances.

HYPERTROPHY - Abnormal growth and hypoplasia; under-development; stunting,

HYPHA - (Plural: Hyphae) One of the threadlike elements of the mycelium; a tubular filament.

HYPO - A prefix denoting a deficiency, lack or less than the normal or desirable amount.

I

IGR - Insect Growth Regulator.

ILLEGAL RESIDUE - Residue that is in excess of a pre-established government-enforced safe level.

IMMINENT HAZARD - A situation which exists when the continued use of a pesticide would likely result in unreasonable, harmful effects on the environment.

IMMUNE - A state of not being affected by disease or poison; exempt from or protected against.

IMPERMEABLE - Not capable of being penetrated. Semi-permeable means permeable to some substances but not to others.

IMPERVIOUS - Hard to penetrate or soak through. The condition of soil when it will not soak up water.

INACTIVE - Not involved in the pesticide action; not reacting chemically with anything.

INCINERATOR - A special high-heat furnace or burner which reduces everything to a non-toxic ash or vapor. Used for disposing of some highly toxic and moderately to slightly toxic pesticides.

INCOMPATIBLE - Not capable of being mixed or used together; in the case of pesticides, the effectiveness of one or more is reduced; or they cause injury to plants or animals.

INCORPORATE INTO SOIL - The mixing of a pesticide into the soil by mechanical means.

INDUCE VOMITING - To make a person or animal throw up.

INERT INGREDIENT - An inactive ingredient. An ingredient in a formulation which has no pesticidal action.

INFECTION - The development and establishment of a pathogen (e.g., a bacterium) in its host which will produce a disease.

INFEST - To be present in number (e.g., insects, mites, nematodes, bacteria, fungi) on the surface matter or in the soil. Do not confuse with infect (infection) that applies only to living, diseased plants or animals.

INFESTATION - Pests that are found in an area or location where they are not wanted.

IN-FURROW - An application to or in a furrow in which a crop is planted.

INGESTED - Taken into the digestive system.

INGREDIENTS - The simplest constituents of the economic poison which can reasonably be determined and reported.

INHALATION - To take air into the lungs; to breathe in.

INHALATION TOXICITY - Poisoning through the respiratory system.

INHIBITOR - A chemical, usually of the regulator type, that prevents or suppresses growth or other physiological processes in plants. 2-amino-1, 2,4-triazole (ATA) is an example of a growth inhibitor.

INJECT - To force a pesticide chemical into a plant, animal, building or other enclosure, or the soil.

INJURIOUS - Harmful. Can or will cause damage.

IN LIEU OF - Instead of. In place of.

INOCULATION - The introduction of an infective agent (the inoculum) into or on living tissues producing a specific disease.

INORGANIC COMPOUNDS - Those compounds lacking carbon.

INSECT - Any of the numerous small invertebrate animals generally having segmented bodies and for the most part belonging to the class *Insecta*, comprising six-legged, usually winged forms.

INSECT GROWTH REGULATOR - A synthetic, organic pesticide which mimics insect hormonal action so that the exposed insect cannot complete its normal development cycle, and dies without becoming an adult.

INSECTICIDE - A substance or mixture of substances intended to prevent or destroy any insects which may be present in any environment.

INSTAR - The form of an insect between molts, numbered to designate the various periods; e.g., the first instar is a stage between the egg and the first molt, etc.

INTEGRATED CONTROL - Utilizing multiple approaches to pest control. When pesticides are required they should exert the least adverse effects on the pests' natural enemies.

INTERMITTENTLY - Starting and stopping from time to time.

INTERNALLY - Inside. See TAKE INTERNALLY.

INTERVAL - Period of time. The time between two applications. The distance between two points.

INTRADERMAL - Within the skin.

INVERSION - When temperature increases with elevation from the ground, an inversion condition exists.

INVERT EMULSION - One in which oil is the continuous phase and water is suspended in it.

IONIC SURFACTANT - One that ionizes or dissociates in water.

IRRITATING - Annoying. Making a person or animal uncomfortable by burning stinging, tickling, making the eyes water, etc.

ISOMERS - Two or more chemical compounds having the same structure but different properties.

J

JET (HYDRAULIC) AGITATOR - A device that keeps a tank mix from settling out of suspension by means of water flowing under pressure (See HYDRAULIC AGITATOR).

JOINTING STAGE - The elongation of the areas between the joints or nodes of a grass or cereal stem to elevate and expose the young seed head.

JOINTLY LIABLE - When two or more persons or companies share legal responsibility for negligence.

JURISDICTION - The extent or range of judicial or other authority.

JUVENILE HORMONE - A specific group of complex organic substances which as hormones, regulate the development of larval characteristics in insects. These are internally secreted by the endocrine glands in insects; however, synthetic chemical analogs are being developed as insecticides.

K

kg. - The abbreviation for kilogram which is a metric measurement of weight equivalent to 1000 grams or approximately 2.2 pounds.

KNAPSACK DUSTER - A duster carried on the back. It is operated by a bellows on top of a cylindrical dust container, the bellows being actuated by a hand lever at the side of the operator.

KNAPSACK SPRAYER - A sprayer that can be strapped on the back and used to apply liquid pesticide chemicals. The attached hose has a nozzle at the tip that can be aimed at the spot to be treated.

L

LABEL - All written, printed or graphic matter on, or attached to the economic poison, or the immediate container thereof, and the outside container or wrapper to the retail package of the economic poison.

LABELING - All information and other written, printed or graphic matter upon the economic poison or any of its accompanying containers or wrappers to which reference is made on the label or in supplemental literature accompanying the economic poison.

LABILE - Easily destroyed.

LACTATING ANIMAL - Any animal that is producing milk.

LACTATION - The period during which an animal produces and secretes milk.

LARVA - The worm-like or grub-like immature or growing stage of an insect. It hatches from an egg and later goes into a resting stage called the pupa. The larva looks very different from the adult. Many insects cause most or all of their damage as larvae. Example: Caterpillar.

LARVICIDE - An insecticide used to kill larvae of insects.

LATENT - Dormant.

LATERAL MOVEMENT - Chemical movement in a plant or in the soil to the side or horizontal movement in the roots or soil layers.

LAY-BY TREATMENTS - Applications of pesticides after the last cultivation.

LC_{50} - A means of expressing the toxicity of a compound present in air as a dust, mist, gas or vapor. It is generally expressed as micrograms per liter as a dust or mist but in the case of a gas or vapor as ppm. The LC_{50} is the statistical estimate of the dosage necessary to kill 50 percent of a very large population of the test species, through toxicity on inhalation under stated condition or by law, the concentration which is expected to cause death in 50 percent of the test animals so treated.

LD_{50} - A common method of expressing the toxicity of a compound. It is generally expressed as milligrams of the chemical per kilogram of body weight of the test animal (mg/kg). An LD_{50} is a statistical estimate of the dosage necessary to kill 50 percent of a very large population of the test species under stated conditions (e.g. single oral dose of aqueous solution), or by law, the dose which is expected to cause death within 14 days in 50 percent of the test animals so treated. If a compound has an LD_{50} of 10 mg/kg it is more toxic than one with an LD_{50} of 100 mg/kg.

LEACHING - Movement of a substance downward or out of the soil in or with water as the result of water movement.

LEGAL RESIDUE - Residue that is within safe levels according to the regulations.

LEGUME - One of a family of plants called Legumenosae. Examples: peas, beans, alfalfa, and soybeans.

LESION - A small diseased or abnormal area on a plant or an animal. It can be a spot, scab, canker, or blister, often caused by fungi, bacteria, viruses or nematodes.

LETHAL - Fatal or deadly.

LIABILITY - Legal responsibility for actions performed.

LICENSING - The issuing of a certificate of permission from a constituted authority to carry on a business, profession, or service.

LIFE CYCLE - The complete succession of developmental stages in the life of an organism.

LIMITATION - Restriction. The most that is allowed.

LINEAR FEET - Running feet. Example: The length of a field is measured in linear feet.

LITER - A unit of volume in the metric system equal to a little more than a quart.

LOCALIZED - Limited to a given area or part. Something that occurs in a small area or on a certain part of a plant or animal.

LODGED - Flattened or bent over and tangled. Refers to hay, grain, corn, and other plants that fall over if a disease, insects, or a storm damages them.

LOW CONCENTRATE SOLUTION - A solution which contains a low concentration or a small amount of active ingredient in a highly refined oil. The solutions are usually purchased as stock sprays meant for use in aeresol generators.

LOW PRESSURE BOOM SPRAYER - A machine which can deliver low to moderate volumes of pesticide at pressures of 30-60 PSI. The sprayers are most often used for field and forage crops, pastures, and rights-of-way.

LOW VOLATILE - A liquid or solid that does not evaporate quickly at normal temperatures.

LOW VOLATILE ESTER - An ester with a high molecular weight and a low vapor pressure.

LOW VOLUME SPRAY - A spray application of 0.5 to 5.0 gallons per acre.

M

MACRO - A combining form meaning large.

MACROSCOPIC - Visible to the naked eye without the aid of a microscope.

MAGGOT - Larval stage of flies.

MAMMALS - Warm-blooded animals that nourish their young with milk; their skin is more or less covered with hair.

MANDIBLES - The first pair of jaws in insects, stout and tooth-like in chewing insects.

MARINE - Of the ocean or sea; having to do with plants and animals which live in, on, or around an ocean or sea.

MATERIAL - A substance; often used to mean a pesticide, pesticide formulation, pesticide chemical, active ingredient, or additive ingredient.

MAXIMUM DOSAGE - The largest amount of a pesticide that is safe to use without resulting in excess residues or damage to whatever is being protected.

MECHANICAL AGITATION - The stirring, paddling or swirling action of a device which keeps a pesticide and any additives thoroughly mixed in the spray tank.

MEDIUM VOLUME SPRAY - A spray application of 5.0 to 50 gallons per acre.

MESA - Mining Enforcement and Safety Administration. All respirators designated for use with pesticides are jointly approved by MESA and NIOSH.

MESH (SCREEN) - Standard screens are used to separate solid particles into size ranges. The mesh is stated in number of openings to each linear inch. The finest screen practical in this work is the 325-mesh which has openings 44 microns in diameter, 1 micron being equivalent to 0.001mm. This screen has over 10,500 openings per square inch. Fine dusting sulfur preferably has 95% of the particles passing a 325-mesh screen. A common range for granular formulations is the 15/30 range. Particles small enough to pass a 60-mesh screen are considered dusts.

METABOLITE - A compound derived in the case of a pesticide by chemical action upon the pesticide within a living organism (plant, insect, higher animal, etc.). The action varies (oxidation, reduction, etc.) and the metabolite may be either more toxic or less toxic than before. The same derivative may in some cases develop upon exposure of the pesticide outside a living organism.

METAMORPHOSIS - The series of changes through which an insect passes in its growth from the egg through the larva and pupa to the adult; complete when the pupa is inactive and does not feed; incomplete when there is no pupa or when the pupa is active and feeds.

METRIC SYSTEM - A system of measurement used by most of the world. Because of its international use and scientific application, the U.S. has stated its intention to gradually adopt the metric system. The units of the metric system are meters (for length), grams (for weight), and liters (for volume).

mg. - The abbreviation for milligram - 1/1000 of a gram.

mg/kg - Used to express the amount of pesticide in milligrams per kilogram of animal body weight to produce a desired effect. 1,000,000 milligrams = 1 kilogram = 2.2 pounds.

MICOPLASMA-LIKE ORGANISMS - Organisms recently discovered to be the cause of many plant diseases formerly attributed to viruses; organisms smaller than bacteria and larger than viruses.

MICRO - A combining form meaning small.

MICROBIAL INSECTICIDE - Bacteria or other tiny plants or animals used to prevent damage by or destroy insects. Example: Milky spore disease of Japanese beetle.

MICROFAUNA - Microscopic forms of animal life.

MICROFLORA - Microscopic forms of plant life.

MICROGRAM - A metric weight measurement equal to 1/1,000,000th of a gram; approximately 28,500,000 micrograms equal 1 ounce.

MICRON - A unit to measure the diameter of spray droplets. Approximately 25,000 µ(microns) equal an inch.

MICROORGANISM - A microscopic animal or vegetable organism.

MICROSCOPIC - Visible under the microscope.

MIGRATING - Moving from place to place. Example: Armyworms in search.

MILDEW - A plant disease characterized by a thin, whitish coating of mycelial growth and spores on the surface of infected plant parts.

MILLIGRAM - A metric weight measurement equal to 1/1,000th of a gram; approximately 28,500 mg equals 1 ounce.

MINERAL SPIRITS - Similar to kerosene. It can be used as a solvent for some pesticides.

MINIMIZE - To reduce to the smallest possible amount. Example: to minimize spray drift or hazard to the environment.

MISCIBLE - Able to be mixed.

MISCIBLE LIQUIDS - Two or more liquids capable of being mixed and which will remain mixed under normal conditions.

MISDIAGNOSE - To make a mistake in deciding what pest has caused the problem.

MIST BLOWER - Spray equipment in which hydraulic atomization of the liquid at the nozzle is aided by an air blast past the source of spray.

MITE - Tiny eight-legged animal with a body divided into two parts. It has no antennae (feelers). During the nymphal stage it has six legs.

MITICIDE - A chemical used to control mites (acarids).

MITOTIC POISON - A chemical that disrupts cell division and resultant growth.

MODE OF ACTION - Manner in which herbicides kill or prevent weed or plant growth.

MOLD - A fungus-caused growth often found in damp or decaying areas or on living things.

MOLLUSCICIDE - A compound used to control slugs and snails which are intermediate hosts of parasites of medical importance to man.

MOLLUSKS - Any of a large family of invertebrate animals, including snails and slugs.

MONITORING SYSTEM - A regular system of keeping track of and checking up on whether or not pesticides are escaping into the environment.

MONOCOT - A seed plant having a single cotyledon (monocotyledon) or leaf; includes grasses, corn, lilies, orchids and palms.

MONOCULTURE - Growing some single crop and not using the land for any other purpose.

MONOPHAGOUS - Limited to a single kind of food, as in the case of the boll weevil which restricts its feeding to the cotton plant.

MORBIDITY - The relative incidence of disease or the state of wasting away.

MORTALITY - Death rate.

MOSAIC - A virus type plant infection showing patchwork of discolored areas on a leaf.

MPH - Miles per hour.

MULCH - A layer of wood chips, dry leaves, straw, hay, plastic strips, or other material placed on the soil around plants to hold moisture in the ground, keep weeds from growing, soak up rain, reduce soil temperatures, or keep fruits and vegetables from touching the ground.

MULTIPURPOSE SPRAY OR DUST - One that controls a wide range of pests.

MUTAGENIC - Capable of producing genetic change.

MYCELIUM - (Plural: Mycelia) - Mass of hyphae constituting the body of a fungus.

N

NAPHTHA - A hydrocarbon fraction of crude oil. When crude is distilled, it is separated into naphtha, kerosene, gas oil, fuel oil, lube oil and residues...with each fraction itself a mixture of many different chemicals. Naphtha is especially valuable as a raw material for both chemicals and gasoline. European ethylene plants, for example, use naphtha as their primary feedstock.

NARROW LEAF SPECIES - Botanically - Those plants classified as mono-cotyledonae; Morphologically - Those plants having narrow leaves and parallel veins. Examples: grasses, sedges, rushes, and onions. Compare to BROADLEAF SPECIES.

NATURAL CONTROL - A control of undesirable pests by natural forces, usually predators and parasites, but may be by pathogens or physical means.

NATURAL ENEMIES - The predators and parasites in the environment which attack pest species.

NECROSIS - Localized death of living tissue, i.e., death of a certain area of a leaf or of a certain area of an organ.

NECROTIC - A term used to describe tissues exhibiting varying degrees of dead areas or spots.

NEGLIGENCE - Failure to do a job or duty; an act or state of neglectful-ness.

NEGLIGIBLE RESIDUE - A tolerance which is set for a food or feed crop which contains a very small amount of pesticide at harvest as a result of indirect contact with a chemical.

NEMATICIDE - A material, often a soil fumigant used to control nematodes infesting roots of crop plants.

NEMATODE - A member of a large group (phylum Nematoda), also known as threadworms, roundworms, etc. Some larger kinds are internal parasites of man and other animals. Nematodes, sometimes called eelworms, injurious to crop plants are slender, free-living, micro-scopic, worm-like organisms in the soil.

NEOPRENE - A synthetic rubber often used to make gloves and boots which offer protection against most pesticides.

NERVOUS SYSTEM - All nerve cells and tissues in animals, including the brain, spinal cord, ganglia, nerves, and nerve centers.

NEUTRAL SOIL - A soil neither acid or alkaline with a pH of 7.0.

NEUTRALIZE - To destroy the effect of or to counteract the properties of something.

NIOSH - National Institute for Occupational Safety and Health. All respirators designated for use with pesticides are jointly approved by NIOSH and MESA.

NITROPHENOL - A synthetic organic pesticide which contains carbon, hydrogen, nitrogen and oxygen.

NO-RESIDUE - As the term applies to pesticides -- the act of registration of an economic poison on the basis of the absence of a residue at time of harvest on the raw agricultural product when the economic poison is used as directed.

NO-TILL - Planting crop seeds directly into stubble or sod with no more soil disturbance than is necessary to get the seed into the soil.

NON-ACCUMULATIVE - Does not build up or store in an organism or in the environment.

NON-CORROSIVE - Opposite of corrosive.

NONIONIC - Chemically inert. A surfactant that does not ionize is classed as nonionic, in contrast to anionic and cationic compounds.

NON-PERSISTENT - Only lasts for a few weeks or less. Example: A pesticide may disappear because it is broken down by light or microorganisms; or it may evaporate.

NON-SELECTIVE - A chemical that is generally toxic to plants or animals without regard to species. A non-selective herbicide may kill or harm all plants.

NON-TARGET ORGANISM - Means a plant or animal other than the one against which the pesticide is applied.

NON-TOXIC - Not poisonous.

NON-VOLATILE - A pesticide chemical that does not evaporate (turn to a gas or a vapor).

NOXIOUS WEED - A weed arbitrarily defined by law as being especially undesirable, troublesome and difficult to control. Definition of the term "noxious weed" will vary according to legal interpretations.

NOZZLES - Devices which control drop size, rate, uniformity, thoroughness, and safety of a pesticide application. The nozzle type determines the ground pattern and safety of a pesticide application. Examples: Flat fan, even flat fan, cone, flooding, off-set, atomizing, broadcast, and solid stream nozzles.

NYMPH - The early stage in the development stage of insects which have no larva stage. It is the stage between egg and adult during which growth occurs in such insects as cockroaches, grasshoppers, aphids, termites.

O

OBLIGATE - Compelled to live under only one type of environment.

OILS - Usually refers to aromatic parafinic oils used as diluents in formulating products as carriers of pesticides or for direct use.

OLEFINS - Reactive, hydrogen-deficient hydrocarbons characterized by one or more "double bond" per molecule. Ethylene and propylene, widely-used "building block" basic chemicals, are olefins.

OLIGOPHAGOUS - Restricted to a few kinds of food. For instance, the common cabbage worm feeds on plants related to the cabbage, such as turnips, mustard and other plants of the crucifer family.

ONCOGENIC - The property to produce tumors (not necessarily cancerous) in living tissues.

OPERATING SPEED - The constant rate at which a pesticide sprayer moves during application; usually measured in miles per hour or feet per minute.

ORAL TOXICITY - Ability of a pesticide chemical to sicken or kill an animal or human if eaten or swallowed.

ORGANIC COMPOUNDS - A large group of chemical compounds that contain carbon.

ORGANIC MATTER - Plant and animal debris or remains found in the soil in all stages of decay. The major elements in organic matter are oxygen, hydrogen, and carbon.

ORGANISM - Any living thing; plant, animal, fungus, bacteria, insect, etc.

ORGANOCHLORINE - Same as chlorinated hydrocarbon.

ORGANOPHOSPHATE - An organic compound containing phosphorus; parathion and malathion are two examples.

ORIFICE - The opening or hole in a nozzle through which liquid material is forced out and broken up into a spray.

ORIGINAL CONTAINER - The package (can, bag, or bottle, etc.) in which a company sells a pesticide chemical. A package with a label telling what the pesticide is and how to use it correctly and safely.

ORNAMENTALS - Plants used to add beauty to homes, lawns, and gardens. They include trees, shrubs, and small colorful plants.

OSMOSIS - The transfer of materials that takes place through a semi-permeable membrane that separates two solutions, or between a solvent and a solution that tends to equalize their concentrations. The walls of living cells are semipermeable membranes and much of the activity of the cells depends on osmosis.

OVER-THE-TOP - A pesticide application over-the-top of a growing plant.

OVICIDES - Pesticides or agents used to destroy insect, mite or nematode eggs.

OXIDATION - To combine with oxygen.

OXIMES - Chemical compounds characterized by the general formula R-CH = NOH.

P

PARASITE - A plant or animal that harms another living plant or animal (the host) by living and feeding on or in it. Some of our worst pests are parasites which cause disease or injury to animals and plants grown by man or to man himself.

PATHOGEN - Any micro-organism which can cause disease. Most pathogens are parasites but there are a few exceptions.

PATHOLOGY - Branch of science of the origin, nature, and course of diseases.

PELLET - A dry formulation of pesticide mixed with other components in discreet particles, usually larger than 10 cubic millimeters.

PENETRANT - Wetting agents that enhance the ability of a liquid to enter into pores of a substrate, to penetrate the surface, are termed penetrating agents or penetrants.

PENETRATION - The ability to get through; the process of entering.

PERCENT BY WEIGHT - A percentage which expresses the active ingredient weight as a part of the total weight of the formulation. Example: one pound of active ingredient added to and mixed with three pounds of inert materials results in a formulation which is 25% pesticide by weight.

PERCENT CONCENTRATION - The weight or volume of a given compound in the final mixture expressed as a percentage.

PERENNIAL - A plant that normally lives for more than two years. Trees and shrubs are perennial plants. Some perennials die back to the roots each winter but new shoots grow again in the spring.

PERSIST - To stay. To remain.

PERSISTENT PESTICIDE - A pesticide chemical (or the metabolites) that remains active in the environment more than one growing season. These compounds often accumulate and are stored in animal and plant tissues.

PEST - By law, forms of plant and animal life and viruses when they exist under circumstances that make them injurious to plants, man, domestic animals, other useful vertebrates, useful invertebrates or other articles or substances.

PESTICIDE (Economic Poison) - As defined under the Federal Insecticide, Fungicide, and Rodenticide Act, economic poison (Pesticide) "means any substance or mixture of substances intended for preventing, destroying, repelling, or mitigating any insects, rodents, nematodes, fungi, or weeds, or any other forms of life declared to be pests; and any substance or mixture of substances intended for use as a plant regulator, defoliant or desiccant."

PESTICIDE TOLERANCE - The amount of pesticide residue which may legally remain in or on a food crop. Federal residue tolerances are established by the Environmental Protection Agency.

PETROCHEMICAL - Generally, a chemical derived from crude oil or natural gas. The term embraces "basic chemicals" and "derivatives", but is difficult to use to precisely designate specific chemical materials. Some of today's "petrochemicals", for example, may someday be profitably derived from coal, oil shale or manufactured gas.

PETROLEUM DISTILLATE - Kerosene.

PETROLEUM OILS - Pesticides used to control insects in plants; they are refined from crude oil.

PETROLEUM PRODUCTS - Anything which contains gasoline, kerosene, oil, or similar products.

pH VALUE - The degree of acidity or alkalinity. The pH scale of 0 to 14 expresses intensity of acidity or alkalinity in the same manner that degrees in the thermometer express intensity of heat. The pH value of 7.0, halfway between 0 and 14, is neither acid nor alkaline. pH values below 7.0 indicate acidity with its intensity increasing as the numbers decrease. Conversely, pH values above 7.0 indicate alkalinity with its intensity increasing as the numbers increase.

PHEROMONES - Chemicals produced by insects and other animals to communicate with and influence the behavior of other animals of the same species. (NOTE: Most Pheromones used today in pest management are synthetic and are used to monitor insect populations.)

PHOTODECOMPOSITION - Destroyed by light.

PHOTOSYNTHESIS - The manufacture of simple sugars by green plants utilizing light as the energy source; a process by which carbohydrates are formed in the chlorophyll containing tissues of plants exposed to light.

PHYSICAL PROPERTIES - Examples are solubility, volatility, inflammability, state of being solid, liquid, or gas.

PHYTOPLANKTON - Microscopic plant life living suspended in water.

PHYTOTOXIC - Injurious to plants.

PISCICIDE - A pesticide used to control fish.

PLANT-DERIVED PESTICIDE - Same as botanical insecticides.

PLANT DISEASES - Harmful conditions or sicknesses which negatively affect plant life; fungi, bacteria, and viruses most often cause plant diseases.

PLANT GROWTH REGULATOR - A substance that alters the growth of plants. The term does not include substances intended solely for use as plant nutrients or fertilizers.

POINT OF DRIP OR RUNOFF - When a spray is applied until it starts to run or drip off the ends of the leaves and down the stems of plants or off the hair or feathers of animals.

POISON - A chemical causing a deleterious effect when absorbed by a living organism (biocide).

POISON CONTROL CENTER - An agency, generally a hospital, which has current information as to the proper first aid techniques and antidotes for all poisoning emergencies.

POLLINATORS - Bees, flies, and other insects which visit flowers and carry pollen from flower to flower in order for many plants to produce fruit, vegetables, buds, and seeds.

POLLUTANT - A harmful chemical or waste material discharged into the water, soil, or atmosphere. An agent that makes something dirty or impure.

POLLUTE - To add an unwanted material (often a pesticide) which may do harm or damage. To render unsafe or unclean or impure by carelessness or misuse.

POLYPHAGOUS - Feeding on a wide range of food species, not necessarily related; for instance, the corn earworm, which, because of its damage to crops other than corn, is called also the tomato fruitworm and the bollworm.

POSTEMERGENCE - After the appearance of a specified weed or crop.

POTABLE - Water suitable for drinking.

POTENCY - The strength of something. Example: How deadly a poison is.

POTENTIATION - The joint action of two pesticides to bring about an effect greater than the sum of their individual effects.

POUR-ON - A pesticide which is poured along the midline of the backs of livestock.

ppb - Parts per billion. A way of expressing amounts of chemicals in foods, plants, animals, etc. One part per billion equals 1 lb. in 500,000 tons.

ppm - Parts per million. A way of expressing amounts of chemicals in foods, plants, animals, etc. One part per million equals 1 lb. in 500 tons.

PRACTICAL KNOWLEDGE - Means the possession of pertinent facts and comprehension together with the ability to use them in dealing with specific problems and situations.

PRECAUTIONS - Warnings. Safeguards.

PRECIPITATE - To settle out. A solid substance that forms in a liquid and sinks to the bottom of the container.

PRECIPITATION - The amount of rain, snow, sleet or hail that falls.

PREDACIDE - A pesticide used to control vertebrate predator pests.

PREDATOR - An insect or other animal that attacks, feeds on and destroys other insects or animals. Predators help to reduce the number of pests which cause disease, damage and destruction.

PRE-DISPOSE TO CHEMICAL INJURY - To increase the chances of damage or harm by careless handling and incorrect use of pesticide chemicals. One must read the label and follow directions carefully to use the pesticide chemicals safely.

PREEMERGENCE - Prior to emergence of the specified weed or crop.

PRE-HARVEST - The time just prior to the picking, cutting, or digging up of a crop.

PRE-HARVEST INTERVAL - See DAYS TO HARVEST.

PRE-PLANT - Application of a pesticide prior to planting a crop.

PRESSURE - The amount of force on a certain area. The pressure of a liquid pesticide forced out of a nozzle to form a spray is measured in pounds per square inch.

PREVAILING - The predominant or general occurence or use.

PRIVATE APPLICATOR - Means a certified applicator who uses or supervises the use of any pesticide which is classified for restricted use for purposes of producing any agricultural commodity on property owned or rented by him or his employer or (if applied without compensation other than trading of personal services between producers of agricultural commodities) on the property of another person.

PRODUCT - A term used to describe a pesticide as it is sold. It usually contains the pesticide chemical plus a solvent and additives.

PROLONGED EXPOSURE - Contact with a pesticide chemical or its residue for a long time.

PROMULGATED - To put into operation (such as a law or rule), or to make known by declaration.

PROPELLANT - Agent in self-pressurized pesticide products that produces the force required to dispense the active ingredient from the container.

PROPER DOSAGE - The right amount, according to label instructions.

PROPERTIES - The characteristics or traits which describe a pesticide.

PROPRIETARY CHEMICAL - A chemical made and marketed by a person having the exclusive right to manufacture and sell it.

PROPYLENE - A colorless gas usually produced as a by-product of ethylene, manufactured from propane, butane, or naphtha. An important ingredient for making isopropanol, butanol, propylene oxide, polypropylene, glycerin, acrylonitrile and phenol.

PROTECTANT - A chemical applied to the plant or animal surface in advance of the pest (or pathogen) to prevent infection or injury by the pest.

PROTECTIVE EQUIPMENT - Means clothing or any other materials or devices that shield agains unintended exposure to pesticides.

PROTOPAMCHLORIDE - An antidote for organophosphate poisoning, *but not for carbamate poisoning.*

PROTOZOA - One celled animals or a colony of similar cells.

psi - Pounds per square inch.

PUBESCENT - Hairy. It affects ease of wetting of foliage; also retention of spray on foliage.

PUPA - The pupa is the resting stage of many insects. The stage between the larva (caterpillar or maggot, etc.) and the adult (butterfly or fly, etc). Some caterpillars spin a silk cocoon before they change to a pupa inside.

PUTREFACTION - The decomposition of proteins by micro-organisms under anaerobic conditions, resulting in the production of incompletely oxidized compounds, some of which are foul smelling.

PYRETHRIN - Either of two liquid esters derived from chrysanthemums, the active ingredient of pyrethrum.

PYRETHROIDS - Natural pyrethrum found in the flowers of plants belonging to the chrysanthemum family. Pyrethrum compounds are found in flowers consisting of four esters, two of which are pyrethrins.

PYRETHRUM - An insecticide made from the dried chrysanthemum flower heads.

PYROPHORIC - Tendency for a material to ignite spontaneously.

Q

QUARANTINE - Regulation forbidding sale or shipment of plants or plant parts, usually to prevent disease, insect, nematode, or weed invasion of an area.

R

RADIOACTIVE - Giving off atomic energy in the form of radiation, such as in alpha, beta or gamma rays.

RADIOISOTOPES - One of a broad class of elements capable of becoming radioactive and giving off atomic energy. Some radioisotopes occur naturally, others are produced artificially. The word is synonymous with radioactive elements and includes tracer elements.

RATE - Rate refers to the amount of active ingredient or acid equivalent of a pesticide applied per unit area (such as 1 acre). Rate is preferred to the occasionally used terms, dosage and application.

RECOMMENDATION - Suggestion or advice from a County Agent, Extension Specialist, or other agricultural authority.

REENTRY INTERVAL - The length of time between the pesticide applications and entry into the field to conduct hand labor.

REGISTERED PESTICIDES - Pesticides approved by the U.S. Environmental Protection Agency for use as stated on the label of the container.

REGISTRATION - Approved by EPA or a state agency for the use of a pesticide as specified on the label.

REGULATED PEST - Means a specific organism considered by a state or federal agency to be a pest requiring regulatory restrictions, regulations, or control procedures in order to protect the host, man and/or his environment.

REGULATORY OFFICIALS - People who work with the federal or state government and enforce rules, regulations, and laws.

REINFESTATION - The return of insects or other pests after they left or were destroyed.

RELATIVE HUMIDITY - The amount of moisture in the air compared to the total amount that the air could hold at that temperature.

REMOTE - Very slight or faint chance; far apart or distant.

REPEATED CONTACT OR INHALATION - To touch or breathe in a pesticide several times over a period of time.

REPELLENT - A compound that is annoying to a certain animal or other organism, causing it to avoid the area in which it is placed.

REPTILES - Animals of the class Vertebrata that are cold-blooded and possess scaly skin. Examples: snakes, turtles, and lizards.

RESIDUAL PESTICIDE - A pesticide chemical that can destroy pests or prevent them from causing disease, damage, or destruction for more than a few hours after it is applied.

RESIDUE - The amount of chemical which remains on the harvested crop.

RESISTANCE - The ability of an organism to suppress or retard the injurious effects of a pesticide.

RESISTANT SPECIES - One that is difficult to kill with a particular herbicide.

RESPIRATION - Breathing.

RESPIRATOR - A face mask used to filter out poisonous gases and dust particles from the air so that a person can breathe and work safely. A person using the most poisonous pesticide chemicals must use a respirator as directed on the pesticide label.

RESPIRATORY TOXICITY - Intake of pesticides through air passages into the lungs. Pesticides may reach the lungs as a vapor or as extremely fine droplets or particles. Most compounds are more toxic by this route. Safety equipment is essential to protect the pesticide handler or applicator from the hazards of respiratory exposure whenever it is recommended on the label.

RESTRICTED USE PESTICIDE - Means a pesticide that is classified for restricted use under the provisions of the Federal Insecticide, Fungicide and Rodenticide Act, as amended.

RESTRICTIONS - Limitations.

REVOCATION - To nullify or withdraw. To revoke a license.

RHIZOME - Underground root-like stem that produces roots and leafy shoots. Examples: the white underground parts of Johnsongrass and horsenettle, the black parts of Russian knapweed.

RODENT - All animals of the order Rodentia, such as rats, mice, gophers, woodchucks or squirrels.

RODENTICIDE - A substance or mixture of substances intended to prevent, destroy, repel or mitigate rodents.

RPAR - Rebuttable Presumption Against Registration. An EPA process whose objectives are to identify pesticide chemicals which present "unreasonable adverse effects on the environment".

rpm - Revolutions per minute.

RUNAWAY PEST - Any pest organisms which enter a new territory where they have no natural enemies, and therefore reproduce with little interference, resulting in a large population which can overrun an area.

RUNOFF - The sprayed liquid which does not remain on the plant. See Point of Drift or Runoff.

RUSSETTING - Rough, brownish markings on leaves, fruit, or tubers. This can be caused by some insects, fungi, pesticide chemicals, or possibly weather.

S

SAFENER - A material added to a pesticide to eliminate or reduce phytotoxic effects to certain species.

SAFETY - The practical certainty that injury will not result from the proper use of a pesticide or implement.

SALT - One of a class of compounds formed when the hydrogen atom of an acid radical is replaced by a metal or metal-like radical. The most common salt is sodium chloride, the sodium salt of hydrochloric acid. Other metal or metal-like salts in food may include phosphorous, calcium, potassium, sodium, magnesium, sulfur, manganese, iron, cobalt, zinc and other metals. They may be present as chlorides, sulfates, phosphates, lactates, citrates or in combination with proteins as in calcium caseinate.

SANITATION - Cleaning up. Keeping gardens, fields, animals, or buildings clean in order to reduce the number of insects and other pest and disease problems. This is one way to cut down on the use of pesticides.

SAPROPHYTE - An organism living upon dead or decaying organic matter.

SCIENTIFIC NAME - The one name used throughout the world by scientists for each animal and plant. These names are made up of two words based on the Greek and Latin languages and are called the genus and species.

SEED BED - Land prepared and used to plant seeds.

SEED PROTECTANT - A chemical applied to seed before planting to prevent disease and insect attack of seeds and new seedlings.

SEED TREATMENT - The application of a pesticide chemical to seeds before planting in order to protect them from injury or destruction by insects, fungi, and other soil pests.

SEGMENT - A ring or subdivision of the body or of an appendage between areas of flexibility associated with muscle attachment.

SEIZURE - To take or impound a crop or animal if it contains more than the allowable pesticide residue. Also the onset of a fit or convulsions.

SELECTIVE PESTICIDE - A chemical that is more toxic to some species (plant, insect, animal microorganisms) than to others.

SENSITIVITY - Susceptible to effects of toxicant at low dosage; not capable of withstanding effects; for example, many broadleaved plants are sensitive to 2,4-D.

SEX ATTRACTANTS - See PHEROMONES.

SHOCK - The severe reaction of the human body to a serious injury which can result in death if not treated (even if the actual injury was not a fatal one).

SHORT TERM PESTICIDE - A pesticide which breaks down almost immediately, after application, into non-toxic by-products.

SIDE-DRESSING - To put fertilizer, or a pesticide in granular form on or in the ground near plants after they have started to grow.

SIGNAL WORDS - Words which must appear on pesticide labels to denote the relative toxicity of the product. The signal words are:"DANGER - POISON"(for highly toxic), "WARNING" (for moderatley toxic), and "CAUTION" (for low-order toxicity). The symbol of the Skull and Crossbones must appear on the labels of highly toxic pesticides along with the words "DANGER-POISON".

SIGN - Some evidence of exposure to a dangerous pesticide; an outward signal of a disease or poisoning in a plant or animal, including people. Compare to SYMPTOM.

SIGNS OF POISONING - Warnings or symptoms of having breathed in, touched, eaten, or drunk a dangerous pesticide that could cause injury or death.

SILVICIDE - A term applied to herbicides used to control undesirable brush and trees, as in wooded areas.

SITE - An area, location, building, structure, plant, animal, or other organism to be treated with a pesticide to protect it from or to reach and control the target pest.

SLIMICIDE - A chemical used to prevent slimy growths, as in wood pulping processes for manufacture of paper and paperboard.

SLURRY - A thick suspension of a pesticide made from wettable powder and water.

SMOKE - Particles of a pesticide chemical between .001 and .1µ(micron) in diameter. The particles are released into the air by burning. See MICRON.

SOFT WATER - Water with few minerals or other chemicals dissolved in it. Soap makes suds easily in soft water.

SOIL APPLICATION - Application of a chemical to the soil rather than to vegetation.

SOIL DRENCH - To soak or wet the surface of the ground with a pesticide chemical. Generally, fairly large volumes of the pesticide preparation are needed to saturate the soil to any depth.

SOIL FUMIGANT - A pesticide that will evaporate quickly. When added to the soil the gas formed kills pests in the soil. Usually a tarpaulin, plastic sheet, or layer of water is used to trap the gas in the soil until it does the job.

SOIL INCORPORATION - Mechanical mixing of the pesticide with the soil.

SOIL INJECTION - Mechanical placement of the pesticide beneath the soil surface with a minimum of mixing or stirring. Common method of applying liquids which change into gases.

SOIL STERILANT - A chemical that prevents the growth of plants, micro-organisms, etc., when present in soil. Soil sterilization may be temporary or relatively permanent, depending on the nature of the chemical being applied.

SOLUBILITY - The ability of a chemical to dissolve in another chemical or solvent; a measure of the amount of substance that will dissolve in a given amount of another substance.

SOLUBLE - Will dissolve in a liquid.

SOLUBLE POWDER - A powder formulation that dissolves and forms a solution in water.

SOLUTION - A preparation made by dissolving a solid, liquid, or gaseous substance into another substance (usually a liquid) without a chemical change taking place. Example: Sugar in water.

SOLVENT - A liquid which will dissolve a substance forming a true solution (liquid in molecular dispersion).

SPACE BOMB - An aerosol spray that can be used in rooms or buildings. A container having a pesticide plus a chemical under pressure which forces the pesticide out as a spray or mist.

SPACE SPRAY - A pesticide forced out of an aerosol container or sprayer as tiny droplets which fill the air in a room or building and destroy insects and other pests.

SPECIES - A group of living organisms which are very nearly alike, are called by the same common name, and can interbreed successfully.

SPECIFIC CATEGORY - Any one of the pesticide use categories designated by FIFRA or by state regulations; a special area (such as agricultural plant pest control or regulatory pest control) requiring state certification for use of Restricted Use Pesticides by commercial applicators.

SPECIFIC GRAVITY - Density. The ratio of the mass (weight) of a material to the mass of an equal volume of water at a specified temperature such as $20°$ C.

SPIDERS - Small animals closely related to insects; they have eight jointed legs, two body regions, no antennae (feelers) and no wings. Spiders are often grouped with mites and ticks.

SPILLAGE - The leaking, running over, or dripping of any pesticide chemical. It should be cleaned up immediately for safety.

SPORE - An inactive form of a micro-organism that is resistant to destruction and capable of becoming active again.

SPOT TREATMENT - A treatment directed at specific plants or areas rather than a general application.

SPRAY - A mixture of a pesticide with water or other liquid applied in tiny droplets.

SPRAY CONCENTRATE - A liquid formulation of pesticide that is diluted with another liquid (usually water or oil) before using.

SPRAY DEPOSIT - The amount of pesticide chemical that remains on a spray surface after the droplets have dried.

SPRAY DRIFT - The movement of airborne spray particles from the intended area of application.

SPREADER - A substance which increases the area that a given volume of liquid will cover on a solid or on another liquid.

STAGE - Any definite period in the development of an insect; e.g., egg stage, caterpillar stage, etc., or in the development of plant; e.g., seedling stage, tillering stage, milk stage, etc.

STAGE OF DEVELOPMENT - A defined period of growth; usually refers to an insect. An insect goes through many changes in its growth, from an egg to an adult; each such change is a stage of development.

STANDARD - Means the measure of knowledge and ability which must be demonstrated as a requirement for certification.

STERILIZE - Treat with a chemical or other agent to kill every living thing in a certain area.

STICKER - A material added to a pesticide to increase tenacity rather than to increase initial deposit.

STOLON - Above ground runners or slender stems that develop roots, shoots, and new plants at the tip of nodes as in the strawberry plant or Bermuda grass.

STOMACH POISON - Pesticide that must be eaten by an insect or other animal in order to kill the animal.

STRUCTURAL PESTS - Pests which attack and destroy buildings and other structures, clothing, stored food, and manufactured and processed goods. Examples: termites, cockroaches, clothes moths, rats, dry rot fungi.

STYLET - A small stiff instrument used by insects to pierce plant or animal tissue for the purpose of feeding.

SUBACUTE TOXICITY - Results produced in test animals of various species by long term exposure to repeated doses or concentrations of a substance.

SUBMERSED PLANT - An aquatic plant that grows with all or most of its vegetative tissue below the water surface.

SUCTION HOSE - A hose through which water is pulled from a pond or stream or spray from the spray tank to the pump.

SUMMER ANNUALS - Plants which germinate in the spring, make most of their growth in the summer, and die in the fall after flowering and seeding.

SUPPLEMENT - Same as adjuvant. Substance added to a pesticide to improve its physical or chemical properties. May be a sticker, spreader, wetting agent, safener, etc., but usually not a diluent.

SUPPRESS - To keep a pest from building up in numbers. To stop a pest from causing injury or destruction.

SURFACE ACTIVE AGENT - A substance that reduces the interfacial tension of two boundary lines. Most pesticide adjuvants may be considered surface active agents. Also known as SURFACTANTS.

SURFACE SPRAY - A pesticide spray which is applied in order to completely cover the outside of the object to be protected.

SURFACE TENSION - Due to molecular forces at the surface, a drop of liquid forms an apparent membrane that causes it to ball up rather than to spread as a film.

SURFACE WATER - Rivers, lakes, ponds, streams, etc., which are located above ground.

SURFACTANT - A material which reduces surface tension between two unlike materials such as oil and water. A spreader or wetting agent.

SUSCEPTIBILITY - Means the degree to which an organism is affected by a pesticide at a particular level of exposure.

SUSCEPTIBLE - Capable of being injured, diseased, or poisoned by a pesticide; not immune.

SUSCEPTIBLE SPECIES - A plant or animal that is affected by moderate amounts of a pesticide.

SUSPENDED - A pesticide use that is no longer legal and the remaining stocks cannot be used. (Note: This order is more severe than CANCELLED). Also, describes particles that are dispersed (or held) in a liquid.

SUSPENSION - A system consisting of very finely divided solid particles dispersed in a liquid.

SWATH - The width of a treated area when a ground rig or spray plane makes one trip across a field.

SYMBIOSIS - The living together in intimate association of two diverse types of organisms.

SYMPTOM - Warning that something is wrong. Any indication of disease or poisoning in a plant or animal. This information is used to figure out what insect, fungus, other pest, or pesticide is causing the disease, damage, or destruction.

SYNDROME - Symptoms chracterizing a particular abnormality.

SYNERGISM - When the effect of two or more pesticides applied together is greater than the sum of the individual pesticides applied separately.

SYNERGIST - A chemical that when mixed with a pesticide increases its toxicity. The synergist may or may not have pesticidal properties of its own.

SYNTHESIS - A coming together of two or more substances to form a new material.

SYSTEMIC - A term used to describe certain pesticides that function by entering and becoming distributed within the plant as opposed to the pesticides that function by contact of the plant's surface.

SYSTEMIC PESTICIDE - A chemical which is translocated within the plant. For example, a systemic insecticide can be applied to the soil, enter the roots of the plant, travel to the leaves and kill insects feeding on the leaves.

T

TAKE INTERNALLY - To eat or swallow.

TANK MIX - The mixture of two or more compatible pesticides in a spray tank in order to apply them simultaneously.

TAPROOT - The main root of some plants. It acts as an anchor and reaches water deep in the soil.

TARGET - The plants, animals, structures, areas, or pests to be treated with a pesticide application.

TECHNICAL MATERIAL - The pesticide as it is manufactured by a chemical company. It is then formulated with other materials to make it usable in wettable powder, dust, liquid, or other form.

TEMPORARY TOLERANCE - A tolerance established on an agricultural commodity by EPA to permit a pesticide manufacturer or his agent time, usually one year, to collect additional residue data to support a petition for a permanent tolerance. In essence, this is an experimental tolerance.

TENACITY - Syn. Adherence. The resistance of pesticide deposit to weathering as measured by retention.

TENACITY INDEX - A ratio of the residual deposit over the initial deposit when subjected to prescribed washing techniques.

TERATOGENESIS - Structural abnormalities of prenatal origin, present at birth or manifested shortly thereafter.

TERMINATE - Finish. Stop.

TERMINAL GROWTH - The tips or ends of a growing plant.

TEST ANIMALS - Laboratory animals, usually rats, fish, birds, mice, or rabbits, used to determine the toxicity and hazards of different pesticides.

THERAPEUTANTS - Remedies for disease, drugs.

THERMAL - Of or pertaining to heat.

THORAX - The second or intermediate region of the insect body bearing the true legs and wings, made up of three rings, named in order, pro-, meso, and metathorax.

THOROUGH COVERAGE - Application of spray or dust where all parts of the plant or area treated is covered.

TICKS - Tiny animals closely related to insects; they have eight jointed legs, two body regions, no antennae (feelers), and no wings. They are bloodsucking organisms and are often found on dogs, cows, or other wild animals. Ticks are often grouped with mites and spiders.

TILLERING STAGE - The development of side shoots from the base of a single stemmed grass or cereal plant.

TIME INTERVAL - The period of time that is required between the last application of a pesticide and harvesting in order to ensure that the legal residue tolerance will not be exceeded.

TOLERANCE - By law, a regulation that establishes the maximum amount of a pesticide chemical that may remain on a raw agricultural commodity.

TOLERANT - Same meaning as resistant. For example, grass is tolerant of 2,4-D to the extent that this herbicide can be used selectively to control broadleaved weed without killing the grass.

TOPICAL APPLICATION - Implies application to the top or to the upper surface of the plant; thus applied from above.

TOXIC - Poisonous; injurious to animals and plants through contact or systemic action.

TOXICANT - An agent capable of being toxic; a poison.

TOXICITY - The natural capacity of a substance to produce injury. Toxicity is measured by oral, dermal and inhalation studies on test animals.

TOXIN - A poison produced by a plant or animal.

TRACER ELEMENT - A radioactive element used in biological and other research to trace the fate of a substance or follow stages in a chemical reaction, such as the metabolic pathway or a nutrient or growth formation in plants or animals. Radioactive elements that have proved useful for tracer work in nutrition research are carbon 14, calcium 45, cobalt 60, strontium 90 and phosphorus 32.

TRADEMARK - Defined as "A word, letter, device, or symbol, used in connection with merchandise and pointing distinctly to the origin or ownership of the article to which it is applied".

TRADE NAME - Same as Brand Name.

TRANSLOCATED PESTICIDE - One that is moved within the plant or animal from the site of entry. Systemic pesticides are translocated.

TRANSPORT - To carry from one place to another, usually in a truck or trailer.

TREATED AREA - A building, field, forest, garden, or other place where a pesticide has been applied.

TRIAL USE - This notation indicates that this material is worthy of trial use on a portion (perhaps 10%) of the grower's acreage. Results may be variable, but the recommendations are the best available based on the present knowledge.

TRIVIAL NAME - A name long commonplace and used everywhere; example, nicotine.

TUBER - An underground stem used for storage of reserve food, ie., irish potato.

TURBIDITY - Suspended matter in water preventing light penetration.

U

ULTRA-HAZARDOUS - A job or activity that is very dangerous.

ULTRA LOW VOLUME SPRAY - A spray application of 0.05 to 0.5 gallons per acre.

ULV - Ultra Low Volume Spray.

UNAUTHORIZED PERSONS - People who have no right doing something because they have not been told or trained to do it.

UNCONTAMINATED - Does not contain hazardous pesticide residues, filth or other undesirable material.

UNDER THE DIRECT SUPERVISION OF - The act or process whereby application of a pesticide is made by a competent person acting under the instruction and control of a certified applicator who is responsible for the actions of that person and who is available if and when needed, even though such certified applicator is not physically present at the time and place the pesticide is applied.

UNDERGROUND WATER - Water and waterways which are below the soil surface; this is where wells get their water.

UNIFORM COVERAGE - The application of a pesticide chemical evenly over a whole area, plant, or animal.

UNINTENTIONALLY - Did not mean to do it, done accidentally.

U.S. BUREAU OF MINES - An agency of the United States Government that tests respirators and gas masks to find ones that can be used with highly poisonous pesticide chemicals.

USDA - United States Department of Agriculture.

V

VAPOR - Gas Steam, mist, fog or fume.

VAPOR DRIFT - The movement of vapors of a volatile chemical from the area of application.

VAPORIZE - To evaporate; to form a gas and disappear into the air.

VAPOR PRESSURE - The property which causes a chemical compound to evaporate. The lower the vapor pressure, the more volatile the compound.

VECTOR - An insect or other animal that carries a disease organism from one host to another. Example: aphids, leafhoppers, and nematodes can transfer viruses from plant to plant.

VERMIN - Pests; usually rats, mice, or insects.

VERTEBRATE - An animal with a bony spinal column. Examples: mammals, fish, birds, snakes, frogs, toads.

VIABILITY - Being alive. Example: A seed is viable if it is capable of sprouting or germinating.

VICTIM - Someone who is injured, poisoned, or hurt in any way.

VIRUS - A sub-microscopic pathogen that requires living cells for growth and is capable of causing disease in plants or animals. Plant viruses are often spread by insects.

VISCOSITY - A property of liquids that determines whether they flow readily or resist flow. Viscosity of liquids usually increases with a decrease in temperature.

VOLATILE - A compound is said to be volatile when it evaporates (changes from a liquid to a gas) at ordinary temperatures on exposure to air.

VOLATILITY - The ability of a solid or liquid to evaporate quickly to ordinary temperatures when exposed to the air.

VOLUME - The amount, mass, or bulk.

VOMITOUS - What comes up when you throw up; matter which is vomited.

W

WAITING PERIOD - See TIME INTERVAL.

WARNING - Beware. Used on labels of pesticide containers having moderately toxic pesticides as defined by the Federal Insecticide, Fungicide and Rodenticide Act.

WARRANTY CLAUSE - A statement on the label limiting the liability of the chemical company.

WATER SOLUBILITY - Capable of being homogenously mixed with water.

WEATHERING - The wearing away of pesticides from the surfaces they were applied to because of rain, snow, ice, and heat.

WEED - A plant that is undesirable due to certain characteristics or its presence in certain areas. A plant growing in a place where it is not wanted.

WEED CONTROL - The process of inhibiting weed growth and limiting weed infestations so that crops can be grown profitably or other operations can be conducted efficiently.

WEED ERADICATION - The elimination of all live plants, plant parts, and seeds of a weed infestation from an area.

WETTABLE POWDER - A solid (powder) formulation which, on addition to water, forms a suspension used for spraying. It is prepared by adding water soluble agents to the formulation.

WETTING AGENT - A compound which reduces surface tension and causes a liquid to contact plant surfaces more thoroughly.

WHORL - The point where leaves or other plant parts unfold or are formed in a circular pattern. Example: A corn plant before the tassel grows out.

WIDE RANGE - Pesticides ability to kill several different kinds of pests.

WILDLIFE - All living things that are not human or domesticated, nor pests, as used here, including birds, mammals, and acquatic life.

WILTING - Drooping of leaves.

WINTER ANNUAL - Plants which germinate in the fall and complete their life cycle by early summer.

X

XEROPHYTE - A plant adapted for growth under dry conditions.

XYLENE - Any of three isomeric hydrocarbons of the benzene series used as solvents.

Y

YELLOWS - One of the various plant diseases, such as *aster yellows*, whose prominent symptom is a loss of green pigment in the leaves.

Z

ZERO TOLERANCE - By law, no detectable amount of the pesticide may remain on the raw agricultural commodity when it is offered for shipment. Zero tolerances are no longer allowed.

ZOOPLANKTON - Microscopic animal life living suspended in water.

APPENDIX

APPENDIX

UNITED STATES FEDERAL PESTICIDE CONTROL OFFICES

ENVIRONMENTAL PROTECTION AGENCY

Office of the DAA for Pesticides
Environmental Protection Agency
401 M Street, S.W.
Washington, DC 20460

EPA REGIONAL PESTICIDE OFFICES

EPA Region I
John F. Kennedy Federal Bldg.
Boston, MA 02203

EPA Region II
26 Federal Plaza, Room 907
New York, NY 10007

EPA Region III
6th and Walnut St., Curtis Bldg.
Philadelphia, PA 19106

EPA Region IV
345 Courtland St., N.E., Room 204
Atlanta, GA 30308

EPA Region V
Federal Office Bldg.
230 South Dearborn Street
Chicago, IL 60604

EPA Region VI
1201 Elm St.
1st International Bldg.
Dallas, TX 75270

EPA Region VII
324 E. 11th Street
Kansas City, MO 64106

EPA Region VIII
1860 Lincoln St., Suite 900
Denver, CO 80295

EPA Region IX
215 Fremont Street
San Francisco, CA 94105

EPA Region X
1200 6th Avenue
Seattle, WA 98101

STATE PESTICIDE CONTROL OFFICES

ALABAMA
Agricultural Chem. & Plant Industry
Department of Agriculture & Industry
Beard Building, P. O. Box 3336
Montgomery, AL 36109

ALASKA
Dept. of Environmental Conservation
P. O. Box 2309
Palmer, AK 99645

ARIZONA
Office of State Chemist
P. O. Box 1586
Mesa, AZ 95201

ARKANSAS
Division of Feeds, Fertilizer
   and Pesticides
State Plant Board
One Natural Resources Drive
Little Rock, AR 72205

CALIFORNIA
Calif. Dept. of Food & Agriculture
Division of Pest Management
1220 N Street, Room A414
Sacramento, CA 95814

COLORADO
Division of Plant Industry
Colo. Dept. of Agriculture
1525 Sherman Street
Denver, CO 80203

CONNECTICUT
Hazardous Materials Management
Dept. of Environmental Protection
State Office Bldg., Capitol Ave.
Hartford, CT 06115

DELAWARE
Delaware State Dept. of Agriculture
P. O. Drawer D
Dover, DE 19901

DISTRICT OF COLUMBIA
Hazardous Chemical Control
D.C. Government/Bureau Consumer
    Health Services
415 12th St., N.W., Room 301
Washington, DC 20004

FLORIDA
Inspection Division
Department of Agriculture
Mayo Building
Tallahassee, FL 32301

GEORGIA
Pesticide Division
Georgia Dept. of Agriculture
Capitol Square
Atlanta, GA 30334

HAWAII
Pesticides Branch
Plant Industry Division
P. O. Box 22159
Honolulu, HI 96822

IDAHO
Pesticide Enforcement
Idaho Dept. of Agriculture
Division of Plant Industries
P. O. Box 790
Boise, ID 83701

ILLINOIS
Bureau of Plant & Apiary Protection
Illinois Dept. of Agriculture
Emmerson Bldg., State Fairgrounds
Springfield, IL 62706

INDIANA
Office of Indiana State Chemist
Department of Biochemistry
Purdue University
West Lafayette, IN 47907

IOWA
Iowa Dept. of Agriculture
Wallace Building
Des Moines, IA 50319

KANSAS
Weed & Pesticide Division
Kansas State Board of Agriculture
1720 South Topeka Avenue
Topeka, KS 66612

KENTUCKY
Division of Pesticides
Capitol Plaza Tower
Frankfort, KY 40601

LOUISIANA
Pesticide & Environmental Programs
Box 44153
Capital Station
Baton Rouge, LA 70803

MAINE
Board of Pesticides Control
Maine Dept. of Agriculture
State House Station #28
Augusta, ME 04333

MARYLAND
Maryland Dept. of Agriculture
Parole Plaza Office Bldg.
Annapolis, MD 21401

MASSACHUSETTS
Division of Food & Drugs
Massachusetts Dept. of Public Health
305 South Street
Jamaica Plain, MA 02130

MICHIGAN
Department of Food & Agriculture
100 Cambridge Street
Boston, MA 02202

MINNESOTA
Agronomy Services Division
State Dept. of Agriculture
90 West Plato Boulevard
St. Paul, MN 55107

MISSISSIPPI
Division of Plant Industry
Mississippi Dept. of Agriculture
and Commerce
P. O. Box 5207
Mississippi State, MS 39762

MISSOURI
Plant Industries Division
Missouri Dept. of Agriculture
P. O. Box 630
Jefferson City, MO 65102

MONTANA
Environmental Management Division
Montana Dept. of Agriculture
Agriculture/Livestock Building
Capitol Station
Helena, MT 59601

NEBRASKA
Bureau of Plant Industry
Department of Agriculture
301 Centennial Mall
Lincoln, NE 68509

NEVADA
Nevada Dept. of Agriculture
350 Capitol Hill Ave.
P. O. Box 11100
Reno, NV 89510

NEW HAMPSHIRE
New Hampshire Dept. of Agriculture
Pesticide Control Division
85 Manchester Street
Concord, NH 03301

NEW JERSEY
Division of Environmental Quality
CN 027
Trenton, NJ 08625

NEW MEXICO
Division of Pesticide Management
New Mexico Dept. of Agriculture
Box 3AQ
Las Cruces, NM 88003

NEW YORK
Bureau of Pesticide Management
Dept. of Environmental Conservation
50 Wolf Road
Albany, NY 12233

NORTH CAROLINA
Food & Drug Protection Division
North Carolina Dept. of Agriculture
P. O. Box 27647
Raleigh, NC 27611

NORTH DAKOTA
North Dakota St. Laboratories Dept.
P. O. Box 937
Bismarck, ND 58505

OHIO
Division of Plant Industry
Ohio Dept. of Agriculture
8995 East Main Street
Reynoldsburg, OH 43068

OKLAHOMA
Plant Industry Division
Oklahoma State Dept. of Agriculture
122 State Capitol Building
Oklahoma City, OK 73105

OREGON
Plant Division
Oregon Dept. of Agriculture
635 Capitol Street, N.E.
Salem, OR 97310

PENNSYLVANIA
Bureau of Plant Industry
Pennsylvania Dept. of Agriculture
2301 North Cameron Street
Harrisburg, PA 17120

RHODE ISLAND
Division of Agriculture
Rhode Island Department
   Environmental Management
83 Park Street
Providence, RI 02903

SOUTH CAROLINA
Regulatory & Public Service Program
Clemson University
Clemson, SC 29631

SOUTH DAKOTA
Division of Ag. Reg. & Inspections
South Dakota Dept. of Agriculture
Anderson Building
Pierre, SD 57501

TENNESSEE
Plant Industries Division
Tennessee Dept. of Agriculture
Box 40627, Melrose Station
Nashville, TN 37204

TEXAS
Agricultural & Environmental
  Sciences Division
P. O. Box 12847
Austin, TX 78711

UTAH
Utah State Agricultural Department
147 North 200 West
Salt Lake City, UT 84103

VERMONT
Plant Industry Division
Vermont Dept. of Agriculture
116 State St./State Office Bldg.
Montpelier, VT 05602

VIRGINIA
Div. of Product & Industry Regulation
VA Dept. of Agriculture & Consumer Serv.
P. O. Box 1163
Richmond, VA 23209

WASHINGTON
Grain & Chemical Division
Washington State Dept. of Agriculture
406 General Administration Bldg.
Olympia, WA 98504

WEST VIRGINIA
Plant Pest Control Division
West Virginia Dept. of Agriculture
State Capitol Bldg.
Charleston, WV 25305

WISCONSIN
Plant Industry Division
Wisconsin Dept. of Agriculture,
  Trade & Consumer Protection
801 West Badger Rd., P. O. Box 8911
Madison, WI 53708

WYOMING
Wyoming Dept. of Agriculture
2219 Carey Avenue
Cheyenne, WY 82002

PUERTO RICO
Analysis & Reg. of Agricultural
  Materials
Division of Laboratory
Department of Agriculture
P. O. Box 10163
Santurce, PR 00908

VIRGIN ISLANDS
Dept. of Conservation and
  Cultural Affairs
Div. of Natural Resources Management
P. O. Box 4340
St. Thomas, VI 00801

ADDRESSES OF CANADIAN PROVINCIAL REGULATORY OFFICES

Newfoundland/Terre-Neuve

Newfoundland Department of Environment
Box 4750
St. John's, Newfoundland
A1C 5T7

Prince Edward Island/
Ile Du Prince Edouard

Prince Edward Island Environmental
   Control Commission
Box 2000
Charlottetown, Prince Edward Island
C1A 7N8

Nova Scotia/Nouvelle-Ecosse

Nova Scotia Environment
Box 2107
Halifax, Nova Scotia
B3J 3B7

New Brunswick/Nouveau-Brunswick

Environmental Services
New Brunswick Environment
Box 6000
Fredericton, New Brunswick
E3B 5H1

Quebec

Etudes Specialisees
Ministere de l'Environment
2360 Chemin Ste-Foy
Ste-Foy, Quebec
G1V 4H2

Ontario

Pesticides Control Section
Ontario Ministry of the Environment
135 St. Clair Avenue West
Suite 100
Toronto, Ontario
M4V 1P5

Manitoba

Technical Services Branch
Manitoba Department of Agriculture
911 Norquay Building
Winnipeg, Manitoba
R3C 0P8

Saskatchewan

Policy, Planning and Research Branch
Saskatchewan Environment
5th Floor, 1855 Victoria Avenue
Regina, Saskatchewan
S4P 3V5

Alberta

Pollution Control Division
Alberta Environment
Edmonton, Alberta
T5K 2J6

British Columbia/Colombie-Britannique

Pesticide Control Branch
Ministry of the Environment
Province of British Columbia
Victoria, British Columbia
V8V 1X5

CANADA PESTICIDE CONTROL OFFICE

Pesticide Division
Plant Products and
   Quarantine Directorate
Agriculture Canada
Ottawa, Ontario
K1A 0C6

Pesticides are most frequently listed by common names in publications. It is sometimes difficult, however, to recognize a particular pesticide by the common name if an individual is familiar with it only by its trade name. This cross reference will be helpful in looking for certain pesticides listed in the following toxicity tables.

TRADE NAMES LISTED ALPHABETICALLY

Trade Name	Common Name	Trade Name	Common Name
A-Rest	ancymidol	Basagran	bentazon
AAtrex	atrazine	Basalin	fluchloralin
Abate	temophos	Basamaize	prynachlor
Acaraben	chlorobenzilate	Basfapon	dalapon
Acaralate	chloropropylate	Baycor	bitertanol
Acti-Dione	cycloheximide	Baygon	propoxur
Advantage	carbosulfan	Bayleton	triadimefon
Agri-Strep	streptomycin	Bayrusil	quinalphos
Agrimycin	streptomycin	Baytan	triadimenol
Agritox	trichloronate	Baytex	fenthion
Agrothion	fenitrothion	Baythion	phoxim
Alanap	naptalam (NPA)	Beam	tricyclazone
Alar	daminozide	Benlate	benomyl
Aldrite	aldrin	Benzac	2,3,6 TBA
Altosid	methoprene	Benzahex	BHC (benzene
Amaze	isofenphos		hexachloride)
Ambush	permethrin	Betanal	phenmedipham
Amex	butralin	Betanex	desmedipham
Amiben	chloramben	Betasan	bensulide
Ammate	ammonium sulfamate (AMS)	Bidrin	dicrotophos
Animert V-101	tetrasul	Bim	tricyclazone
Ansar	MSMA	Biotrol	bacillus thuringiensis
Ansar 157	ammonium methanearsonate (AMA)	Black Leaf 40	nicotine
		Bladafume	sulfotep
Antor	diethatyl ethyl	Bladex	cyanazine
Antracol	propineb	Bolero	benthiocarb
Antu	antu	Bolstar	sulprofos
Aqualin	acrolein	Bomyl	bomyl
Arasan	thiram	Botran	DCNA (dicloran)
Arbotect	thiabendazole	Bravo	chlorothalonil
Aresin	monolinuron	Brestan	triphenyltin acetate
Aspon	ASP-51	Brominal	bromoxynil
Asulox	asulam	Brom-O-Gas	methyl bromide
Atlacide	sodium chlorate	Bromofume	ethylenedibromide (EDB)
Atlas A	sodium arsenite	Buctril	bromoxynil
Avadex	diallate	Butoxone	2,4-DB
Avadex BW	triallate	Butyrac	2,4-DB
Avenge	difenzoquat	Bux	bufencarb
Azak	terbutol	Bux	metalkamate
Azodrin	monocrotophos	CF 125	chlorflurecol
BAAM	amitraz	COCS	copper oxychloride
Balan	benefin	Caddy	cadmium chloride
Banvel	dicamba	Calo-gran	calomel
Baron	erbon	Can-Trol	MCPB

Trade Name	Common Name	Trade Name	Common Name
Caparol	prometryn	Diphacin	diphacinone
Caragard	terbumeton	Dipterex	trichlorfon
Carbamult	promecarb	Diquat	diquat
Carbyne	barban	Dithane D-14	nabam
Carzol	formetanate	Dithane M-22	maneb
	hydrochloride	Dithane M-45	mancozeb
Casoron	dichlobenil	Dithane Z-78	zineb
Castrix	crimidine	Dosanex	metoxuron
Ceredon	benquinox	Dow General	dinoseb (DNPB)
Chemform	bordeaux mixture	Dowcide	PCP (pentachloro phenol)
Chipco Turf	mecoprop (MCPP)	Dowfume	methyl bromide
Chlorocide	chlorbenside	Dowpon	dalapon
Cidial	phenthoate	Dozer	fenuron - TCA
Ciodrin	crotoxyphos	Drinox-H34	heptachlor
Co-Ral	coumaphos	Du-Ter	triphenyltin hydroxide
Co-Rax	warfarin	Dual	metolachlor
Cobex	dinitroamine	Dursban	chlorpyrifos
Compound 1080	sodium carbonate	Dybar	fenuron
Copper-Count N	copper ammonium	Dyfonate	fonofos
	fluroacetate	Dylox	trichlorfon
Cotoran	fluometuron	Dymid	diphenamide
Counter	terbufos	Dyrene	anilazine
Crag Nemacide	dazomet	Ekamet	etrimphos
Croneton	ethiophencarp	Ekatin	thiometon
Curacron	profenfos	Elgetol 30	DNOC
Cyanogas	calcium cyanide	Endothal	endothall
Cyclon	hydrocyanic acid (HCN)	Endrex	endrin
Cycocel	chlormequat chloride	Eptam	EPTC
Cyflee	cythioate	Eptapur	buturon
Cygon	dimethoate	Eradex	chinothionat
Cyolane	phosfolan	Ethanox	ethion
Cyprex	dodine	Ethyl Guthion	azinphos-ethyl
Cython	malathion	Etrofol	CPMC
Cytrolane	mephosfolan	Etrofolan	MIPC
Daconate	MSMA plus surfactant	Euparen	dichlofluanid
Dacthal	DCPA	Euparen M	tolyfluanid
Dasanit	fensulfothion	Evik	ametryn
Dechlorane	mirex	Evital	norflurazon
Ded-Weed	silvex	Faneron	bromofenoxim
De-Fend	dimethoate	Far-Go	triallate
Delan	dithianon	Fenac	fenac
Delnav	dioxathion	Ferbam	ferbam
Demosan	chloroneb	Ficam	bendiocarb
Dessicant	arsenic acid	Florocid	sodium fluoride
Devrinol	napropamide	Folcidin	cypendazole
Di-Syston	disulfoton	Folimat	omathoate
Dibrom	naled	Fruitone	1-napthalenacetic acid
Dicurane	chlorotoluron	Fumarin	coumafuryl
Dieldrite	dieldrin	Fumazone	dibromo chlorpropane
Difolatan	captafol	Fundal	chlordimeform
Dimecron	phosphamidon	Funginex	triforine
Dimilin	diflubenzuron	Fungo 50	methyl thiophanate
Dinofen	dinobuton	Furadan	carbofuran
Dipel	__bacillus__ __thuringiensis__	Furloe	chlorpropham (CIPC)

Trade Name	Common Name	Trade Name	Common Name
Fusilade	fluazifop-butyl	Morocide	binapacryl
Galecron	chlordimeform	Murfotox	mecarbam
Gardona	tetrachlorvinphos	Mylone	dazomet, DMTT
Gardoprim	terbuthylazine	Nabac	hexachlorophene
Garlon	triclopyr	Neguvon	trichlorfon
Gatnon	benzthiazuron	Nem-A-Tak	fosthieton
Gesaran	metoprotryn	Nemacur	fenamiphos
Glyodin	glyodin	Nemagon	dibromochloropropane
Goal	oxyflurofen	Neo-Pynamin	tetramethrin
Gramoxone	paraquat	Nexion	bromophos
Guthion	azinphos-methyl	Norex	chloroxuron
Hanane	dimefox	Nortron	ethofumisate
Herban	norea	Nudrin	methomyl
Hoelon	diclofop methyl	Nuvanol N	iodofenphos
Hydrol	allyxcarb	OMPA	schradan
Hyvar	bromacil	Omite	propargite
Igran	terbutryn	Ordram	molinate
Imidan Prolate	phosmet	Orthene	acephate
Imugan	chloraniformethane	Orthocide	captan
IPC	propham	Orthorix	lime-sulfur
Karathane	dinocap	Outfox	cyprazine
Karmex	diuron	PMAS	PMA
Kelthane	dicofol	Paarlan	isopropalin
Kepone	chlordecone	Padan	cartap
Kerb	pronamide	Panogen	mema
Kloben	neburon	Paracide	para-dichlorobenzene
Koban	MF-344	Paradow	para-dichlorobenzene
Kocide	copper hydroxide	Parnon	parinol
Korlan	ronnel	Parzate	nabam
Krenite	fosamine ammonium	Patoran	metobromuron
Kryocide	cryolite	Pencal	calcium arsenate
Kuron	silvex	Pentac	dienochlor
Lannate	methomyl	Perthane	ethylan
Lasso	alachlor	Phaltan	folpet
Lesan	fenaminosulf	Phosdrin	mevinphos
Lexone	metribuzin	Phostoxin	aluminum phosphide
Lintox	lindane		(phosphine)
Lironion	difenoxuron	Phosfon	chlorphonium
Lorox	linuron	Phosvel	leptophos
Lorsban	chlorpyrifos	Picfume	chloropicrin
MH-30	maleic hydrazide (MH)	Pik-Off	glyoxime
Manzate D	maneb	Pipron	piperalin
Marlate	methoxychlor	Pirimor	pirimicarb
Matacil	aminocarb	Pival	pindone
Mertect	thiabendazole	Planavin	nitralin
Mesurol	mercaptodimethur	Plantvax	oxycarboxin
Meta-Systox R	oxydemeton-methyl	Plictran	cyhexatin
Milogard	propazine	Polybor-Chlorate	sodium chlorate borate
Mobilawn	dichlofenthion	Polyram	metiram
Mocap	ethoprop	Potablan	monalide
Modown	bifenox	Pounce	permethrin
Monitor	methamidophos	Pramitol	prometone
Monobor-Chlorate	sodium chlorate borate	Prefix	chlorthiamid
Morestan	oxythioquinox	Preforan	fluorodifen

Trade Name	Common Name	Trade Name	Common Name
Prefox	ethiolate	Systox	demeton
Princep	simazine	Talon	brodifacoum
Probe	methazole	Tandex	karbutilate
Prodan	sodium fluosilicate	Telone	dichloropropene
Prowl	pendimethalin	Telvar	monuron
Pydrin	fenvalerate	Temik	aldicarb
Pynamin	allethrin	Tenoran	chloroxuron
Pyramin	pyrazon	Terraclor	PCNB
Quintar	dichlone	Terrazole	terrazole
Rabon	tetrachlorvinphos	Tersan 1991	benomyl
Racumin	coumatetralyl	Tersan SP	chloroneb
Rad-E-Cate	sodium cacodylate	Thimet	phorate
Rad-E-Cate 25	cacodylic acid	Thiocron	amidithion
Ramrod	propachlor	Thiodan	endosulfan
Randox	CDAA	Thuricide	bacillus thuringiensis
Raticate	norbormide	Tiguvon	fenthion
Redion V-18	tetradifon	Tillam	pebulate
Ridomil	metalaxyl	TOK	nitrofen
Rodex	warfarin	Tolban	profluralin
Ro-Neet	cycloate	Tomorin	coumachlor
Ronilan	vinclozolin	Topsin	thiophanate
Ronstar	oxadiazon	Topsin-M	thiophanate-methyl
Roundup	glyphosate	Torak	dialifor
Rout	bromacil, diuron	Tordon	picloram
Rovral	iprodione	Treflan	trifluralin
Rozol	chlorophacinone	Triangle	copper sulfate
Ruelene	crufomate	Trithion	carbophenothion
Sancap	dipropetryn	Tronabor	borax
Semeron	desmetryne	Truban	terrazole
Sencor	metribuzin	Trysben	trichlorobenzoic acid
Sevin	carbaryl	Tupersan	siduron
Shoxin	norbormide	Urox	monuron-TCA
Sinbar	terbacil	Vacor	pyriminol
Sinox	DNOC	Valone	PMP
Snip	dimetilan	Vapam	SMDC
Sodium TCA	TCA	Vapona	DDVP, dichlorvos
Sonar	fluridone	Vapotone	TEPP
Soprabel	acid lead arsenate	Vegadex	CDEC
Soyex	fluorodifen	Velpar	hexazinone
Spectracide	diazinon	Vendex	fenabutin oxide
Spergon	chloranil	Vernam	vernolate
Spike	tebuthiuron	Vidden D	dichloropropane-
Stam	propanil		dichloropropene
Stik	(NAA)	Vikane	sulfuryl flouride
Strobane T	toxaphene	Vitavax	carboxin
Suffix	benzoylprop ethyl	Vydate L	oxamyl
Sumithion	fenitrothion	Warbex	famphur
Supracide	methidathion	Zectran	mexacarbate
Surflan	oryzalin	Zelio	thallium sulfate
Sutan	butylate	Zerlate	ziram
Synthrin	resmethrin	Zolone	phosalone
		Zorial	norflurazon

This list is not intended to be all inclusive. An excellent reference for additional pesticides and information is the annual <u>Farm Chemicals Handbook</u> published by Meister Publishing Co., 37841 Euclid Ave., Willoughby, Ohio 44094. Some products listed have been or are about to be discontinued by their manufacturers, but are still included for historical interest and reference.

Table 1

Toxicity Values of Insecticides and Acaricides

Common Name	Trade Name	Producer	Chemical Class	Acute Oral LD$_{50}$	Dermal Toxicity
acephate	Orthene	Chevron Chemical	Organic Phosphate	866-945	- -
acid lead arsenate	Soprabel	Several	Inorganic Salt	10-50**	- -
aldicarb	Temik	Union Carbide	Methylcarbamoyl	7**	* *
aldrin	Aldrite and Others	Shell Chemical Co.	Chlorinated Hydrocarbon	55*	*
allethrin	Pynamin	M. G. King Co.	Botanical	>920	- -
allyxcarb	Hydrol	Bayer	Carbamate	90-99*	- -
amidithion	Thiocron	CIBA-GEIGY	Organic Phosphate	600-660	- -
aminocarb	Matacil	Bayer Mobay	Carbamate	30**	*
amitraz	BAAM	Tuco (Upjohn)	Formamidine	553	*
ASP-51	Aspon	Stauffer Chemical	Organic Phosphate	1440	- -
azinphos-ethyl	Ethyl Guthion	Bayer	Organic Phosphate	17.5**	*
azinphos-methyl	Guthion	Bayer Mobay	Organic Phosphate	13-16**	*
bacillus thuringiensis	Biotrol Dipel Thuricide	Abbott Labs TH Agr. & Nutr. Co. Sandoz	Microbial	>15,000	- -

*Moderately Toxic   **Highly Toxic

Table 1

Toxicity Values of Insecticides and Acaricides

Common Name	Trade Name	Producer	Chemical Class	Acute Oral LD_{50}	Dermal Toxicity
bendiocarb	Ficam	Fison Corp.	Carbamate	143*	– –
BHC (benzene hexachloride)	Benzahex	Hooker Chemical	Chlorinated Hydrocarbon	600–1250	– –
binapacryl	Morocide	American Hoechst	Dinitrophenyl Bentenoate	421*	– –
bomyl	Bomyl	Hopkins Ag. Chem. Co.	Organic Phosphate	31**	* *
bromophos	Nexion	Celamerck GMBH & Co.	Organic Phosphate	3750–7706	– –
bufencarb	Bux	Chevron Chemical	Carbamate	170*	– –
carbaryl	Sevin	Union Carbide	Carbamate	500–850	– –
carbofuran	Furadan	FMC Corporation	Carbamate	11**	– –
carbophenothion	Trithion	Stauffer Chemical	Organic Phosphate	32**	* *
carbosulfan	Advantage	FMC Corporation	Carbamate	209*	– –
cartap	Padan	Takeda	Dimethylamino	250*	– –
chinothionat	Eradex	Bayer	Trithio Carbonate	3400	– –
chlorbenside	Chlorocide	Boots Co. Ltd.	Chlorinated Sulfide	>3000	– –
chlordane	Many	Velsicol	Chlorinated Hydrocarbon	457–590*	*
chlordecone	Kepone	Allied Chem. Corp.	Chlorinated Ketone	114–140*	– –
chlordimeform	Fundal Galecron	CIBA-GEIGY Nor-Am Chemical	Formamidine	127–352*	– –

*Moderately Toxic   **Highly Toxic

Table 1

Toxicity Values of Insecticides and Acaricides

Common Name	Trade Name	Producer	Chemical Class	Acute Oral LD$_{50}$	Dermal Toxicity
chlorfenvinphos	Sapecron Supona	CIBA-GEIGY Shell Chemical	Organic Phosphate	10-39**	* *
chlormephos	Dotan	Murphy Chemical	Organic Phosphate	7**	* *
chlorobenzilate	Acaraben	CIBA-GEIGY	Chlorobenzilate	960	- -
chloropropylate	Acaralate	CIBA-GEIGY	Dichlorobenzilate	5000	- -
chlorpyrifos	Dursban Lorsban	Dow Chemical	Organic Phosphate	97-276*	- -
copper aceto arsenite	- - -	Los Angeles Chemical	Inorganic Salt	22**	- -
coumaphos	Co-Ral	Mobay	Organic Phosphate	56-230*	*
CPMC	Etrofol	Bayer	Carbamate	648	*
crotoxyphos	Ciodrin	Shell Chemical	Organic Phosphate	125*	*
crufomate	Ruelene	Dow Chemical	Organic Phosphate	770	- -
cryolite	Kryocide	Pennwalt Corp.	Inorganic Salt	>10,000	- -
cyhexatin	Plictran	Dow Chemical	Tricyclohexyl Hydroxystanane	540	- -
cythioate	Cyflee	American Cyanamid	Organic Phosphate	160*	- -
DDT	Many	Montrose Chemical Co.	Chlorinated Hydrocarbon	250*	- -
DDVP	Vapona	Shell Chemical Co.	Organic Phosphate	56-80*	* *
demeton	Systox	Mobay	Organic Phosphate	2.5-12**	* *

*Moderately Toxic   **Highly Toxic

409

Table 1

Toxicity Values of Insecticides and Acaricides

Common Name	Trade Name	Producer	Chemical Class	Acute Oral LD$_{50}$	Dermal Toxicity
dialifor	Torak	BFC Chemicals	Organic Phosphate	43-53**	- - -
diazinon	Spectracide	CIBA-GEIGY	Organic Phosphate	300-400*	- - -
dichlofenthion	Mobilawn	Mobil Chemical	Organic Phosphate	270*	- - -
dichlorvos	Vapona	Shell Chemical Co.	Organic Phosphate	56-80*	* *
dicofol	Kelthane	Rohm & Haas	Trichloroethanol	809	- - -
dicrotophos	Bidrin	Shell Chemical Co.	Organic Phosphate	22**	*
dieldrin	Dieldrite	Shell Chemical Co.	Chlorinated Hydrocarbon	60*	*
dienochlor	Pentac	Hooker Chemical Co.	Cyclopentadiene	3160	- - -
diflubenzuron	Dimilin	TH Agr. & Nutr. Co.	Urea	>4640	- - -
dimefox	Hanane	Murphy Chemical Co.	Organic Phosphate	1-2**	* *
dimethoate	Cygon De-Fend	American Cyanamid BASF Wyandotte Corp.	Organic Phosphate	320-380*	*
dimethrin	- - -	MGK Co.	Carboxylate	>15**	- - -
dimetilan	Snip	CIBA-GEIGY	Carbamate	>50*	- - -
dinobuton	Dinofen	Murphy Chemical Co.	Isopropylcarbonate	140*	*
dinocap	Karathane	Rohm & Haas	Nitrophenol	980	- - -
dioxathion	Delnav	Hercules Chemical	Organic Phosphate	110*	*

*Moderately Toxic

**Highly Toxic

Table 1

Toxicity Values of Insecticides and Acaricides

Common Name	Trade Name	Producer	Chemical Class	Acute Oral LD$_{50}$	Dermal Toxicity
disulfoton	Di-Syston	Mobay	Organic Phosphate	2.6-12.5**	* *
DNOC	Elgetol 30	Bayer	Nitrophenol	20-50**	* *
endosulfan	Thiodan	FMC Corporation	Dioxathiepinoxide	30-110**	*
endrin	Endrex	Shell Chemical Velsicol	Chlorinated Hydrocarbon	7-15**	* *
EPN	- - -	E. I. duPont	Organic Phosphate	14-42**	* *
ethion	Ethanox	Rhone-Poulenc	Organic Phosphate	208*	*
ethiophencarp	Croneton	Bayer	Carbamate	411*	*
ethoprop	Mocap	Rhone-Poulenc	Organic Phosphate	61.5*	*
ethylan	Perthane	Rohm & Haas	Dichloroethane	8170	- -
etrimphos	Ekamet	Sandoz	Organic Phosphate	1800	- -
famphur	Warbex	American Cyanamid	Organic Phosphate	36-62**	*
fenabutin oxide	Vendex	Shell Chemical	Hexaxis Distannoxane	263*	- -
fenitrothion	Agrothion Sumithion	American Cyanamid Sumitomo Chemical	Organic Phosphate	500*	*
fensulfothion	Dasanit	Mobay	Organic Phosphate	2-10**	* *
fenthion	Baytex Tiguvon	Mobay	Organic Phosphate	250-300*	*
fenvalerate	Pydrin	Shell Chemical	Benzene Acetate	451*	- -

*Moderately Toxic   **Highly Toxic

Table 1

Toxicity Values of Insecticides and Acaricides

Common Name	Trade Name	Producer	Chemical Class	Acute Oral LD_{50}	Dermal Toxicity
fonofos	Dyfonate	Stauffer Chemical	Organic Phosphate	8-17.5**	* *
formetanate hydrochloride	Carzol	Nor-Am	Hydrochloride	20**	- - -
gamma BHC (lindane)	Several	Hooker Chemical	Chlorinated Hydrocarbon	88-125*	*
heptachlor	Drinox-H34	Velsicol	Chlorinated Hydrocarbon	40-188**	* *
iodofenphos	Nuvanol N	CIBA-GEIGY	Organic Phosphate	2100	- - -
isofenphos	Amaze	Mobay	Organic Phosphate	28**	* *
lead arsenate	- - -	FMC Corporation	Inorganic Salt	10-50**	- - -
leptophos	Phosvel	Velsicol	Organic Phosphate	52.8*	- - -
----	Lethane 384	Rohm & Haas	Thiocyanate	90*	- - -
lindane	Lintox	Hooker Chemical	Chlorinated Hydrocarbon	88-125*	- - -
malathion	Cython	American Cyanamid	Organic Phosphate	1375	- - -
mecarbam	Murfotox	Murphy Chemical Co.	Organic Phosphate	36**	*
mephosfolan	Cytrolane	American Cyanamid	Organic Phosphate	8.9**	* *
mercaptodimethur	Mesurol	Mobay	Carbamate	45**	- - -
metalkamate	Bux	Chevron Chemical	Carbamate	170*	- - -
methamidophos	Monitor	Mobay	Organic Phosphate	18.9**	* *

*Moderately Toxic   **Highly Toxic

Table 1

Toxicity Values of Insecticides and Acaricides

Common Name	Trade Name	Producer	Chemical Class	Acute Oral LD$_{50}$	Dermal Toxicity
methidathion	Supracide	CIBA-GEIGY	Organic Phosphate	65*	– –
methomyl	Lannate Nudrin	E. I. duPont Shell Chemical	Carbamate	17**	– –
methoprene	Altosid	Zoecon	Growth Regulator	34,600	– –
methoxychlor	Marlate	E. I. duPont	Chlorinated Hydrocarbon	6,000	– –
methyl parathion	Many	Monsanto	Organic Phosphate	9–25**	*
mevinphos	Phosdrin	Stauffer Shell Chemical	Organic Phosphate	3.7–12**	* *
mexacarbate	Zectran	Dow Chemical Co.	Carbamate	19**	*
MIPC	Etrofolan	Bayer	Carbamate	485*	*
mirex	Dechlorane	Allied Chemical	Butapentalene	306*	*
monocrotophos	Azodrin	Shell Chemical	Organic Phosphate	20**	*
naled	Dibrom	Chevron Chemical	Organic Phosphate	430*	*
nicotine	Black Leaf 40	Chemical Formulators	Pyridine	50–60*	*
omathoate	Folimat	Bayer	Organic Phosphate	50**	*
oxamyl	Vydate L	E. I. duPont	Thiooxamimidate	5.4–37**	– –
oxydemeton-methyl	Meta-Systox R	Mobay	Organic Phosphate	56–65*	*
oxythioquinox	Morestan	Mobay	Quinoxalinone	2500–3000	*

*Moderately Toxic   **Highly Toxic

413

Table 1

Toxicity Values of Insecticides and Acaricides

Common Name	Trade Name	Producer	Chemical Class	Acute Oral LD_{50}	Dermal Toxicity
parathion	Many	Bayer Monsanto	Organic Phosphate	13**	**
paris green	- - -	Los Angeles Chemical	Arsenical	22**	- - -
permethrin	Ambush Pounce	ICI Americas FMC Corp.	Synthetic Pyrethroid	>4,000	- - -
phenthoate	Cidial	Bayer	Organic Phosphate	300-400*	- - -
phorate	Thimet	American Cyanamid	Organic Phosphate	2-4**	**
phosalone	Zolone	Rhone-Poulenc	Organic Phosphate	120*	*
phosfolan	Cyolane	American Cyanamid	Organic Phosphate	8.9**	**
phosmet	Imidan Prolate	Stauffer Chemical	Organic Phosphate	300*	- - -
phosphamidon	Dimecron	Chevron Chemical	Organic Phosphate	20-22.4**	*
phoxim	Baythion	Bayer	Organic Phosphate	1845	- - -
pirimicarb	Pirimor	Plant Protection Ltd.	Carbamate	147*	- - -
profenfos	Curacron	CIBA-GEIGY	Organic Phosphate	400*	*
promecarb	Carbamult	Nor-Am	Carbamate	74-90*	- - -
propargite	Omite	Uniroyal Chemical	Propynyl Sulfite	2200	- - -
propoxur	Baygon	Mobay	Carbamate	95-104*	- - -
pyrethrum	- - -	FMC Corporation MGK Company	Botanical	1500	- - -
quinalphos	Bayrusil	Bayer	Organic Phosphate	66*	*

*Moderately Toxic   **Highly Toxic

Table 1

Toxicity Values of Insecticides and Acaricides

Common Name	Trade Name	Producer	Chemical Class	Acute Oral LD_{50}	Dermal Toxicity
resmethrin	Synthrin	FMC Corporation	Carboxylate	4240	- - -
ronnel	Korlan	Dow Chemical Co.	Organic Phosphate	1740	*
rotenone	- - -	FMC Corporation	Botanical	132*	*
ryania	- - -	Penick & Co.	Botanical	1200	- - -
sabadilla	- - -	Prentiss Drug	Botanical	4000	- - -
schradan	OMPA	Murphy Chemical Co.	Organic Phosphate	9**	* *
sodium fluoaluminate	Kryocide	Pennwalt Corporation	Inorganic Salt	>10,000	- - -
sodium fluoride	Florocid	Allied Chemical Co.	Inorganic Salt	75-150*	- - -
sodium fluosilicate	Prodan	Tamogan Ltd.	Inorganic Salt	125*	- - -
sulfotep	Bladafume	Bayer	Organic Phosphate	7-10**	* *
sulprofos	Bolstar	Mobay Chem. Corp.	Organic Phosphate	107*	*
temophos	Abate	American Cyanamid	Organic Phosphate	2030	- - -
TEPP	Vapotone	Miller Chemical	Organic Phosphate	1.2-2**	* *
terbufos	Counter	American Cyanamid	Organic Phosphate	3.5**	* *
tetrachlorvinphos	Gardona Rabon	Shell Chemical	Organic Phosphate	4000-5000	- - -
tetradifon	Redion V-18	Phillips-Duphar	Sulfone	>14,700	- - -

*Moderately Toxic   **Highly Toxic

415

Table 1

Toxicity Values of Insecticides and Acaricides

Common Name	Trade Name	Producer	Chemical Class	Acute Oral LD_{50}	Dermal Toxicity
tetramethrin	Neo-Pynamin	FMC Corporation	Carboyxlate	>20,000	- -
tetrasul	Animert V-101	Phillips-Duphar	Diphenyl Sulfide	10,800	- -
thiometon	Ekatin	Sandoz Ltd.	Organic Phosphate	100-120*	*
toxaphene	Strobane T	Hercules Chemical	Chlorinated Hydrocarbon	69*	*
trichlorfon	Dylox Dipterex Neguvon	Bayer	Organic Phosphate	450-500*	- -
trichloronate	Agritox	Bayer	Organic Phosphate	37.5**	* *

*Moderately Toxic   **Highly Toxic

416

Table 2

Toxicity Values of Herbicides and Plant Growth Regulators

Common Name	Trade Name	Producer	Chemical Class	Acute Oral LD50	Dermal Toxicity
acrolein	Aqualin	Shell Chemical Co.	Propenal	46**	S
alachlor	Lasso	Monsanto	Acetanilide	1800	- -
allyl alcohol	- -	Dow Chemical Shell Chemical	Alcohol	64*	* *
ametryn	Evik	CIBA-GEIGY	Triazine	1110	- -
amitrole	Several	Union Carbide	Triazole	25,000	- -
ammonium methanearsonate (AMA)	Ansar 157	Cleary Vineland	Arsonate	600	- -
ammonium sulfamate (AMS)	Ammate	E. I. duPont	Sulfamate	3900	- -
ancymidol	A-Rest	Lilly (Elanco)	Pyrimidine Methanol	4500	*
arsenic acid	Dessicant	Penwalt Corp.	Orthoarsenic Acid	48-100**	* *
asulam	Asulox	Rhone-Poulenc	Carbamate	>5000	- -
atrazine	AAtrex	CIBA-GEIGY	Amino Triazine	3080	- -
barban	Carbyne	Velsicol	Carbanilate	1350	- -
benefin	Balan	Lilly (Elanco)	Dinitro-toluidine	>10,000	- -
bensulide	Betasan	Stauffer Chemical	Sulfonamide	1082	- -
bentazon	Basagran	BASF Wyandotte Corp.	Benzothiadiazin	1100	M

*Moderately Toxic   **Highly Toxic   M - Mild Skin Irritation   S - Severe Skin Irritation

Table 2

Toxicity Values of Herbicides and Plant Growth Regulators

Common Name	Trade Name	Producer	Chemical Class	Acute Oral LD_{50}	Dermal Toxicity
benthiocarb	Bolero	Chevron Chemical Co.	Carbamate	1903	- -
benzoylprop ethyl	Suffix	Shell Chemical Co.	Aminoproprionate	1555	- -
benzthiazuron	Gatnon	Bayer	Methylurea	1280	- -
bifenox	Modown	Rhone-Poulenc	Nitro-Benzoate	6400	- -
borax	Tronabor	Kerr-McGee	Borate	2660-5190	- -
bromacil	Hyvar Rout Faneron	E. I. duPont Hopkins CIBA-GEIGY	Methyluracil	5200	M
bromofenoxim			Dinitrophenyl	1217	- -
bromoxynil	Brominal Buctril	Union Carbide Rhone-Poulenc	Benzonitrile	190*	- -
butralin	Amex	Union Carbide	Dinitrobenzenamine	12,600	- -
buturon	Eptapur	BASF Wyandotte Corp.	Urea	5800	- -
butylate	Sutan	Stauffer Chemical	Carbamate	4659	- -
cacodylic acid	Rad-E-Cate 25	Vineland Chemical	Arsinic Acid	700	- -
calcium arsenate	Pencal	Pennwalt	Inorganic Salt	298*	M
CDAA	Randox	Monsanto	Chloraceteamide	750	M
CDEC	Vegadex	Monsanto	Carbamate	850	M
chloramben	Amiben	Union Carbide	Benzoic Acid	5620	- -

*Moderately Toxic   **Highly Toxic   M - Mild Skin Irritation   S - Severe Skin Irritation

Table 2

Toxicity Values of Herbicides and Plant Growth Regulators

Common Name	Trade Name	Producer	Chemical Class	Acute Oral LD$_{50}$	Dermal Toxicity
chlorflurecol	GF 125	EM Industries	Carboxylate	6800	- -
chlormequat chloride	Cycocel	American Cyanamid	Methyl Ammonium Chloride	570	- -
chlorotoluron	Dicurane	CIBA-GEIGY	Dimethylurea	>10,000	- -
chloroxuron	Tenoran Norex	CIBA-GEIGY Nor-Am	Dimethylurea	3700	- -
chlorphonium	Phosfon	Mobil Chemical	Phosphonium Chloride	178*	- -
chlorpropham (CIPC)	Furloe	PPG Industries	Carbamate	3800	- -
chlorthiamid	Prefix	Shell Chemical	Dichlorothiobenzamide	757	- -
copper sulfate	Triangle	Phelps Dodge	Inorganic Compound	470*	S
cyanazine	Bladex	Shell Chemical	Triazine	334*	- -
cycloate	Ro-Neet	Stauffer Chemical	Carbamate	3160-4100	- -
cyprazine	Outfox	Gulf	Triazine	1200	- -
2,4-D	Many	Many	Phenoxyacetic Acid	300-1200*	- -
2,4-DB	Butyrac Butoxone	Union Carbide Rhone-Poulenc	Phenoxybutyric Acid	700	- -
dalapon	Baspafon Dowpon	BASF Wyandotte Corp. Dow Chemical	Dichloroproprionic Acid	970	M
daminozide	Alar	Uniroyal	Dimethyl Hydrazide	8400	- -
dazomet	Mylone	Stauffer Chemical	Thiadiazine	320-500*	- -

*Moderately Toxic   **Highly Toxic   M - Mild Skin Irritation   S - Severe Skin Irritation

Table 2

Toxicity Values of Herbicides and Plant Growth Regulators

Common Name	Trade Name	Producer	Chemical Class	Acute Oral LD_{50}	Dermal Toxicity
DCPA	Dacthal	Diamond Shamrock	Terepthalic Acid	3000	- - -
desmedipham	Betanex	Nor-Am	Carbamate	>10,250	- - -
desmetryne	Semeron	CIBA-GEIGY	Triazine	1390	- - -
diallate	Avadex	Monsanto	Carbamate	395*	M
dicamba	Banvel	Velsicol	Benzoic Acid	1040	- - -
dichlobenil	Casoron	TH Agr. & Nutr. Co.	Benzonitrile	3160	- - -
diclofop methyl	Hoelon	American Hoechst	Methyl Proponoate	679	- - -
diethatyl ethyl	Antor	BFC Chemicals	Ethyl Ester	2300	- - -
difenoxuron	Lironion	CIBA-GEIGY	Dimethylurea	7750	- - -
difenzoquat	Avenge	American Cyanamid	Pyrazolium Methyl Sulfate	470*	- - -
dinitroamine	Cobex	U. S. Borax	Dinitroaniline	3700	* *
dinoseb (DNPB)	Dow General	Dow Chemical	Dinitrophenol	46-60**	M
diphenamide	Dymid	Lilly (Elanco) Tuco (Upjohn)	Diphenyl Acetamide	1000	- - -
dipropetryn	Sancap	CIBA-GEIGY	Triazine	5000	- - -
diquat	Diquat	Chevron Chemical	Pyridilium	400-440*	S
diuron	Karmex Rout	E. I. duPont Hopkins	Dimethylurea	3400	- - -

*Moderately Toxic   **Highly Toxic   M - Mild Skin Irritation   S - Severe Skin Irritation

Table 2

Toxicity Values of Herbicides and Plant Growth Regulators

Common Name	Trade Name	Producer	Chemical Class	Acute Oral LD$_{50}$	Dermal Toxicity
DMTT	Mylone	Stauffer Chemical	Thiadiazine Thione	640	M
DNOC	Sinox	Hopkins FMC Corp.	Dinitrophenol	20-50**	*
DSMA	Many	Diamond Shamrock Vineland	Methanearsonate	>3150	- - -
endothall	Endothal	Pennwalt Corp.	Dicarboxylic Acid	51*	S
EPTC	Eptam	Stauffer Chemical	Carbamate	1630	- - -
erbon	Baron	Dow Chemical	Dichlorpropionate	1000-3500	M
ethiolate	Prefox	Gulf	Carbamate	542	- - -
ethofumisate	Nortron	BFC Chemicals	Methane Sulfonate	5650	- - -
fenac	Fenac	Union Carbide	Phenylacetic Acid	1780	- - -
fenuron	Dybar	E. I. duPont	Dimethylurea	6400	M
fenuron - TCA	Dozer	Hopkins	Phenylurea	5700	M
fluazifop-butyl	Fusilade	ICI Americas	Butyl Propionate	1272	- - -
fluchloralin	Basalin	BASF Wyandotte Corp.	Dinitroaniline	1550	- - -
fluometuron	Cotoran	CIBA-GEIGY	Urea	8900	- - -
fluorodifen	Preforan Soyex	CIBA-GEIGY Nor-Am	Diphenyl Ether	9000	- - -
fluridone	Sonar	Lilly (Elanco)	Pyridinone	>10,000	- - -

*Moderately Toxic **Highly Toxic M - Mild Skin Irritation S - Severe Skin Irritation

Table 2

Toxicity Values of Herbicides and Plant Growth Regulators

Common Name	Trade Name	Producer	Chemical Class	Acute Oral LD_{50}	Dermal Toxicity
fosamine ammonium	Krenite	E. I. duPont	Carbamoylphosphonate	24,000	- - -
glyoxime	Pik-Off	CIBA-GEIGY	Dioxime	1175	- - -
glyphosate	Roundup	Monsanto	Phosphonoglycine	4300	- - -
hexazinone	Velpar	E. I. duPont	Triazine	1690	- - -
isopropalin	Paarlan	Lilly (Elanco)	Dinitroanaline	5000	- - -
karbutilate	Tandex	FMC Corp.	Carbamate	3000	- - -
linuron	Lorox	E. I. duPont	Phenylurea	1500	M
maleic hydrazide (MH)	MH-30	Uniroyal	Pyridazinone	3900-6950	- - -
MCPA	Many	Many	Oxyacetic Acid	700-800	- - -
MCPB	Can-Trol	Rhone-Poulenc	Phenoxybutyric Acid	680	- - -
mecoprop (MCPP)	Chipco Turf Herb.	Rhone-Poulenc	Propanoic Acid	930	- - -
methazole	Probe	Velsicol	Oxadiazolidinedione	1350	- - -
metobromuron	Patoran	CIBA-GEIGY	Methylurea	3000	M
metolachlor	Dual	CIBA-GEIGY	Acetamide	2780	- - -
metoprotryn	Gesaran	CIBA-GEIGY	Aminotriazine	>5000	- - -
metoxuron	Dosanex	Sandoz Ltd.	Dimethylurea	3200	- - -

*Moderately Toxic **Highly Toxic M - Mild Skin Irritation S - Severe Skin Irritation

Table 2

Toxicity Values of Herbicides and Plant Growth Regulators

Common Name	Trade Name	Producer	Chemical Class	Acute Oral LD_{50}	Dermal Toxicity
metribuzin	Sencor Lexone	Chemagro E. I. duPont	Triazine	1930	- -
molinate	Ordram	Stauffer Chemical	Carbothioate	501–720	- -
monalide	Potablan	Schering	Dimethylvaleramide	>4000	- -
monolinuron	Aresin	American Hoechst	Methylurea	2250	- -
monuron	Telvar	E. I. duPont	Dimethylurea	3600	M
monuron-TCA	Urox	Hopkins	Dimethylurea-TCA	2300	- -
MSMA	Ansar Daconate	Diamond-Shamrock	Methanearsonate	700	- -
napropamide	Devrinol	Stauffer Chemical	Proprionamide	>5000	- -
1-napthalenacetic acid (NAA)	Fruitone Stik	Union Carbide FMC Corp.	Acetamide	1000	- -
naptalam (NPA)	Alanap	Uniroyal Chemical	Pthalic Acid	8000	- -
neburon	Kloben	E. I. duPont	Methylurea	>11,000	M
nitralin	Planavin	Shell Chemical	Dinitroanaline	>2000	- -
nitrofen	TOK	Rohm & Haas	Nitrophenol Ether	2630	M
norea	Herban	Hercules	Dimethylurea	2000	- -
norflurazon	Evital Zorial	Sandoz Ltd.	Pyridazinone	>8000	- -
oryzalin	Surflan	Lilly (Elanco)	Dinitroanaline	>10,000	- -

*Moderately Toxic   M - Mild Skin Irritation   S - Severe Skin Irritation

**Highly Toxic

423

Table 2

Toxicity Values of Herbicides and Plant Growth Regulators

Common Name	Trade Name	Producer	Chemical Class	Acute Oral LD_{50}	Dermal Toxicity
oxadiazon	Ronstar	Rhone-Poulenc	Oxadiazolin	>8000	---
oxyflurofen	Goal	Rohm & Haas	Nitrobenzene	>5000	---
paraquat	Gramoxone	Chevron Chemical	Bipyridilium	150*	M
PCP (pentachloro phenol)	Dowcide	Dow Chemical	Pentachlorophenol	50-140*	M
pebulate	Tillam	Stauffer Chemical	Carbamate	1120	---
pendimethalin	Prowl	American Cyanamid	Dinitrobenzamine	1250	---
phenmedipham	Betanal	Nor-Am	Carbanilate	>8000	---
picloram	Tordon	Dow Chemical	Picolinic Acid	8200	---
profluralin	Tolban	CIBA-GEIGY	Dinitrotoluidine	2200	---
prometone	Pramitol	CIBA-GEIGY	Triazine	2980	---
prometryn	Caparol	CIBA-GEIGY	Triazine	3750	---
pronamide	Kerb	Rohm & Haas	Dichlorbenzamide	8350	---
propachlor	Ramrod	Monsanto	Acetanilide	1200	M
propanil	Stam	Rohm & Haas	Phenylanilide	1384	---
propazine	Milogard	CIBA-GEIGY	Triazine	5000	---
propham	IPC	PPG Industries	Carbanilate	5000	---

*Moderately Toxic   **Highly Toxic   M - Mild Skin Irritation   S - Severe Skin Irritation

Table 2

Toxicity Values of Herbicides and Plant Growth Regulators

Common Name	Trade Name	Producer	Chemical Class	Acute Oral LD$_{50}$	Dermal Toxicity
prynachlor	Basamaize	BASF Wyandotte Corp.	Acetanilide	1777	M
pyrazon	Pyramin	BASF Wyandotte Corp.	Pyridazone	3300	M
siduron	Tupersan	E. I. duPont	Phenylurea	750	M
silvex	Ded-Weed Kuron	TH Agr. & Nutr. Co. Dow Chemical	Propionic Acid	650	– – –
simazine	Princep	CIBA-GEIGY	Triazine	5000	– – –
SMDC	Vapam	Stauffer Chemical	Carbamate	820	– – –
sodium arsenite	Atlas A	Los Angeles Chemical	Inorganic Arsenical	10–50**	* *
sodium cacodylate	Rad-E-Cate	Vineland Chemical	Organic Salt	700	– – –
sodium chlorate	Atlacide	Rhone-Poulenc	Inorganic Salt	1200	M
sodium chlorate borate	Monobor-chlorate Polybor-chlorate	U. S. Borax	Mixture of Salts	4300	– – –
2,4,5-T	Many	Several	Phenoxyacetic Acid	500	– – –
2,3,6 TBA (trichlorobenzoic acid)	Benzac Trysben	E. I. duPont Union Carbide	Benzoic Acid	1644	– – –
TCA	Sodium TCA	Dow Chemical	Acetic Acid	500	M
tebuthiuron	Spike	Lilly (Elanco)	Dimethylurea	644	– – –
terbacil	Sinbar	E. I. duPont	Methyluracil	>5000	– – –
terbumeton	Caragard	CIBA-GEIGY	Methoxytriazine	485*	– – –

*Moderately Toxic   **Highly Toxic   M – Mild Skin Irritation   S – Severe Skin Irritation

Table 2

Toxicity Values of Herbicides and Plant Growth Regulators

Common Name	Trade Name	Producer	Chemical Class	Acute Oral LD_{50}	Dermal Toxicity
terbuthylazine	Gardoprim	CIBA-GEIGY	Triazine	2160	– – –
terbutol	Azak	Hercules Chemical	Carbamate	34,000	– – –
terbutryn	Igran	CIBA-GEIGY	Triazine	2400-2980	– – –
triclopyr	Garlon	Dow Chemical	Pyridinyloxyacetic Acid	630	– – –
trifluralin	Treflan	Lilly (Elanco)	Dintroanaline	3700	– – –
triallate	Avadex BW Far-Go	Monsanto	Carbamate	1675-2165	– – –
– – – –	Ureabor	Occidental Chemical	Mixture	2710	– – –
vernolate	Vernam	Stauffer Chemical	Carbamate	1780	– – –

Moderately Toxic   **Highly Toxic   M - Mild Skin Irritation   S - Severe Skin Irritation

Table 3

Toxicity Values of Fungicides and Plant Disease Chemicals

Common Name	Trade Name	Producer	Chemical Class	Acute Oral LD$_{50}$	Dermal Toxicity
anilazine	Dyrene	Mobay	Triazin-Amine	>5000	- -
benomyl	Benlate Tersan 1991	E. I. duPont	Carbamate	>10,000	- -
benquinox	Ceredon	Bayer	Benzoquinone	100*	- -
bitertanol	Baycor	Mobay	Triazole-Ethanol	>5000	- -
bordeaux mixture	Chemform	Chemical Formulators	Inorganic Compound	300*	M
cadmium chloride	Caddy	W. A. Cleary Corp. Vineland Chemicals	Inorganic Salt	88*	- -
calomel	Calo-gran	Mallinckrodt	Inorganic Salt	210*	- -
captafol	Difolatan	Chevron Chemical	Dicarboximide	6200	S
captan	Orthocide	Chevron Chemical	Dicarboximide	9000	- -
carboxin	Vitavax	Uniroyal	Carboxanilide	3820	- -
chloraniformethane	Imugan	Bayer	Trichloroethane	2500	- -
chloranil	Spergon	Uniroyal	Benzoquinone	4000	M
chloroneb	Demosan Tersan SP	E. I. duPont	Methoxybenzene	>11,000	- -
chlorothalonil	Bravo	Diamond Shamrock	Phthalonitrile	>10,000	M
copper ammonium carbonate	Copper-Count N	Mineral Research Corp.	Inorganic Salt	- - -	M
copper hydroxide	Kocide	Kocide Chemical	Inorganic Compound	1000	M

*Moderately Toxic **Highly Toxic M - Mild Skin Irritation S - Severe Skin Irritation

427

Table 3

Toxicity Values of Fungicides and Plant Disease Chemicals

Common Name	Trade Name	Producer	Chemical Class	Acute Oral LD$_{50}$	Dermal Toxicity
copper oxychloride	COCS	Rhone-Poulenc	Inorganic Salt	700-800	M
cycloheximide	Acti-Dione	Tuco (Upjohn)	Antibiotic	2**	S
cypendazole	Folcidin	Bayer	Benzimidazole	2500	- - -
DCNA (dicloran)	Botran	Tuco (Upjohn)	Nitroaniline	>5000	M
dichlone	Quintar	Hopkins	Naphthoquinone	1520	M
dichlofluanid	Euparen	Bayer	Phenylsulfamide	1000	*
dinocap	Karathane	Rohm & Haas	Nitrophenol	980	- - -
dithianon	Delan	Celamerk	Dicarbonitrile	610	- - -
dodine	Cyprex	American Cyanamid	Guanidine Acetate	1000	M
fenaminosulf	Lesan	Mobay	Sulfonate	75*	M
ferbam	Ferbam	FMC Corp.	Carbamate	>17,000	M
fixed coppers	Many	Several	Inorganic Compound	3000-6000	M
folpet	Phaltan	Chevron Chemical Stauffer Chemical	Phthalimide	>10,000	M
formaldehyde (formalin)	- - -	Several	Methanal	800	M
- - -	Fore	Rohm & Haas	Carbamate	>7500	- - -
glyodin	Glyodin	Agway Inc.	Inidizoline Acetate	4600-7600	- - -

*Moderately Toxic   **Highly Toxic   M - Mild Skin Irritation   S - Severe Skin Irritation

Table 3

Toxicity Values of Fungicides and Plant Disease Chemicals

Common Name	Trade Name	Producer	Chemical Class	Acute Oral LD_{50}	Dermal Toxicity
hexachlorophene	Nabac	Kalo Lab., Inc.	Trichlorophenol	320*	--
iprodione	Rovral	Rhone-Poulenc	Imidazolidine-carboxamide	3500	--
lime-sulfur	Orthorix	Chevron Chemical	Calcium Polysulfides	low	S
mancozeb	Dithane M-45	Rohm & Haas	Carbamate	>8000	--
maneb	Dithane M-22 Manzate D	Rohm & Haas E. I. duPont	Carbamate	6750	M
mema	Panogen	Shell Chemical	Methoxyethylmercury Acetate	25**	--
metalaxyl	Ridomil	CIBA-GEIGY	Methyl-Ester	669	--
methyl thiophanate	Fungo 50	Mallinckrodt	Thioallophanate	9700	--
metiram	Polyram	FMC Corporation	Carbamate	>10,000	--
MF-344	Koban	Mallinckrodt	Thiadiazole	4000	--
nabam	Dithane D-14	Rohm & Haas	Carbamate	395*	M
oxycarboxin	Plantvax	Uniroyal	Carboxanilide	2000	--
parinol	Parnon	Elanco (Lilly)	Pyridine Methanol	5000	M
PCNB	Terraclor	Olin Corporation	Pentachloro-nitrobenzene	200*	M
piperalin	Pipron	Elanco (Lilly)	Dichlorobenzoate	2500	--
PMA	PMAS	W. A. Cleary Corp.	Phenylmercury Acetate	22**	--
propineb	Antracol	Bayer	Dithiocarbamate	8500	M

*Moderately Toxic   **Highly Toxic   M – Mild Skin Irritation   S – Severe Skin Irritation

Table 3

Toxicity Values of Fungicides and Plant Disease Chemicals

Common Name	Trade Name	Producer	Chemical Class	Acute Oral LD$_{50}$	Dermal Toxicity
- - -	Quintar	Hopkins	Dichloronapthoquinone	1300	- - -
streptomycin	Agrimycin Agri-Strep	Merck Chem. Div. Charles Pfizer	Antibiotic	9000	M
terrazole	Terrazole Truban	Olin Mathieson Mallinckrodt	Thiadiazole	2000	- - -
thiabendazole	Mertect Arbotect	Merck Chem. Div. Hopkins	Benzimidazole	3100	- - -
thiophanate	Topsin	Pennwalt	Carbamate	>15,000	- - -
thiophanate-methyl	Topsin-M	Pennwalt	Carbamate	7500	- - -
thiram	Arasan Tersan	E. I. duPont	Disulfide	780	M
tolyfluanid	Euparen M	Bayer	Dimethyl-sulfamide	>1000	M
triadimefon	Bayleton	Mobay	Dimethylbutanone	400*	- - -
triadimenol	Baytan	Mobay	Triazole-ethanol	700	- - -
tricyclazone	Bim, Beam	Lilly (Elanco)	Benzothiazole	250*	- - -
triforine	Funginex	Celamerck	Formamide	>16,000	- - -
triphenyltin acetate	Brestan	American Hoechst	Triphenyltin Acetate	140-491*	- - -
triphenyltin hydroxide	Du-Ter	TH Ag & Nutr. Co.	Hydroxide	208-245*	- - -
vinclozolin	Ronilan	BASF Wyandotte Corp.	Oxazolidinedione	>10,000	- - -
zineb	Dithane Z-78 Parzate	Rohm & Haas FMC Corp.	Dithiocarbamate	>5200	M
ziram	Zerlate	Pennwalt	Dithicarbamate	1400	M

*Moderately Toxic   **Highly Toxic   M - Mild Skin Irritation   S - Severe Skin Irritation

Table 4

Toxicity Values of Rodenticides and Predacides

Common Name	Trade Name	Producer	Chemical Class	Acute Oral LD_{50}
antu	Antu	Several	Naphthylthiourea	2**
arsenic trioxide	- - -	Several	Inorganic Salt	13**
barium carbonate	- - -	Sherwin Williams	Inorganic Salt	630-750
brodifacoum	Talon	ICI Americas	Hydroxybenzopyron	0.27**
chlorophacinone	Rozol	Chempar	Dioxyindane	20.5**
coumachlor	Tomorin	CIBA-GEIGY	Hydroxy-Coumarin	900-1200
coumafuryl	Fumarin	Union Carbide	Hydroxy-Coumarin	25**
coumatetralyl	Racumin	Bayer	Hydroxy-Coumarin	25**
crimidine	Castrix	Bayer	Methylprimidine	1-2**
diphacinone	Diphacin	Velsicol	Indandione	15**
norbormide	Raticate Shoxin	Several	Dicarboximide	5**
pindone	Pival	Motomco	Indandione	50*
PMP	Valone	Motomco	Indandione	50*
pyriminol	Vacor	Rohm & Haas	Nitrophenyl urea	>500
sodium fluoroacetate	Compound 1080	Aceto Chemical	Fluoroacetate	1-4**
strychnine sulfate	- - -	Several	Inorganic Salt	30-60**

*Moderately Toxic   **Highly Toxic

431

Table 4

Toxicity Values of Rodenticides and Predacides

Common Name	Trade Name	Producer	Chemical Class	Acute Oral LD_{50}
thallium sulfate	Zelio	Bayer	Inorganic Salt	16**
warfarin	Rodex Co-Rax	Hopkins Prentiss Drug Velsicol	Hydroxy-Coumarin	1**
zinc phosphide	- - -	Hooker Chemicals	Inorganic Salt	45.7**

*Moderately Toxic   **Highly Toxic

Table 5

Toxicity Values of Nematicides

Common Name	Trade Name	Producer	Chemical Class	Acute Oral LD_{50}	Dermal Toxicity
aldicarb	Temik	Union Carbide	Carbamate	7**	* *
carbofuran	Furadan	FMC	Carbamate	11**	- -
dazomet	Crag Nemacide Mylone	Stauffer Chemical Hopkins Ag Chemical	Isothiocyanate	320-500*	- -
dibromo chlorpropane	Fumazone Nemagon	Dow Chemical Shell Chemical	Dibromochloro Propane	170-300*	*
dichlofenthion	Mobilawn	Mobil Chemical	Organic Phosphate	270*	- -
dichloropropane - dichloropropene	Vidden D	Dow Chemical	Mixture	250-500*	*
dichloropropene	Telone	Dow Chemical	Dichloropropene	250-500*	*
ethylenedibromide (EDB)	Bromofume	Dow Chemical	Ethylenedibromide	146*	S
fenamiphos	Nemacur	Mobay	Phosphoramidate	8.1-9.6**	* *
fosthieton	Nem-A-Tak	American Cyanamid	Organic Phosphate	4.7**	* *
methomyl	Lannate Nudrin	E. I. duPont Shell Chemical	Thioacetimide	17**	- -
methyl bromide	Dowfume	Dow Chemical	Bromomethane	20**	* *
SMDC	Vapam	Stauffer	Dithiocarbamate	820	- -

*Moderately Toxic   **Highly Toxic   M - Mild Skin Irritation   S - Severe Skin Irritation

Table 6

Toxicity Values of Commodity or Space Fumigants

Common Name	Trade Name	Producer	Chemical Class	Acute Vapor PPM Per Million Parts Air
aluminum phosphide (phosphine)	Phostoxin	Degesch America, Inc.	Inorganic	0.3**
calcium cyanide	Cyanogas	American Cyanamid	Cyanide	5**
carbon disulfide	- - -	FMC Corp. PPG Industries Stauffer Chemical	Inorganic	200*
carbon tetrachloride	- - -	Dow Chemical FMC Corp. PPG Industries Stauffer Chemical	Inorganic	300*
chloropicrin	Picfume	Dow Chemical	Nitromethane	20**
ethylene dibromide (EDB)	Bromofume Dowfume	Dow Chemical	Organic	200*
ethylene dichloride (EDC)	- - -	Several	Organic	1000
ethylene oxide (ETO)	- - -	Several	Organic	500
hydrocyanic acid (HCN)	Cyclon	- - -	Inorganic gas	10**
methyl bromide	Bromo Gas	Several	Bromoethane	200*
paradichloro benzene	Paracide Paradow	Allied Chemical Dow Chemical PPG Industries	Chlorobenzene	75**
sulfuryl flouride	Vikane	Dow Chemical	Flouride	5**

*Moderately Toxic   **Highly Toxic

INDEX

COMMON PESTICIDE CLASSIFICATIONS, NAMES, AND TRADE NAMES